This book
the las

# The IMA Volumes in Mathematics and Its Applications

## Volume 7

*Series Editors*
George R. Sell   Hans Weinberger

# Institute for Mathematics and Its Applications
# IMA

The **Institute for Mathematics and Its Applications** was established by a grant from the National Science Foundation to the University of Minnesota in 1982. The IMA seeks to encourage the development and study of fresh mathematical concepts and questions of concern to the other sciences by bringing together mathematicians and scientists from diverse fields in an atmosphere that will stimulate discussion and collaboration.

The IMA Volumes are intended to involve the broader scientific community in this process.

Hans Weinberger, Director
George R. Sell, Associate Director

## IMA Programs

1982–1983 Statistical and Continuum Approaches to Phase Transition
1983–1984 Mathematical Models for the Economics of Decentralized Resource Allocation
1984–1985 Continuum Physics and Partial Differential Equations
1985–1986 Stochastic Differential Equations and Their Applications
1986–1987 Scientific Computation
1987–1988 Applied Combinatorics
1988–1989 Nonlinear Waves

## Springer Lecture Notes from the IMA

*The Mathematics and Physics of Disordered Media*
Editors: Barry Hughes and Barry Ninham
(Lecture Notes in Mathematics, Volume 1035, 1983)

*Orienting Polymers*
Editor: J. L. Ericksen
(Lecture Notes in Mathematics, Volume 1063, 1984)

*New Perspectives in Thermodynamics*
Editor: James Serrin
(Springer-Verlag, 1986)

*Models of Economic Dynamics*
Editor: Hugo Sonnenschein
(Lecture Notes in Economics, Volume 264, 1986)

George Papanicolaou
Editor

# Random Media

With 14 Illustrations

Springer-Verlag
New York Berlin Heidelberg
London Paris Tokyo

George Papanicolaou
Institute for Mathematics and
Its Applications
Minneapolis, Minnesota 55455

AMS Classification: 73 B 35

Library of Congress Cataloging in Publication Data
Papanicolaou, George.
Random media
(The IMA volumes in mathematics and its applications)
Bibliography: p.
1. Stochastic processes. 2. Random fields.
3. Probabilities. 4. Mathematical physics. I. Title.
II. Series.
QC20.7.S8P37 1987 519.2 87-4923



Printed and bound by R.R. Donnelley & Sons, Harrisonburg, Virginia
Printed in the United States of America.

9 8 7 6 5 4 3 2 1

ISBN 0-387-96524-6 Springer-Verlag New York Berlin Heidelberg
ISBN 0-540-96524-6 Springer-Verlag Berlin Heidelberg New York

# The IMA Volumes in Mathematics and Its Applications

## Current Volumes:

**Volume 1:** Homogenization and Effective Moduli of Materials and Media
Editors: Jerry Ericksen, David Kinderlehrer, Robert Kohn, and J.-L. Lions

**Volume 2:** Oscillation Theory, Computation, and Methods of Compensated Compactness
Editors: Constantine Dafermos, Jerry Ericksen, David Kinderlehrer, and Marshall Slemrod

**Volume 3:** Metastability and Incompletely Posed Problems
Editors: Stuart Antman, Jerry Ericksen, David Kinderlehrer, and Ingo Müller

**Volume 4:** Dynamical Problems in Continuum Physics
Editors: Jerry Bona, Constantine Dafermos, Jerry Ericksen, and David Kinderlehrer

**Volume 5:** Theory and Applications of Liquid Crystals
Editors: Jerry Ericksen and David Kinderlehrer

**Volume 6:** Amorphous Polymers and Non-Newtonian Fluids
Editors: Constantine Dafermos, Jerry Ericksen and David Kinderlehrer

**Volume 7:** Random Media
Editor: George Papanicolaou

## Forthcoming Volumes:

*1985–1986: Stochastic Differential Equations and Their Applications*
Percolation Theory and Ergodic Theory of Infinite Particle Systems
Hydrodynamic Behavior and Interacting Particle Systems
Stochastic Differential Systems, Stochastic Control Theory and Applications

*1986–1987: Scientific Computation*
Computational Fluid Dynamics and Reacting Gas Flows
Numerical Algorithms for Modern Parallel Computer Architectures

D
A
B
C
E

CONTENTS

# FOREWORD

This IMA Volume in Mathematics and its Applications

**RANDOM MEDIA**

represents the proceedings of a workshop which was an integral part of the 1984-85 IMA program on STOCHASTIC DIFFERENTIAL EQUATIONS AND THEIR APPLICATIONS We are grateful to the Scientific Committee:

Daniel Stroock (Chairman)
Wendell Fleming
Theodore Harris
Pierre-Louis Lions
Steven Orey
George Papanicolaou

for planning and implementing an exciting and stimulating year-long program.

We especially thank George Papanicolaou for organizing a workshop which produced fruitful interactions between mathematicians and scientists from both academia and industry.

George R. Sell

Hans Weinberger

# PREFACE

During September 1985 a workshop on random media was held at the Institute for Mathematics and its Applications at the University of Minnesota. This was part of the program for the year on Probability and Stochastic Processes at IMA. The main objective of the workshop was to bring together researchers who work in a broad area including applications and mathematical methodology. The papers in this volume give an idea of what went on and they also represent a cross section of problems and methods that are currently of interest.

A look at the Table of Contents shows that one-dimensional problems (Lyapunov indices, density of states, localization, etc.) received considerable attention. This is understandable since they lead to tractable mathematical problems and since there has been rapid progress in recent years. There is considerable progress in several dimensional problems as well, in particular on localization of cases in multimensional random media. This topic was discussed at length.

A related volume on THE MATHEMATICS AND PHYSICS OF DISORDERED MEDIA contains the proceedings of an IMA workshop held in 1983 and appears as Volume 1035 in the Springer Lecture Notes in Mathematics series.

George Papanicolaou

# STABLE CONVERGENCE AND ASYMPTOTIC CAPACITY MEASURES[1]

by

J. R. Baxter and N. C. Jain

University of Minnesota

Minneapolis, MN 55455

(1) This work was supported in part by the National Science Foundation.

## 1. Introduction

Let $B_t$ denote Brownian motion taking values in $R^d$, $d \geq 3$. Let $D_n$ be a sequence of closed sets in $R^d$. Let $\tau_n$ denote the first hitting time of $D_n$. We study conditions under which the stopping times $\tau_n$ have, in a certain sense , a weak limit, denoted by T. We will refer to this type of convergence as stable convergence. The limit T of the $\tau_n$ is a randomized stopping time, and will turn out to be an "exponential killing time", where the rate h is a function of position in $R^d$. If we denote the sample space of standard Brownian motion by C (path space) then we will consider our randomized stopping times to be defined on $C\times[0,1]$, where [0,1] means the unit interval with Lebesgue measure. $\tau_n \to T$ stably simply means that $\tau_n|_A$ converges in distribution to $T|_{A\times[0,1]}$ for each measurable subset A of C. The sets $D_n$ are to be thought of as consisting of many small bodies, which become ever smaller and more densely distributed as $n \to \infty$. The function h represents a "limiting capacity density" for these bodies, as will be made clear.

One reason for studying this type of problem is to obtain limit theorems for solutions of the diffusion equation on $D_n^c$ with Dirichlet boundary conditions on $\partial D_n$, as specified in (1)-(3) below.

Let f be a bounded Borel function on $R^d$ and let $u_n(x,t) = E^x\left[f(B_t)\chi_{\{\tau(n)>t\}}\right]$. Then $u_n$ satisfies:

(1.1) $\partial u_n(x,t)/\partial t = (1/2)\Delta u_n(x,t)$ for $t>0$, $x\epsilon D_n^C$,

(1.2) $u_n(x,t) = 0$ for $t>0$, $x\epsilon\partial D_n$.

(1.3) $u_n(x,0) = f(x)$ for $x\epsilon D_n^C$.

We may think of f as the initial distribution for diffusing matter with density $u_n$ under the condition that any matter which touches the body $D_n$ is adsorbed and passes out of the solution. If the body $D_n$ is finely divided, one hopes that the actual density $u_n$ is close in some sense to a limiting density associated with the sequence of bodies $D_n$. First a particular case of the general situation will be described.

Let $D_i(n)$, $i=1,\dots,k(n)$, be compact sets, for each $n=1,2\dots$ . Let there exist a sequence $\rho_n>0$, $\rho_n\to 0$, such that diameter$(D_i(n)) < \rho_n$ for each i and each n. Let $D_n = D_1(n)\cup\dots\cup D_{k(n)}(n)$. Let $\lambda_i(n)$ denote the classical equilibrium measure on $D_i(n)$, and let $\lambda(n) = \lambda_1(n) + \dots + \lambda_{k(n)}(n)$. Suppose the sequence $\lambda(n)$ is bounded in total mass, and that there exists a finite measure $\lambda$ such that:

(1.4) $\lambda(n) \to \lambda$ vaguely as $n\to\infty$,

(1.5) $\Sigma^*\langle\lambda_i(n),\lambda_j(n)\rangle \to \langle\lambda,\lambda\rangle$ as $n\to\infty$,

where $\Sigma^*$ denotes the sum over all i and j with $i\neq j$, and for any measures $\mu$ and $\nu$,

(1.6) $\langle\mu,\nu\rangle = \int\int |x-y|^{-d+2}\mu(dx)\nu(dy)$, the classical energy inner product.

Intuitively, we may say that (1.4) specifies a limiting density for the bodies $D_i(n)$, and that (1.5) requires that these bodies be rather evenly spread out.

In [3] we prove

**Theorem 1.1** If the limit measure $\lambda$ above has bounded density h with respect to Lebesgue measure m on $R^d$, then: given any bounded Borel set A in $R^d$, $\varepsilon>0$ and $t_0<\infty$, there exists an integer $n_0$ such that for all $n\geq n_0$

(1.7) $m(\{x\colon x\epsilon A \text{ and } \sup_{0\leq t\leq t_0}|u_n(x,t) - u(x,t)| > \varepsilon\}) < \varepsilon$, where u is defined by

(1.8) $u(x,t) = E^x[f(B_t)\chi_{\{T>t\}}]$,

T being the randomized stopping time corresponding to the multiplicative functional $\exp(-\int_{[0,t]} h(B_s)ds)$, so that T is the exponential killing mentioned earlier.

Here for simplicity we have denoted an expectation involving a randomized stopping time (in this case T) by the same symbol as an ordinary expectation, although strictly speaking T is independent of the usual sample space of Brownian motion.

The limit u in theorem 1 satisfies:

(1.9) $\partial u(x,t)/\partial t = (1/2)\Delta u(x,t) - h(x)u(x,t)$ for $t>0$, $x \epsilon R^d$,

with the same initial condition as before, namely

(1.10) $u(x,0) = f(x)$ for $x \epsilon R^d$.

Theorem 1.1 is a generalization of the convergence result of Papanicolaou and Varadhan proved in [12]. Other approaches to problems of this type are given in [6], [7], [8], [10], [11], [13], [14], [15].

In [3] we also show

**Theorem 1.2** If the limit measure $\lambda$ above has bounded density h with respect to Lebesgue measure m on $R^d$, then from any subsequence n(k), we can extract another subsequence n(k(i)), such that, for m-a.e. x in $R^d$, $\tau_{n(k(i))}$ converges stably to T with respect to $P^x$, where T is the randomized stopping corresponding to exponential killing with rate h.

Theorem 1.2 could be rephrased as an "in measure" convergence result analogous to theorem 1.1, by defining a topology for stable convergence, but the a.e. convergence for subsequences formulation seems simpler in this case. Of course, theorem 1.2 implies that for any probability measure $\nu$ that is absolutely continuous with respect to Lebesgue measure, $\tau_n$ converges stably to T with respect to $P^\nu$.

Theorem 1.2 implies theorem 1.1 via another result of [3], which we state as:

**Theorem 1.3** Let $T_n$ be arbitrary randomized stopping times that converge stably to a limit T with respect to some probability $P^\nu$. Let t be fixed. For each n, let $\gamma_n$ be the distribution of $B_t|\{T_n>t\}$ with respect to $P^\nu$, and let $\gamma$ be the corresponding distribution of $B_t|\{T>t\}$. Assume that

$P^{\nu}(T=t)=0$. Then $\gamma_n$ converges to $\gamma$ in total variation norm.

Theorem 1.3 expresses the fact that the continuity of the transition density of Brownian motion allows one to deduce norm convergence for distributions on the state space from stable convergence on the sample space. It should be noted that stable convergence also gives information about other quantities, for example it shows the weak convergence of the distributions of the $B_{T_n}$.

The proof of theorem 1.2 uses the fact, proved in [2], that the space of randomized stopping times is compact with respect to stable convergence. This reduces the proof of theorem 1.2 to showing that every limit point of the sequence $\tau_n$ is the same, in other words, to a uniqueness argument.

In a later paper [5], a new result was obtained, which is given below as theorem 1.4. To state theorem 1.4 we need the notion of the classical potential associated with a measure $\mu$, which we write as Pot $\mu$, defined by

(1.11) $\quad \text{Pot } \mu\,(x) = k\int |x-y|^{-d+2}\mu(dy)$, where $k=\int_0^\infty e^{-1/2t}\,dt/[2\pi t]^{d/2}$.

For any measure $\mu$, and any function f, we write $f\mu$ to mean the measure $\gamma$ defined by $\gamma(A) = \int_A f\,d\mu$. Thus Pot $\chi_B\,\mu$ denotes the potential of the measure $\chi_B\,\mu$. We denote Lebesgue measure on $R^d$ by m. If $\mu$ has a density h with respect to Lebesgue measure we may write Pot h for Pot $\mu$. The capacity of a compact set W will be denoted by c(W). A property which is true except on a polar set in $R^d$ is said to hold quasi-everywhere (q.e.). The <u>additive functional</u> $A_t$ associated with a potential g is defined as usual to be the additive functional such that $E^x[A_\infty] = g(x)$ (so that $g\circ B_t + A_t$ is a <u>martingale</u>). Of course if g=Pot h then $A_t=\int_{[0,t]} h\circ B_s\,ds$. If $\lambda$ is a measure on $R^d$ such that Pot $\chi_K\lambda$ is bounded on $R^d$ for each bounded ball K, then we can define an additive functional $A_t$ associated with $\lambda$, even if Pot $\lambda$ is $\infty$ on $R^d$: we set

(1.12) $\quad A_t=\lim_{n\to\infty} A_t(n)$,

where $A_t(n)$ is the additive functional associated with Pot $\chi_{K(n)}\lambda$, for any sequence of balls K(n) such that $K(n)\uparrow R^d$.

**Definition 1.1** Let $D_n$ be any sequence of closed sets in $R^d$. The total capacity measure for the sequence $D_n$ is defined to be the minimal measure $\lambda$ such that $\limsup_{n\to\infty} c(D_n \cap W) \le \lambda(W)$ for every compact set $W$ in $R^d$. It is shown in [5] that $\lambda$ exists. If every subsequence of $D_n$ has the same total capacity measure, we will say that the sequence $D_n$ has a limiting capacity measure $\lambda$.

We can now state

**Theorem 1.4** Let $D_n$ be a sequence of closed sets with limiting capacity measure $\lambda$, such that Pot $\chi_K \lambda$ is bounded on $R^d$ for each bounded ball $K$. Let $\tau_n$ be the first hitting time of $D_n$. Let $A_t$ be the additive functional associated with $\lambda$, let $F_t$ be the multiplicative functional $\exp(-A_t)$, and let $T$ be the randomized stopping time associated with $F$. Let $\nu$ be any finite measure on $R^d$ that does not charge polar sets. Then from any subsequence $n(k)$, we can extract another subsequence $n(k(i))$, such that, for $\nu$-a.e. $x$ in $R^d$, $\tau_{n(k(i))}$ converges stably to $T$ with respect to $P^x$.

Theorem 1.4 implies theorem 1.2, because, as shown in [5], any sequence $D_n$ satisfying conditions (1.4) and (1.5) has a limiting capacity measure $\lambda$. Of course, the existence of a limiting capacity measure is a much weaker hypothesis than conditions (1.4) and (1.5).

In the present paper we will give some applications of theorems 1.3 and 1.4. We will consider the convergence of stopped distributions (section 3), homogenization for Brownian motion with a nonanticipating drift (section 4), and convergence of solutions of the Dirichlet problem (section 5).

## 2. Notations and Auxiliary Results.

Let $X=(\Omega,\mathcal{M},\mathcal{M}_t,X_t,P)$, $0\le t<\infty$, be a stochastic process, where

(2.1) $(\Omega,\mathcal{M},P)$ is a probability space, $(\mathcal{M}_t)$ an increasing right continuous family of $\sigma$-algebras contained in $\mathcal{M}$, and

(2.2) $X_t:\Omega\to E$ is $\mathcal{M}_t$-measurable for each $t<\infty$, $E$ being a topological space with topology $\mathcal{T}$ and Borel $\sigma$-algebra $\mathcal{E}$. We assume that $\mathcal{T}$ has a countable base. We will write $X_t$ as $X(t)$ or $X(t,\omega)$ where convenient.

$X(\cdot,\omega)$ is assumed to be right continuous and have left limits for P-a.e. $\omega$, and the process X is assumed to be quasi-left-continuous.

Let $\mathbf{B}_1$ denote the Borel sets of [0,1] and let $\mathbf{B}$ denote the Borel sets of $[0,\infty]$.

A <u>randomized $\mathcal{M}_t$-stopping time</u> T is a map $T : \Omega \times [0,1] \to [0,\infty]$ such $T(\omega,\cdot)$ is left continuous and nondecreasing, $T(\omega,0) = 0$, and T is an $\mathcal{M}_t \times \mathbf{B}_1$-stopping time in the ordinary sense. An ordinary stopping time $\tau$ can be regarded as a randomized stopping time T defined by $T(\omega,a) = \tau(\omega)$ for $a>0$, $T(\omega,0)=0$. Associated with T is the <u>stopping time probability measure</u>

$F : \Omega \times \mathbf{B} \to [0,1]$, defined by

(2.3) $$F(\omega,[0,t]) = \sup\{a : T(\omega,a) \le t\}.$$

Then

(2.4) $$F(\cdot\,,[0,t]) \text{ is } \mathcal{M}_t\text{-measurable for each } t.$$

We will sometimes write $F(\cdot\,,(t,\infty])$ as $F((t,\infty])$ or $F_t$. In our application $F_t$ will turn out to be a multiplicative functional, but this property will not be explicitly used.

We can recover T from F by

(2.5) $$T(\omega,a) = \inf\{t : F(\omega,[0,t]) \ge a\}.$$

Given any F for which (2.4) holds, T defined by (2.5) is a randomized stopping time, and (2.3) holds. Thus the notions of randomized stopping time and stopping time measure are equivalent.

Probabilities and expectations involving a randomized stopping time should clearly use the probability $P \times m_1$, where $m_1$ is Lebesgue measure on [0,1]. However, we will usually only write P explicitly, since the meaning is clear from the context.

**Definition 2.1** If $T_n$, T are randomized stopping times with stopping time measures $F_n$, F, respectively, P is a probability on $(\Omega,\mathcal{M})$, and $\mathcal{G}$ is a $\sigma$-algebra contained in $\mathcal{M}$, then we say that <u>$T_n$ converges stably to T, with respect to P and $\mathcal{G}$</u>, if for every $A \in \mathcal{G}$, $T_n|_{A\times[0,1]}$ converges in distribution to $T|_{A\times[0,1]}$, with respect to $P\times m_1$.

For a general discussion of stable convergence in other contexts, see [1].

We note that $T_n \to T$ stably with respect to P and $\mathcal{G}$ if and only if

(2.6) $E[Yf \circ T_n] \to E[Yf \circ T]$,

for all $Y \in L^1(\Omega, \mathcal{G}, P)$ and all $f \in \mathbf{C}([0,\infty])$. Furthermore, it is enough to check (2.6) on any dense sets in $L^1(\Omega, \mathcal{G}, P)$ and $\mathbf{C}([0,\infty])$.

We will normally assume that

(2.7) $\mathcal{G}$ is countably generated,

and

(2.8) $X_t$ is $\mathcal{G}$-measurable mod P, for every t.

Because of (2.7), we have [2]:

**Theorem 2.1** From any sequence $T_n$ of randomized stopping times we can extract a stably convergent subsequence with limit T.

Because of (2.8), we also have [2]:

**Theorem 2.2** If $T_n$ converges stably to T, and $T<\infty$, P-a.e., then $X_{T_n}$ converges in distribution to $X_T$.

Let

(2.9) $\mathcal{F}_t^0 = \sigma(X_s: s \le t)$, $\mathcal{F}_t = (\mathcal{F}_t^0)_+$, $\mathcal{F} = \sigma(X_s: 0 \le s < \infty)$.

For most applications we can take $\Omega = C$, the space of continuous functions $\omega$ from $[0.\infty)$ to $R^d$, $X_t(\omega) = B_t(\omega) \equiv \omega(t)$, and let $\mathcal{M}_t = \mathcal{F}_t$, $\mathcal{G} = \mathcal{M} = \mathcal{F}$. However, for section 4 it seems convenient to be able to work in a slightly more general setting, so we will not restrict ourselves to the canonical version of Brownian motion. Instead, we shall follow [9] and define the process X to be a d- dimensional Brownian motion if the following conditions hold:

(2.10) $E = R^d$;

(2.11) The process X has stationary independent increments, and $X_{t+s} - X_s$ is normal with mean 0 and variance $\delta_{ij} t$ for all t;

(2.12) $X_0$ is independent of the set of process increments;

(2.13) $X(\cdot,\omega)$ is continuous for P-a.e. $\omega$.

**Theorem 2.3** Let $X = (\Omega, \mathcal{M}, X_t, P)$ be a d-dimensional Brownian motion in the sense of (2.10)-(2.13). Let $\mathcal{F}, \mathcal{F}_t$ be defined by (2.9). Let $T_n$ and T be randomized $\mathcal{F}_t$-stopping times, such that $T_n \to T$ stably with respect to P and $\mathcal{F}$. Let t be fixed. Let Y be any integrable $\mathcal{F}$-measurable function.Define signed Borel measures $\gamma_n$ and $\gamma$ by

(2.14) $\int g d\gamma_n = E[Yg\circ X_t \chi_{\{T_n>t\}}]$, $\int g d\gamma = E[Yg\circ X_t \chi_{\{T>t\}}]$,

for all bounded Borel g on $R^d$. Assume

(2.15) $P(T=t) = 0$.

Then

(2.16) $\gamma_n \to \gamma$ in total variation norm as $n\to\infty$.

The proof of theorem 2.3 is the same as that of theorem 1.3.

**Theorem 2.4** Let $X=(\Omega,\mathbf{M},X_t,P)$ be a d-dimensional Brownian motion in the sense of (2.10)-(2.13). Let $D_n$ be a sequence of closed sets with limiting capacity measure $\lambda$, such that Pot $\chi_K\lambda$ is bounded on $R^d$ for each bounded ball K. Let $\sigma_n$ be the first hitting time of $D_n$. Let $\varphi:\Omega\to C$ be the map defined by $\varphi(\omega)(t)=X_t(\omega)$. Let $A_t$ be the additive functional on C associated with $\lambda$, let $G_t = \exp(-A_t\circ\varphi)$, let G be the stopping time measure defined by $G_t$, and let S be the randomized stopping time associated with G. Let $\nu$ be the distribution of $X_0$. Assume that $\nu$ does not charge polar sets. Then $\sigma_n$ converges stably to S with respect to P and $\mathbf{M}$.

Of course, if $\lambda$ has a density h with respect to Lebesgue measure m on $R^d$ then $G_t$ is just given by

(2.17) $$G_t(\omega) = \exp(-\int_{[0,t]} h(\varphi(\omega)(s))ds) = \exp(-\int_{[0,t]} h\circ X_s(\omega)ds).$$

**Proof.** Let $\mathbf{F}_t$ and $\mathbf{F}$ be defined by (2.9). Let $\tau_n$ be the first hitting time of $D_n$ on C. Let $F_t = \exp(-A_t)$, let F be the stopping time measure defined by $F_t$, and let T be the randomized stopping time on C associated with F. Clearly $\sigma_n=\tau_n\circ\varphi$ and $S=T\circ\varphi_1$, where $\varphi_1:\Omega\times[0,1]\to C\times[0,1]$ is defined by $\varphi_1(\omega,a)=(\varphi(\omega),a)$. By theorem 1.4, $\tau_n\to T$ stably on C with respect to $P^\nu$ and the canonical $\sigma$-field on C. Since $P\varphi^{-1}=P^\nu$, we have easily that $\sigma_n\to S$ stably with respect to P and $\mathbf{F}$.

If $Y\in L^1(\Omega,\mathbf{M},P)$ and $f\in\mathbf{C}([0,\infty])$, let $Z=E[Y|\mathbf{F}]$. Then $E[Yf\circ\sigma_n] = E[Zf\circ\sigma_n]$ $\to E[Zf\circ S] = E[Yf\circ S]$, and the theorem follows by (2.6).

By making the necessary definitions one could of course formulate and prove a result fully analogous to theorem 1.4 for a general version of Brownian motion equipped with a family of probabilities $P^x$.

**3. Stopped Distributions**

Let a sequence $D_n$ of closed sets be given, with limiting capacity measure $\lambda$. Let $\tau_n$ be the first hitting time of $D_n$. Suppose that $\lambda$ satisfies the condition of theorem 1.4, i.e. that Pot $\chi_K\lambda$ is bounded on $R^d$ for every bounded ball K. Let $A_t$ the additive functional associated with $\lambda$, $F_t$ the multiplicative functional $\exp(-A_t)$, and T the randomized stopping time whose stopping time measure F is defined by $F_t$. Let $\nu$ be a measure such that $\tau_n$ converges stably to T with respect to $P^\nu$. For example, $\nu$ might be any probability that does not charge polar sets, by theorem 1.4.

Let $\gamma_{n,t}$ denote the distribution of $B_t$ restricted to $\{\tau_n>t\}$, with respect to $P^\nu$. By theorem 1.3 we know that $\gamma_{n,t}$ converges in total variation norm to $\gamma_t$, where $\gamma_t$ is the distribution of $B_t$ restricted to $\{T>t\}$, with respect to $P^\nu$. Let $u_n(\cdot,t)$, $u(\cdot,t)$ denote the densities of $\gamma_{n,t}$, $\gamma_t$ with respect to Lebesgue measure m. Then $u_n$ has a smooth version satisfying equations (1.1), (1.2) for $t>0$. Fix t, and fix $s<t$. Clearly

(3.1) $$u_n(x,t) = E^x\left[u_n(B_{t-s},s)\chi_{\{\tau_n>t-s\}}\right],$$

and

(3.2) $$u(x,t) = E^x\left[u(B_{t-s},s)\chi_{\{T>t-s\}}\right].$$

Since $u_n(x,s)$ is uniformly bounded in x and n for fixed $s>0$, and since $\gamma_{n,s}$ converges in norm to $\gamma_s$, $u_n(\cdot,s)$ converges in $L^2(m)$-norm to $u(\cdot,s)$. Let $u_n(x,\cdot,t)$, $u(x,\cdot,t)$ denote the densities with respect to Lebesgue measure m of the distribution of $B_t$ restricted to the sets $\{\tau_n>t\}$, $\{T>t\}$, with respect to $P^x$. By theorem 1.4 and the argument just given, if we let $U_{n,t}$ denote the map $R^d \to L^2(m)$ that sends x to $u_n(x,\cdot,t)$, and define $U_t$ similarly sending x to $u(x,\cdot,t)$, then $U_{n,t-s}$ converges to $U_{t-s}$ in $\mu$-measure, as a map from any bounded subset of $R^d$ into a normed space, where $\mu$ is any bounded measure that does not charge polar sets. Hence, by (3.2)

(3.3) $u_n(\cdot,t)$ converges in $\mu$-measure to $u(\cdot,t)$ on bounded subsets of $R^d$.

This argument also shows that if $x \in R^d$ happens to be such that $\tau_n$

converges stably to T with respect to $P^x$, then

(3.4) $\quad u_n(x,t) \to u(x,t)$ as $n \to \infty$.

These results indicate how we can obtain convergence results for densities $u_n$ of the "unstopped material" at time t. Slightly more careful use of theorem 1.3 gives the type of convergence described in theorem 1.1 (see [3], theorem 2.2, corollary). We wish now to consider now the convergence of the "stopped material", by which we mean the distribution $\psi_{n,t}$ of $B_{\tau_n}$ restricted to $\{\tau_n \le t\}$, with respect to $P^\nu$. Let $\psi_t$ denote the distribution of $B_T$ restricted to $\{T \le t\}$, with respect to $P^\nu$. Then

**Theorem 3.1** For each $t \ge 0$, $\psi_{n,t}$ converges weakly to $\psi_t$ as $n \to \infty$.

**Proof.** Fix $t \ge 0$. Since $A_t$ is continuous, $F_t$ is continuous. Thus $P(T=t)=0$. As noted above, $\gamma_{n,t}$ converges to $\gamma_t$ in total variation norm. In particular, $B_t$ restricted to $\{\tau_n > t\}$ converges in distribution, with respect to $P^\nu$, to $B_t$ restricted to $\{T > t\}$. From the definition of stable convergence, clearly $\tau_n \wedge \tau$ converges stably to $T \wedge t$. By theorem 2.2, $B_{\tau_n \wedge t}$ converges in distribution to $B_{T \wedge t}$, with respect to $P^\nu$. Theorem 3.1 follows at once.

We note the $\psi_{n,t}$ can be expressed in terms of the time integral of the normal derivative of $u_n$ on the suface $\partial D_n$, and an analytical proof of theorem 3.1 is therefore possible, but the present argument seems much simpler.

## 4. Brownian Motion with a General Drift

Let us consider Brownian motion controlled by a nonanticipating drift, that is, let $X_t$ denote the process on C which satisfies the stochastic differential equation

(4.1) $\quad dX_t = dB_t - v_t dt,$

where $v_t$ is a bounded progressively measurable function on the sample space, and $(C, \mathcal{M}, \mathcal{M}_t, B_t, P^x)$ is the canonical Brownian process.

Let $D_n$, $\lambda$, $\nu$, $\tau_n$, $A_t$, T be as defined at the beginning of section 3. If we consider $P = P^\nu$ on C, then the process $X_t$ on C has initial distribution $\nu$ with respect to P. Let a sequence of stopping times $\sigma_n$ be defined on C, so that $\sigma_n$ is the first hitting time of $D_n$ by the process $X_t$. Define $\varphi:C\to C$ by $\varphi(\omega)(t)=X_t(\omega)$. Then $\tau_n\circ\varphi=\sigma_n$. As in section 2, let S be the randomized stopping time associated with the stopping time measure G, where $G_t=\exp(-A_t\circ\varphi)$, so that $S=T\circ\varphi_1$.

**Theorem 4.1** $\sigma_n$ converges stably to S with respect to $P^\nu$ and $\mathcal{M}$.

**Proof.** To study the limit of the $\sigma_n$, we will consider the $X_t$ process under a new probability Q, defined by

(4.2) $\quad Q = H_tP$ on $\mathcal{M}_t$, where $H_t = \exp(\int_{[0,t]} v_s dB_s - (1/2)\int_{[0,t]} v_s^2 ds)$.

By Girsanov's theorem, $(C,\mathcal{M},\mathcal{M}_t,X_t,Q)$ is a martingale, and its quadratic variation process is $\delta_{ij}t$, so by Levy's theorem $(C,\mathcal{M},X_t,Q)$ is a d-dimensional Brownian motion process as defined in section 2. It follows from theorem 2.4 that $\sigma_n$ converges stably to S with respect to Q and $\mathcal{M}$, and the theorem follows at once from the remark after (2.6) and the fact that P has a density with respect to Q on each $\mathcal{M}_t$.

## 5. The Dirichlet Problem

In this section we will note how to deduce convergence results for solutions of the Dirichlet problem from the theorems of section 1. It should be emphasized that other approaches exist, particularly variational techniques, as in [7],[8]. Further references are given in [7].

Let $D_n$, $\lambda$, $\tau_n$, T be as defined at the beginning of section 3. Let $\nu$ be probability measure that does not charge polar sets. Let U be a bounded open set in $R^d$ and let $\sigma$ denote the first exit time of U. Let g be a bounded Borel function on $\partial U$. Let $f_n$,f be defined by

(5.1) $\quad f_n(x,t) = E^x[g\circ B_\sigma \chi_{\{\tau_n>\sigma\}}]$, $f(x,t) = E^x[g\circ B_\sigma \chi_{\{T>\sigma\}}]$.

Clearly $f_n$ is the solution of the generalized Dirichlet problem on $U\cap D_n^c$, with boundary data equal to 0 on $\partial D_n$ and equal to g on $\partial U-D_n$. We now assume that

(5.2) $\quad P^\nu(T=\sigma) = 0$.

This will be true if $\lambda(\partial U)=0$. More generally, for any Borel set A in $R^d$,

$$(5.3) \qquad P^{\nu}(B_T \in A) \leq \int_A \text{Pot}\ \nu\ d\lambda.$$

This follows from the definition of T when $\lambda$ has a density, and in general by approximating $\lambda$, and using theorem 2.2.

Under assumption (5.2), it is possible to prove the analog of theorem 1.3 with the quantity of that theorem replaced by $\sigma$ [4]. Thus, if we denote the distribution of $B_\sigma$ restricted to $\{\tau_n > \sigma\}$ with repect to $P^x$ by $\pi_n(x)$, and denote the distribution of $B_\sigma$ restricted to $\{T > \sigma\}$ by $\pi(x)$, then considered as maps into the Banach space of measures with total variation norm $\pi_n(x) \to \pi(x)$ whenever $\tau_n \to T$ stably with respect to $P^x$, and hence $\pi_n$ converges to $\pi$ in $\nu$-measure, by theorem 1.4. It follows that

(5.4) $\qquad f_n$ converges to f in $\nu$-measure.

The same argument also shows that if $\tau_n \to T$ stably with respect to x for some $x \in U$ then

(5.5) $\qquad f_n(x) \to f(x)$ .

Of course, if $\lambda$ has a density h with respect to Lebesgue measure m then we can interpret f as the solution of the Dirichlet problem on U for the operator $\Delta$-h with boundary data g.

## References


1. D. J. Aldous, G. K. Eagleson, On mixing and stablity of limit theorems, Annals of Prob. 6 (1978), 325-331
2. J. R. Baxter, R. V. Chacon, Compactness of stopping times, Z. Wahr. und Verw. Gebiete 40 (1977), 169-181.
3. J. R. Baxter, R. V. Chacon, N. C. Jain, Weak limits of stopped diffusions, to appear in T. A. M. S..
4. J. R. Baxter, G. Dal Maso, Umberto Mosco, Stopping times and $\Gamma$-convergence, preprint.
5. J. R. Baxter, N. C. Jain, Asymptotic capacities for finely divided bodies and stopped diffusions, preprint.
6. D. Cioranescu, F. Murat, Un terme etrange venu d' ailleurs, in Nonlinear Partial Equations and their Applications, Volume II, edited by H. Brezis and J. L. Lions, 98- 138, Pitman, London 1982.
7. G. Dal Maso, Umberto Mosco, Wiener's criterion and $\Gamma$-convergence, preprint.
8. G. Dal Maso, Umberto Mosco, A variational Wiener criterion and energy decay estimates for relaxed Dirichlet problems, preprint.

9. J. L. Doob, Classical Potential Theory and Its Probabilistic Counterpart, Springer-Verlag, New York 1984.
10. E. Ya. Hruslov, The method of orthogonal projections and the Dirichlet problem in domains with a fine grained boundary, Math. USSR Sb. 17 (1972) 37-59.
11. M. Kac, Probabilistic methods in some problems of scattering theory, Rocky Mountain J. Math. 4 (1974) pp. 511-538.
12. G. C. Papanicolaou, S. R. S. Varadhan, Diffusions in regions with many small holes, in Stochastic Differential Systems-Filtering and Control, edited by B. Grigelionis, Lecture Notes in Control and Information Sciences 25, 190-206, Springer-Verlag, New York 1980.
13. J. Rauch, M. Taylor, Potential and scattering theory on wildly perturbed domains, J. Funct. Anal. 18 (1975), 27-59.
14. S. Weinryb, Etude asymptotique de l'image par des mesures de $R^3$ de certains ensembles aleatoires lies a la courbe Brownienne, Rapport interne no. 122 de l'Ecole Polytechnique, Feb. 1985.
15. S. Weinryb, Image par une mesure de $R^3$ de l'intersection de deux saucisses de Wiener independantes temps locaux d'intersection relatifs a cette mesure, Rapport interne de l' Ecole Polytechnique, Sept. 1985.

# Effective Equation and Renormalization for a Nonlinear Wave Problem with a Random Source.

By

Luis L. Bonilla

Departamento de Fisica Teórica,
Universidad de Sevilla, Apdo. Correos 1065 Sector Sur
Sevilla, Spain.

## 1. Introduction.

We shall explain how to derive the effective equation for the expected value of the solution of a given nonlinear random partial differential equation. Our method is a generalization of the classical smoothing for linear wave propagation in random media,[1], to nonlinear problems. We also show that the effective equation can be derived from a path integral formulation used by Parisi and Sourlas in the case of a random nonlinear Laplace equation,[2]. They represented the moment-generating functional of the solution as a functional integral over commuting and anticommuting fields. For many problems of physical interest, the meaning of the functional integral, and thus the meaning of the effective equation, is established by a limiting process which involves renormalization. As an illustration we analyze a nonlinear random wave equation, [3]. Our exposition here is based in our joint work with S. Venakides, [3], where a more detailed analysis can be found.

The following is the simplest scalar random wave equation having the same nonlinearity as that of compressional sound waves in an isotropic solid:

$$\partial^2 u(\mathbf{x},t)/\partial t^2 - c^2\Delta u(\mathbf{x},t) + \lambda|\nabla u(\mathbf{x},t)|^2 = f(\mathbf{x},t), \quad \mathbf{x}\in\mathbb{R}^3,\ t\in\mathbb{R}, \tag{1}$$

$$\langle f(\mathbf{x},t)\rangle = 0, \tag{2}$$

$$\langle f(\mathbf{x},t)f(\mathbf{x}',t')\rangle = 2F\,\delta(\mathbf{x}-\mathbf{x}')\,\delta(t-t'). \tag{3}$$

Here u represents the divergence of the displacement vector, c is the speed of the wave, $\lambda$ is a small real number, and f is a white noise source. The expected value $\langle u(\mathbf{x},t)\rangle$ obeys a nonlinear wave equation with nonlocal terms (see Section 5). The most noticeable feature of this effective equation is that the nonlocal part of its linear term is complex. This means that the scattering due to the noise source produces attenuation of the sound waves. The attenuation becomes very strong at high frequencies.

The content of this paper is as follows. We explain the classical smoothing method and our extension to nonlinear equations in Section 2. In Section 3 we show how to obtain the same effective equation from the path integral formalism. Section 4 is devoted to establishing the meaning of the functional integral by means of renormalization. In particular, we show that (1) is renormalizable in three space dimensions. Section 5 contains our results for (1), and Section 6 contains some final comments.

## 2. Smoothing method for nonlinear equations.

Let us consider the following equation for $u$:

$$Lu + \lambda N(u,u) = h + f, \qquad 0 < |\lambda| \ll 1. \tag{4}$$

Here $L$ is a linear deterministic operator, $N(.,.)$ is a deterministic symmetric bilinear form, $h$ is a given function and $f$ is a random source of known statistics (not just gaussian white noise) satisfying, without loss of generality:

$$\langle f \rangle = 0.$$

To keep the presentation simple we have avoided to consider higher nonlinearities in (4) or the case of a random medium, in which $L$ and $N$ contain noisy terms. The method explained below can be easily modified to cover those cases. Our purpose is to derive an equation for the expected value $\langle u \rangle$ of the solution of (4) in terms of known quantities such as $L$, $N$, $h$ and the moments of the noise $f$. Such an equation is called an effective equation of (4). The essence of its derivation can be explained as follows:

We first solve (4) for $u$:

$$u = U[h+f]. \tag{5}$$

We then take the expected value:

$$\langle u \rangle = \langle U[h+f] \rangle \equiv V[h]. \tag{6}$$

Here $V[h]$ is a map which depends on the moments of the noise $f$. Equation (6) is now solved for $h$ to give

$$\begin{aligned} h = V^{-1}[\langle u \rangle] &= -H + \mathbf{L}\langle u \rangle + \mathbf{N}\langle u \rangle, \quad \text{or} \\ \mathbf{L}\langle u \rangle + \mathbf{N}\langle u \rangle &= h + H. \end{aligned} \tag{7}$$

**Remark.** The amount of explicit information on the solution $u$ provided by (5), (6), and (7) decreases in the same order, since (5) gives $u$ exactly, (6) gives only $\langle u \rangle$ and (7) requires to be solved to yield $\langle u \rangle$. In practice though, (5) and (6) are derived iteratively, through schemes which are valid only over a restricted range of the independent variables of the problem. They should be thought of only as asymptotic approximations to the corresponding unknown exact solutions. The approximate effective equation, in spite of being the one which gives the least amount of information on $u$ (in fact presumably because of it) turns out, in several cases, to have the widest range of validity. This is not hard to rationalize if the calculation is performed in the following way: Assuming the exact effective equation exists and has the form (7) with $\mathbf{L}$ and $\mathbf{N}$ undetermined, we iterate it to find a series expression for $\langle u \rangle$. We find another series for $\langle u \rangle$ by iterating (5) and then taking the expected value. We determine $\mathbf{L}$ and $\mathbf{N}$ by identifying the two series. In this calculation the iterations are only used in finding $\mathbf{L}$ and $\mathbf{N}$ and not $u$ or $\langle u \rangle$. For example, in the theory of linear wave propagation in a random medium, the iterative solution does not give a good approximation of $\langle u \rangle$ for short wavelengths. On the other hand, the approximate effective equation can be solved by other means (e.g. in this case, the method of characteristics) to yield $\langle u \rangle$ with greater accuracy, [1]. Thus, the derivation of effective equations is well motivated.

The above derivation is also valid if the operation $\langle . \rangle$ is replaced by some projector $P$ with the properties $LP = PL$, $PN(P.,P.) = N(P.,P.)$ In such a case we obtain instead of (7) an equation for $Pu$ which we may call the *effective equation with respect to the projection P.* The above derivation is called the *smoothing method* if it is used to derive an effective equation with respect to $P = \langle . \rangle$. Having described the smoothing method in general we now give a practical realization of it. Let us write (4) in the following form:

$$u = \beta - \lambda L^{-1} N(u,u), \tag{8a}$$

$$\beta = L^{-1}h + L^{-1}f. \tag{8b}$$

Here $L^{-1}$ is the inverse of the operator $L$, (which we assume invertible). Upon iterating (8a) we find the solution of (4) as an asymptotic series in $\lambda$, of which we just write the first few terms:

$$u = U[h+f] = \beta - \lambda L^{-1}N(\beta,\beta) + \lambda^2 L^{-1}N(\beta, L^{-1}N(\beta,\beta)) + O(\lambda^3).$$

We now apply the expected value operation to both sides of this expression and use $\langle \beta \rangle = L^{-1}h$. The result is

$$\langle u \rangle = \langle U[h+f] \rangle \equiv V[h] = L^{-1}h - \lambda L^{-1}N(L^{-1}h, L^{-1}h) - \lambda L^{-1}\langle N(L^{-1}f, L^{-1}f)\rangle + 2\lambda^2 L^{-1}N(L^{-1}h, L^{-1}N(L^{-1}h, L^{-1}h)) + 2\lambda^2 L^{-1}N(L^{-1}h, L^{-1}\langle N(L^{-1}f, L^{-1}f)\rangle) + 4\lambda^2 L^{-1}\langle N(L^{-1}f, L^{-1}N(L^{-1}h, L^{-1}f))\rangle + O(\lambda^3). \tag{9}$$

Equation (9) is now solved for $h$ by iteration to give

$$L\langle u\rangle + \lambda N(\langle u\rangle, \langle u\rangle) - 4\lambda^2 \langle N(L^{-1}f, L^{-1}N(L^{-1}f, \langle u\rangle))\rangle - 4\lambda^3 \langle N(L^{-1}N(L^{-1}f, \langle u\rangle), L^{-1}N(L^{-1}f, \langle u\rangle))\rangle - 8\lambda^3 \langle N(L^{-1}f, L^{-1}N(\langle u\rangle, L^{-1}N(L^{-1}f, \langle u\rangle)))\rangle + O(\lambda^4) = h - \lambda\langle N(L^{-1}f, L^{-1}f)\rangle + 2\lambda^2 \langle N(L^{-1}f, L^{-1}N(L^{-1}f, L^{-1}f))\rangle + O(\lambda^3). \tag{10}$$

We have not displayed explicitly the $O(\lambda^3)$-term which does not depend on $\langle u\rangle$. Equation (10) is the approximate effective equation which can be written in the form (7) by identifying $\mathcal{L}$, $\mathcal{N}$ and $H$ as follows:

$$\mathcal{L}v = Lv - 4\lambda^2 \langle N(L^{-1}f, L^{-1}N(L^{-1}f, v))\rangle + O(\lambda^3), \tag{11a}$$

$$\mathcal{N}v = \lambda N(v,v) - 8\lambda^3 \langle N(L^{-1}f, L^{-1}N(v, L^{-1}N(L^{-1}f, v)))\rangle - 4\lambda^3 \langle N(L^{-1}N(L^{-1}f, v), L^{-1}N(L^{-1}f, v))\rangle + O(\lambda^4), \tag{11b}$$

$$H = -\lambda\langle N(L^{-1}f, L^{-1}f)\rangle + 2\lambda^2 \langle N(L^{-1}f, L^{-1}N(L^{-1}f, L^{-1}f))\rangle + O(\lambda^3). \tag{11c}$$

Before we find the approximate effective equation for our problems (1) and (2), we shall describe how to recover the effective equation from a more general formalism.

## 3. Functional integrals and effective equations.

It is convenient to work in this section with the random wave equation (1) instead of using a generic notation as in the previous section.

**A. The moment-generating functional.** We define the moment generating functional as the expected value of $\exp[\int J(x)u(x)dx]$ with respect to the distribution function of the noise $f$,

$$\Phi[J,h] = N_1 \int \exp\{\int [J(x) u_{f+h}(x) - \tfrac{1}{4F}\,(f(x))^2]\, dx\}\; Df(x), \tag{12a}$$

$$1/N_1 = \int \exp\{\tfrac{1}{4F} \int [f(x)]^2\, dx\}\; Df(x), \tag{12b}$$

$$\Phi[0,h] = 1. \tag{12c}$$

Here $u_{f+h}(x)$ is the solution of the random wave equation (1) for a given right side $f + h$. $x$ is an abbreviation for $(\mathbf{x}, t)$. The moments of $u_{f+h}$ with respect to the noise $f$ are given by the functional derivatives

$$\langle u_{f+h}(x_1) \ldots u_{f+h}(x_N) \rangle = \frac{\delta^N \Phi[J = 0,h]}{\delta J(x_1) \;\ldots\ldots\; \delta J(x_N)}\,. \tag{13}$$

Let us assume that there is an unique solution of (1) for a given $f + h$. Proceeding formally we can transform (1) as follows,

$$\Phi[J,h] = N_1 \int \exp\{\int Ju - f^2/4F\}\; \delta(\Box u + \lambda(\nabla u)^2 - f - h)\;\; \det M(x)\; Du\, Df$$

$$= N_1 \int \exp\{\int [Ju - f^2/4F \; + iv(\Box u + \lambda(\nabla u)^2 - h \; - f)]\}\, \det M(x)\; Du \;\; Dv Df.$$

(14)

We have used that the delta function is the Fourier transform of 1. M(x) is the operator $\Box + 2\lambda \nabla u(x) \cdot \nabla$, which is the functional derivative of the left side of (1) with respect to $u(x)$. The integration with respect to $f$ can be performed explicitly by completing the square in the argument of the exponential in (14):

$$\Phi[J,h] = N_2 \int \exp\{\int [Ju - ihv - Fv^2 + iv[\Box u + \lambda(\nabla u)^2]]\, dx\}\, \det M(x)\; Du\; Dv.$$

$N_2$ is a normalization factor chosen so that (12c) holds true. We write the generating functional $\Phi[J,h]$ as:

$$\Phi[J,h] = Z[J, -ih, 0, 0\;]/Z[0, -ih, 0, 0\;]. \tag{15}$$

where the functional $Z[\mathbf{J}]$, $\mathbf{J} \equiv (J_1, J_2, J_3, J_4)$, is defined by the path integral

$$Z[\mathbf{J}] = N \int \exp\{\Sigma_{1\le j\le 4}\int J_j(x)U_j(x)dx + \mathcal{A}[\mathbf{U}]\} \Pi_{1\le j\le 4} DU_j(x), \tag{16}$$

$$\mathcal{A}[\mathbf{U}] = \int \{iU_2(x)[\Box U_1(x) + \lambda(\nabla U_1(x))^2] - F(U_2(x))^2$$

$$+U_3(x)[\Box + 2\lambda\nabla U_1(x)\cdot\nabla]U_4(x)]dx, \tag{17}$$

$$Z[\mathbf{0}] = 1. \tag{18}$$

To derive (16)-(18) from our previous formula for $\Phi$, we have used the exact gaussian quadrature:

$$\det M(x) = \int \exp\{\int U_3(x)M(x)U_4(x)dx\} DU_3(x) DU_4(x).$$

Here $U_3(x)$ and $U_4(x)$ are anticommuting fields (see [4], pp. 439-442). In $Z$ (which we will also call moment-generating functional) and in the *action integral* $\mathcal{A}$, we have used the notation $U_1 = u$, $U_2 = v$. The external sources $J_3$ and $J_4$ are anticommuting fields, so that the products $J_jU_j$, j=3,4, are ordinary commuting scalar functions. The anticommuting functions have an auxiliary character and they will not appear in our final results for the effective equation. The functional integral in (16) has to be calculated by dividing the action integral in two parts: a part which is quadratic in the fields $U_j$ and a part which contains higher nonlinearities. Z becomes:

$$Z[\mathbf{J}] = N \exp\{\lambda\mathcal{A}_1[\delta/\delta J_1, \delta/\delta J_2, \delta/\delta J_3, \delta/\delta J_4]\} Z_0[\mathbf{J}], \tag{19a}$$

$$\mathcal{A}_1[\mathbf{U}] = \int\{iU_2[\nabla U_1]^2 + 2U_3\nabla U_1\cdot\nabla U_4\}dx, \tag{19b}$$

$$Z_0[\mathbf{J}] = \int \exp\{\int[iU_2\Box U_1 - FU_2^2 + U_3\Box U_4 + \Sigma_{1\le j\le 4}\int J_jU_j]dx\} \Pi_{1\le j\le 4} DU_j$$

$$= \exp\{\int[-FJ_1\Box^{-2}J_1 - iJ_1\Box^{-1}J_2 + J_3\Box^{-1}J_4]dx\}. \tag{19c}$$

Here $\Box^{-1}$ and $\Box^{-2}$ are the Green's functions of the D'Alembertian operator and of its square, respectively. The path integral (19c) is gaussian and it has been evaluated explicitly. The functional (19a) can be calculated as an asymptotic series in powers of $\lambda$ by expanding the exponential. The result can be represented graphically by means of Feynman diagrams, [4]. Let us now describe the connection with the approximate effective equation of Section 2.

**B. The vertex functional and the effective equation.** The effective equation can be written in a compact form by using functional methods. To see this, we take the functional Legendre transformation of $\ln Z[\mathbf{J}]$. We define the dual variables

$$u_i(x) = \delta \ln Z[\mathbf{J}]/\delta J_i(x), \qquad i = 1,2,3,4. \tag{20}$$

and obtain the transformed functional

$$\Gamma[\mathbf{u}] = \int \sum_{1\leq j\leq 4} J_i(x)u_i(x)dx - \ln Z[\mathbf{J}]. \tag{21}$$

In (21) the J's are obtained as the solutions of equations (20). The u's are nonrandom fields dual to the J's, not to be confused with the random solution of equation (1). $\Gamma[\mathbf{u}]$ is called the vertex functional.

As the Legendre transform is an involution, it follows that

$$\delta\Gamma[\mathbf{u}]/\delta u_j(x) = J_j(x), \qquad j = 1,2,3,4. \tag{22}$$

Let us set $J_i = 0$, $i=1,3,4$, and $J_2=-i\hbar$ in (22). We shall prove below that the solution of these equations for $j=1,3,4$ is $u_i=0$, $i=2,3,4$, for any function $u_1(x)$. On the other hand, equations (12a) and (15) imply that $u_1(x)\equiv\langle u_{f+h}(x)\rangle$, so that

$$i\, \delta\Gamma[\langle u\rangle,0,0,0]/\delta u_2(x) = h(x), \tag{23}$$

is an effective equation for $\langle u\rangle$. That it is the same one which we obtained by the smoothing method is proven by comparing the two asymptotic representations (9) and $\langle u(x)\rangle=\delta Z[0, -i\hbar, 0,0]/\delta J_1(x)$, with $Z[\mathbf{J}]$ as in (19), [3]. To derive (23) from $\langle u\rangle=\delta Z[0,-ih,0,0]/\delta J_1$ is the same as solving (9) for h in order to find (10).

**C. Symmetries.** By examining the action integral (17) we find the following symmetry relations:

$$\mathcal{A}[\mathbf{U};F,\lambda] = \mathcal{A}[U_1+aU_3+\hat{a}U_4, U_2, U_3-iU_2\hat{a}, U_4+iU_2a; F,\lambda], \tag{24}$$

$$\mathcal{A}[\mathbf{U};F,\lambda] = \mathcal{A}[\nu U_1, \nu^{-1}U_2, \nu^{\alpha}U_3, \nu^{-\alpha}U_4; \nu^2F, \nu^{-1}\lambda]. \tag{25}$$

Here $\nu>0$ and $\alpha$ are real numbers, and $a$ and $\hat{a}$ are infinitesimal parameters which are independent and anticommuting. We have displayed explicitly the dependence of the action integral on the parameters $\lambda$ and F. (24) is called a global supersymmetry; supersymmetry because it intertwines commuting and anticommuting fields, and global because $a$ and $\hat{a}$ don't depend on x. Local supersymmetries (in which $a$ and $\hat{a}$ depend on x) are behind the dimensional reduction of Parisi and Sourlas, [2], for random equations with polynomial

nonlinearities which do not involve derivatives. For (1) the gradients in the nonlinear term exclude the possibility of local supersymmetries and dimensional reduction.
By changing variables in the functional integral Z[**J**] according to $U_1 \to U_1 + aU_3 + \hat{a}U_4$, $U_2 \to U_2$, etc. (as in (24)) and using the invariance of the action integral, we find

$$Z[\mathbf{J}] = Z[J_1, J_2 + i\hat{a}J_3 - iaJ_4, J_3 + aJ_1, J_4 + \hat{a}J_1]. \tag{26}$$

Letting first $a$ and then $\hat{a}$ tend to zero in (26), we find the following *Ward identities*:

$$\int \{ iJ_4(x)\, \delta \ln Z[\mathbf{J}]/\delta J_2(x) - J_1(x)\, \delta \ln Z[\mathbf{J}]/\delta J_3(x) \} dx = 0, \tag{27a}$$

$$\int \{ iJ_3(x)\, \delta \ln Z[\mathbf{J}]/\delta J_2(x) + J_1(x)\, \delta \ln Z[\mathbf{J}]/\delta J_4(x) \} dx = 0. \tag{27b}$$

By taking functional derivatives in (27a-b) and then letting $J_i=0$, $i=1,3,4$, and $J_2=-ih$, we find that

$$\delta \ln Z[0,-ih,0,0]/\delta J_j(x) = 0, \qquad \text{for } j=2,3,4.$$

The left side of this expression is equal to $u_j$, $j=2,3,4$, the solution of equation (22). Thus the supersymmetry yields a proof that $u_j=0$, $j=2,3,4$ for $J_i=0$, $i=1,3,4$, and $J_2=-ih$, no matter what the field $u_1(x)$ is.
The scaling symmetry (25) yields the following identity:

$$\left\{ \int dx \{ J_1 \delta/\delta J_1 - J_2 \delta/\delta J_2 + \alpha [J_3 \delta/\delta J_3 - J_4 \delta/\delta J_4] \} + 2F\partial/\partial F - \lambda \partial/\partial\lambda \right\} \ln Z[\mathbf{J}; F, \lambda] = 0 .$$

We have obtained this relation in the following way: By using (25), we derive a similar relation for $\ln Z[\mathbf{J}; F, \lambda]$. Then we differentiate it with respect to $\nu$ and set $\nu=1$. The result is the previous identity.
As $\alpha$ is arbitrary, it follows that:

$$\int dx\, [J_3 \delta/\delta J_3 - J_4 \delta/\delta J_4] \ln Z[\mathbf{J}; F, \lambda] = 0, \tag{28a}$$

and thus

$$\left\{ \int dx \{ J_1 \delta/\delta J_1 - J_2 \delta/\delta J_2 \} + 2F\partial/\partial F - \lambda \partial/\partial\lambda \right\} \ln Z[\mathbf{J}; F, \lambda] = 0 . \tag{28b}$$

The homogeneity relation (28a) implies that all functional derivatives of lnZ (evaluated at $J_i=0$, i=3,4, and $J_1, J_2$ arbitrary) which contain a different number of operators $\delta/\delta J_3(x)$ and $\delta/\delta J_4(x)$ are identically zero. (28b) is a homogeneity relation for the coefficients of the expansion of the functional derivatives of lnZ in powers of $\lambda$, [3].

**Remark.** By using the same derivation as before, we can find a functional integral for $\exp(\int J(x)u(x)dx)$ (or for its average with respect to some noise), where u(x) is the unique solution of any given differential equation (random or not). The supersymmetry (24), with $\mathbf{a}\equiv\mathbf{0}$, is always present in such a functional integral, as it just says that M(x),the argument of the determinant in the path integral, is the derivative of the original equation with respect to u(x).

The supersymmetry (24) with $\mathbf{a}\not\equiv\mathbf{0}$, is a consequence of the self-adjointness of the wave operator, and thus it is not a general feature. The gaussian character of the noise and the monomial nonlinearity in (1) are the causes of the scaling symmetry (25).

## 4. Establishing the meaning of the path integral: Renormalization.

If we try to evaluate the functional Z[**J**] as indicated in the previous section we run into trouble. As we did not expose in detail the perturbation theory for Z[**J**], we shall use the formulae from the smoothing method to illustrate the problems which appear. We indicated that the same effective equation could be obtained by either means. Let us consider the $O(\lambda^2)$ correction to the linear operator $L = \Box$ in equation (10) as an example:

$$\langle N(L^{-1}f, L^{-1}N(L^{-1}f,\langle u\rangle))\rangle =$$
$$\int \nabla\Box^{-1}(x-y)\langle f(y)\cdot\nabla\Box^{-1}(x-z)\,\nabla\Box^{-1}(z-s)f(s)\rangle\cdot\nabla\langle u(z)\rangle dy\,dz\,ds$$
$$= 2F\int \nabla\Box^{-1}(x-y)\cdot\nabla\Box^{-1}(x-z)\,\nabla\Box^{-1}(z-y)\cdot\nabla\langle u(z)\rangle\,dy\,dz. \qquad (29)$$

If we find the Fourier transform of both sides of this equation, the result is:

$$-\langle u(k)\rangle\, F\, c^{-5}\mathbf{k}\cdot\int\mathbf{q}\ \ \mathbf{q}\cdot(\mathbf{k}-\mathbf{q})\,(q^2)^{-2}(k-q)^{-2}d^3\mathbf{q}\,d\Omega/(2\pi)^4. \qquad (30a)$$

Here $q^2 = \mathbf{q}^2c^2-\Omega^2$. The integral (30) diverges logarithmically for large values of $|\mathbf{q}|$. This can be visualized by regularizing the integral over $\mathbf{q}$: We integrate $|\mathbf{q}|$ only from 0 to $1/\varepsilon$, $\varepsilon \ll 1$. We can evaluate explicitly (30) by subtituting $(q^2)^{-2}(k-q)^{-2}$ in it by the Feynman identity:

$$(q^2)^{-2}(k-q)^{-2} = 2\int_0^1 ds_1 \int_0^1 ds_2\, s_1 \delta(1-s_1-s_2)\,[s_1 q^2 + s_2 (k-q)^2]^{-3}.$$

The result is

$$i\,F\,c^{-5}\,\mathbf{k}^2\{c^2\,\mathbf{k}^2/k^2 + \ln k^2/4 - 2\ln\varepsilon + 23/6\}\,\langle u(k)\rangle/12\pi^2. \qquad (30b)$$

Upon regularizing in a similar way the terms of the asymptotic series in powers of $\lambda$ for the effective equation , or more generally those for $\Gamma[\mathbf{u};\lambda]$, we find, [3], that:

1. For $\varepsilon\neq 0$, all the terms in the expansion of $\Gamma[\mathbf{u};\lambda]$ are finite for smooth fields $\mathbf{u}$.

2. As $\varepsilon\to 0$ the coefficients of infinitely many terms in the expansion diverge. At a given order in $\lambda F/c^7$ (the dimensionless expansion parameter) only four of these coefficients are independent. They appear in the expansion of

$$\delta\Gamma[\mathbf{0};\lambda]/\delta u_2,\quad \delta^2\Gamma[\mathbf{0};\lambda]/\,\delta u_1\delta u_2,\quad \delta^2\Gamma[\mathbf{0};\lambda]/\,\delta u_2\delta u_2,\quad \text{and}$$
$$\delta^3\Gamma[\mathbf{0};\lambda]/\,\delta u_1\,\delta u_1\delta u_2$$

in powers of $\lambda$. By forming appropriate linear combinations of these coefficients, we can obtain those of the divergent parts of all the other functional derivatives of $\Gamma[\mathbf{u};\lambda]$. In this situation, the moment generating functional $Z[\mathbf{J};\lambda]$ is called renormalizable:

$Z[\mathbf{J};\lambda]$ is renormalizable if:

a. Infinitely many coefficients in the expansion of $\Gamma[\mathbf{u};\lambda]$ or $Z[\mathbf{J};\lambda]$ in powers of $\lambda$ diverge as $\varepsilon\to 0$.

b. At any given order in the dimensionless expansion parameter (loop expansion), only a finite number of divergent coefficients (the same number for all orders) are independent.

3. If $\mathbf{x}\varepsilon\mathbb{R}^n$, $n<3$, only finitely many terms in the expansion diverge and $Z[\mathbf{J};\lambda]$ is called superrenormalizable. For $n>3$, there is an infinite number of divergent terms in the expansion and the number of independent coefficients grows with the order in $\lambda$. $Z[\mathbf{J};\lambda]$ is then called nonrenormalizable. At $n=3$ the parameter $\lambda$ is dimensionless when we use 1 as the unit of inverse distance. This is a general feature of renormalizable path integrals.

At this point it is convenient to go back to Equation (1) in order to determine what the origin of the divergencies in our equations is. We notice that the white noise $f(\mathbf{x},t)$ in (1) is not a meaningful mathematical object. A thorough analysis would require to reinterpret the white noise in equation (1), so that integrations which contain it make sense. Instead of doing this, we hold to the moment-generating functional as the primary object of our interest and try to find a satisfactory expression for it. This will mean finding a definition of $Z[\mathbf{J};\lambda]$ such that the expansion of this moment-generating functional does not contain divergent coefficients. That this definition makes sense from the point of view of the stochastic equation will not be investigated here.

To understand the origin of the divergencies in our expansions we go back to Equation (29), for example. According to (23), a similar term can be found in the expansion of $\Gamma[\mathbf{u};\lambda]$. We notice that, because of the delta correlated noise, the Green's functions in (29) have overlapping singularities. We thus face the problem of constructing the product of generalized functions whose singular supports overlap. By properly defining the product of generalized functions whose singular supports overlap as a generalized function, we rid of the divergencies, [5,4,3]. It turns out that defining appropriately the product of generalized functions is equivalent to modifying appropriately $Z[\mathbf{J};\lambda]$, [5]. We shall explain below how to modify the action integral in $Z[\mathbf{J};\lambda]$, and therefore the original equation (1), so that no divergent terms are present in the expansions for the new $Z[\mathbf{J};\lambda]$. This modification consists of inserting certain constants in the coefficients of (1), thereby originating the name renormalization.

In [3] we have proven that:

"Let us define the modified action integral

$$\mathcal{A}[\mathbf{U};F,c^2,\lambda] = \int (iU_2[\Box U_1+\lambda(\nabla U_1)^2]- F(U_2)^2+U_3[\Box+2\lambda\nabla U_1\cdot\nabla]U_4]dx \;+\int [(\zeta_1-1)U_2 -ic^2(\zeta_2-1)U_2\Delta U_1+ i\,\lambda(\zeta_3-1)U_2(\nabla U_1)^2 -F(\zeta_4-1)(U_2)^2 - c^2(\zeta_2-1)U_3\Delta U_4 +2\lambda(\zeta_3-1)U_3\nabla U_1\cdot\nabla U_4]dx. \tag{31}$$

The constants $\zeta_j$ have the following expansion in $\lambda$:

$$\zeta_j(\lambda,\varepsilon) = 1+\sum_{1\leq m\leq\infty} A_{j,m}(\varepsilon)\,\lambda^{2m-\delta(j,1)},\quad \delta(j,1)\equiv\delta_{j,1}. \tag{32}$$

We can choose the coefficients $A_{j,m}(\varepsilon)$ in (32) so that the perturbation theory for $Z[\mathbf{J};F,c^2,\lambda]$ does not contain divergent terms as $\varepsilon\to 0$. The coefficients in (32) diverge as $\varepsilon\to 0$".

**Remark 1**. The perturbation theory for (31) still uses (19c) as the unperturbed moment-generating functional. The action $\mathcal{A}_1[\mathbf{U};F,c^2,\lambda]$ now includes an infinite number of terms due to the constants (32). The extra terms in (31) with respect to (17) are called *counterterms*. There is a one-to-one correspondence between the counterterms in (31) and the number of independent divergencies above mentioned.

**Remark 2.** The choice of counterterms is not unique, and in fact, different expressions for the path integral result from different choices. Let us consider any two set of counterterms which eliminate the divergencies. We can obtain the expression for the renormalized functional Z corresponding to one of them from that of the other. To do this we have to add appropiate new counterterms of the same form as those in (31) to the renormalized action. We shall use as our choice of counterterms the one that just eliminates the divergent terms in the expansion of the path integral. This is called *minimal subtraction* (see next Section: compare (30b) with (33) below).

**Remark 3.** The renormalization procedure builds in a new symmetry in the renormalized functionals related to the arbitrariness in the choice of counterterms. This new symmetry is the *Renormalization Group*, which can be used to recover the entire renormalized functional and to find asymptotic expresions thereof, [4,5].

## 5. Results for Equation (1).

After renormalization, the Fourier transform of the effective equation of (1) is

$$[-\omega^2 + \mathbf{k}^2c^2(1+iF\lambda^2c^{-7}[\mathbf{k}^2c^2/k^2+23/6+\ln(k^2/4)]/12\pi^2)]\,\langle u(\mathbf{k},\omega)\rangle + \lambda\int \mathbf{q}\cdot(\mathbf{k}-\mathbf{q})\langle u(q)\rangle\,\langle u(k-q)\rangle\,dq/(2\pi)^4 + O(\lambda^3) = h(q). \tag{33}$$

The most noticeable feature of this equation is the presence of the factor i. For waves of small amplitude, A, we can ignore the nonlinearities. Let us assume that there is no external source, h=0. Then we can obtain the following dispersion relation between the wavenumber $\mathbf{k}$ and the frequency $\omega$ of a given plane wave solution $\langle u(\mathbf{x},t)\rangle = A\exp[i(\mathbf{k}.\mathbf{x}-\omega t)$,

$$-\omega^2 + \mathbf{k}^2c^2(1+iF\lambda^2c^{-7}[\mathbf{k}^2c^2/k^2+23/6+\ln(k^2/4)]/12\pi^2) = 0.$$

For $\lambda^2 \ln k^2$ small this relation yields

$$\mathbf{k}^2c^2 = \omega^2 + (1+i)\lambda(Fc^{-7}/2)^{1/2}\omega[1+29iF\lambda^2c^{-7}/6 + iF\lambda^2c^{-7}\ln\lambda\omega]. \tag{34}$$

The imaginary part of $|\mathbf{k}|$ is the attenuation coefficient. The noisy source scatters the sound so that there is an increasing attenuation of the wave as the frequency increases.

## 6. Final comments.

1. The effective equation of a given linear or nonlinear random problem can be derived equivalently by using the smoothing method or the Parisi and Sourlas formalism. In both cases one is assuming that the original random problem has an unique solution for each value of the external source. The functional formalism, though, allows for a more compact expression of the effective equation. It also gives, in principle, expressions for the higher moments of the solution of the random problem.

2. Those problems in which white noise is present may need renormalization, i.e. reinterpretation, of the moment-generating functional, at least in some space dimensions.

3. In all problems with an unique solution, we can find a global supersymmetric transformation which leaves invariant the moment-generating functional.

4. The lowest order approximation to the effective equation is the one given by the saddle-point method applied to the path integral $Z[\mathbf{J};\lambda]$, [4]. In random problems this approximation usually corresponds to ignore the noise completely. Successive approximations in this method are analogous to the Born series in wave propagation problems, in which the direct solution of the approximate effective equation yields better results. This suggests that the effective equation as given by (23) may be a more appropriate starting point for the analysis. After all there is no reason to think a priori that the effects of the noise are small.

### References.

**[1]** J. B. Keller. Wave propagation in random media. Procs. Sympos. in Appl. Math., vol. **13**, pp. 227-246. AMS, Providence, R.I., 1962.

**[2]** G. Parisi & N. Sourlas. Random magnetic fields, supersymmetry, and negative dimensions. Phys. Rev. Lett. **43**, 744-745, (1979).

**[3]** L.L. Bonilla & S. Venakides. Effective equations and renormalization. In preparation, (1985).

**[4]** C. Itzykson & J.B. Zuber. **Quantum Field Theory.** MacGraw Hill, N.Y., 1980.

**[5]** N.N. Bogoliubov & D.V. Shirkov. **Introduction to the Theory of Quantized Fields.** 3rd. edition. Wiley, N.Y., 1980.

**Acknowledgements.** I wish to thank Prof. Joseph B. Keller for his support during my stay at Stanford University when part of this work was completed. I am also indebted to Profs. Hans Weinberger and George Sell for their hospitality and support during my stay at the I.M.A..

THE SPECTRUM OF BACKSCATTER FOR PULSE REFLECTION FROM A RANDOMLY STRATIFIED MEDIUM

R. Burridge[†], G. Papanicolaou[††], and B. White[†††]

[†]Courant Institute of Mathematical Sciences, New York University
251 Mercer St., New York, N. Y. 10012
Present address: Schlumberger-Doll Research Center
Old Quarry Rd., Ridgefield, CT 06877
[††]Courant Institute of Mathematical Sciences, New York University
251 Mercer St., New York, N.Y. 10012
[†††]Exxon Research and Engineering Co., Corporate Research Science Laboratories,
Route 22 E., Annandale N.J. 08801

Abstract

We consider the problem of a plane wave pulse normally incident on a plane stratified acoustic or elastic half space with material parameters which are stationary random functions of position. For the elastic case we study SH waves, so that the problem is completely one dimensional. It is assumed that the wave length is large compared to the typical size of an inhomogeneity, and that sufficient time has elapsed for the pulse to have travelled through many inhomogeneities. More specifically, let $\epsilon$ be the ratio of the size of a typical inhomogeneity to a wavelength characteristic of the incident pulse. It is assumed that $\epsilon \ll 1$. We consider a subsection, or "time window", of the backscattered signal of width $O(1/\epsilon)$, the order of magnitude of the pulse duration, but centered at a large time $\tau/\epsilon^2$, where $\tau$ is $O(1)$. Then this section of the backscattered process is approximately stationary and Gaussian, with power spectral density $S_\tau(\omega)$ given by

$$S_\tau(\omega) = |\hat{f}(\omega)|^2 \frac{1}{\tau} \mu(\sqrt{\alpha\tau}\,\omega)$$

where $\hat{f}$ is the Fourier transform of the pulse shape, $\alpha$ is a constant which may be computed from the statistics of the random medium, and $\mu$ is a universal function, which we characterize.

## 1. Introduction

We consider the problem of a plane wave pulse normally incident on a plane-stratified acoustic or elastic half space with material parameters (density and compressibility) which are stationary random functions of position. For the elastic case we study SH waves, so that our problem is completely one dimensional. It is assumed that the wave length is large compared to the typical size of an imhomogeneity, and that sufficient time has elapsed for the pulse to have traversed many inhomogeneities. Then provided propagation distance is not too large, effective medium theory is applicable: the wave sees an effectively constant medium, and there is little backscatter. In this paper we go beyond effective medium theory to compute backscatter, and to compute it on a time scale too long for effective medium theory to be valid.

The time scale we consider is that for which the phenomenon of localization[1] occurs. More precisely, a wave travelling with the propagation speed indicated by effective medium theory will have, on our time scale, reached a certain depth. A time-harmonic wave, of frequency corresponding to the center frequency of our pulse, will be exponentially attenuated as this depth becomes large on our scale, in accord with the well-known theory of localization, *i.e.*, by the mechanism of random multiple scattering. Our theory, however, predicts the transient properties of the time varying reflected signal rather than, as is more usual in this regime, the steady state response to a time harmonic excitation.

It is readily apparent that the reflected signal is not statistically stationary in time, since it must ultimately decay to zero as the wave energy is completely reflected from the half space, as is guaranteed by localization theory. We show, however, that the statistics are slowly varying, so that the process is approximately stationary over an appropriately chosen "time window" which is not too long. Furthermore, we show that the statistics of the reflection process can be expressed in terms of a universal function, which does not depend on the statistical details of the random medium.

To explain succinctly our main result, let the time scale be chosen so that it takes a unit of time to traverse, on average, a typical inhomogeneity, when travelling with the effective propagation speed of the random medium. Let $\epsilon$ be the ratio of the size of a typical inhomogeneity to a wavelength characteristic of the incident pulse. Our fundamental assumption is that the wave is low frequency, *i.e.*, that $\epsilon$ is small ( note that we do <u>not</u> assume that fluctuations in the material parameters of the random medium are small). Then it takes $O(1/\epsilon)$ time to traverse a distance comparable to a wavelength, and it is on this time scale that the reflection process is approximately stationary. We center a time window of length $O(1/\epsilon)$ at a time $\tau/\epsilon^2$, far into the record of the reflected process. Then the power spectral density , $S_\tau(\omega)$, of this appropriately scaled section of the process is given approximately by

$$S_\tau(\omega) = |\hat{f}(\omega)|^2 \frac{1}{\tau} \mu(\sqrt{\alpha\tau}\ \omega) \tag{1}$$

where $\hat{f}$ is the Fourier transform of the pulse shape, $\alpha$ is a scale constant computed from the statistics of the random medium. and $\mu$ is a universal function of frequency, $\omega$, which can be determined by the solution of a partial differential equation given below ( equation (51) ). Thus we show universal behavior: the statistics of the medium enter only through the scale parameter $\alpha$.

Equation (1) demonstrates the slow decay of the backscatter with time, as well as a shift to lower frequencies at later times. These phenomena have been noted by other authors, *e.g.* Richards and Menke[2], who performed Monte Carlo simulations.

This paper is a condensation, with some simplification, of results in Burridge, Papanicolaou, and White[3]. The Monte Carlo simulations of Sheng, Zhang, White and Papanicolaou[4] suggest a generalization of the present theory which removes the low frequency assumption. In section 2, we formulate the problem in terms of stochastic partial differential equations, and show how the problem is reformulated in the frequency domain. This reformulation is crucial, since it results in stochastic ordinary differential equations, for which a much wider and more rigorous variety of asymptotic techniques is available. Our asymptotic method for the limit theorem of section 4 is an extension of that in Papanicolaou and Kohler[5]. First, however, the transformation of section 3 is used to reduce the dimensionality of the differential equations. In [3], it is argued in a non-rigorous way that the limit process is Gaussian, a property supported by the Monte-Carlo simulations of Sheng, Zhang, White, and Papanicolaou[6]. This fact, coupled with the power spectral density of equation (1), gives a complete and universal characterization of the canonical backscatter process.

## 2. Problem Formulation and Scaling.

Let the density, $\rho$, and the compressibility, $1/K$, of the random half-space be stationary random functions of the single depth variable z for z>0. Then the Euler equations, for pressure, p, and normal particle velocity, w, are

$$\begin{aligned} &\rho w_t + p_z = 0 \\ &\frac{1}{K} p_t + w_z = 0 \quad . \end{aligned} \tag{2}$$

We define the effective parameters $\rho_0$ and $K_0$ that are consistent with effective medium theory

$$\rho_0 = E[\rho]$$
$$K_0 = (E[1/K])^{-1} \quad . \tag{3}$$

That is, provided the propagation distance is not too large, a low frequency wave will propagate approximately as if it were in a constant medium with parameters $\rho_0, K_0$, and hence propagation speed

$$C_0 = \sqrt{K_0/\rho_0} \quad . \tag{4}$$

For simplicity, we consider a wave incident from a homogeneous half-space, $z<0$, with constant material parameters $\rho_0, K_0$, and striking the interface at $z=0$ at time $t=0$. The Green's function for this problem is then given by the initial conditions

$$\left.\begin{aligned} w &= (l_0/2)\delta(t - z/C_0) \\ p &= (l_0/2)\rho_0 C_0 \delta(t - z/C_0) \end{aligned}\right\} \quad \text{for } t<0 \tag{5}$$

where the parameter $l_0$ is a typical size of the random inhomogeneities, and the normalizer $(l_0/2)$ is chosen for later convenience. We nondimensionalize by first setting

$$\rho = \rho_0(1+\mu(z/l_0))$$
$$\frac{1}{K} = \frac{1}{K_0}(1+\nu(z/l_0)) \tag{6}$$

so that $E[\mu]=E[\nu]=0$, and then setting

$$z' = z/l_0 \qquad p' = p/(\rho_0 C_0{}^2)$$
$$t' = C_0 t/l_0 \qquad w' = w/C_0 \tag{7}$$

to obtain, after dropping primes

$$(1+\mu(z))w_t + p_z = 0$$
$$(1+\nu(z))p_t + w_z = 0 \tag{8}$$

with the initial conditions

$$\left.\begin{aligned} w&=\frac{1}{2}\delta(t-z) \\ p&=\frac{1}{2}\delta(t-z) \end{aligned}\right\} \qquad \text{for } t<0 \qquad (9)$$

We define the down-going wave, A(t,z), and the up-coming wave, B(t,z), as well as the auxilliary processes m(z) and n(z) by

$$\begin{aligned} A&=w+p \qquad & m&=(\mu+\nu)/2 \\ B&=w-p \qquad & n&=(\mu-\nu)/2 \end{aligned} \qquad (10)$$

so that A, B satisfy

$$\begin{aligned} (1+m)A_t+A_z+nB_t&=0 \qquad & A&=\delta(t-z) \text{ for } t<0 \\ (1+m)B_t-B_z+nA_t&=0 \qquad & B&=0 \text{ for } t<0 \ . \end{aligned} \qquad (11)$$

Thus the wave is completely down-going for negative times (we take z positive downward), and we are interested in the up-coming wave at z=0, t>0, *i.e*, in B(t,0) for t>0. In particular, we shall be interested in this Green´s function convolved with a pulse which is slowly-varying on this time scale.

We define the Fourier transforms

$$\begin{aligned} \hat{A}(\omega,z)&=\int_{\infty}^{\infty} e^{i\omega t} A(t,z)\, dt \\ \hat{B}(\omega,z)&=\int_{\infty}^{\infty} e^{i\omega t} B(t,z)\, dt \ . \end{aligned} \qquad (12)$$

Then $\hat{A},\hat{B}$ satisfy

$$\begin{aligned} \hat{A}_z&=i\omega(\hat{A}+m\hat{A}+n\hat{B}) \qquad & \hat{A}(\omega,0)&=1 \\ \hat{B}_z&=-i\omega(\hat{B}+n\hat{A}+m\hat{B}) \qquad & \hat{B}(\omega,L)&=0 \ . \end{aligned} \qquad (13)$$

The boundary condition for $\hat{A}$ follows from the initial condition in equation (11), and the fact that A=0 for z<0, t>0. The boundary condition for $\hat{B}$ is imposed instead of a radiation condition at z=∞; we impose instead a cutoff for the half-space at z=L, and later take limits as L→∞. In the time domain, this cutoff can have no effect for any fixed finite time as long as L is sufficiently large,because of hyperbolicity of equation (2).

We express the hypothesis that the pulse is broad compared to the size of the inhomogeneities by introducing a small parameter $\epsilon \ll 1$, and denoting the pulse shape by $f(\epsilon t)$. Then the backscatter, $g(t)$, can be obtained by multiplying the transform, $\hat{f}(\omega/\epsilon)/\epsilon$, of $f(\epsilon t)$, by the transform, $\hat{B}$, of the Green's function, and inverse transforming. We insert a factor of $1/\sqrt{\epsilon}$ to anticipate the smallness of the backscatter. Then the process of interest is

$$g(t) = \frac{1}{2\pi\sqrt{\epsilon}} \int_{\infty}^{\infty} e^{-i\omega t} \left(\frac{1}{\epsilon} \hat{f}(\epsilon\omega)\right) \hat{B}(\omega,0)\, d\omega \quad . \tag{14}$$

We next introduce the scaling for the approximately stationary sub-sections, or "time windows" of g. We consider times of order $O(1/\epsilon)$ near a central time $\tau/\epsilon^2$, where $\tau$ is $O(1)$. That is, we let $t=\tau/\epsilon^2+\sigma/\epsilon$, where $\sigma$ is $O(1)$, and consider g as a stochastic process with time parameter $\sigma$, when $\tau$ is held fixed. Thus the process within this time window is, upon substitution in (14) and a change of variables

$$\begin{aligned} g_\tau^\epsilon(\sigma) &= g(\tau/\epsilon^2+\sigma/\epsilon) \\ &= \frac{1}{2\pi\sqrt{\epsilon}} \int_{\infty}^{\infty} e^{-i\omega\tau/\epsilon - i\omega\sigma} \hat{f}(\omega)\, \hat{B}(\epsilon\omega,0)\, d\omega \quad . \end{aligned} \tag{15}$$

It is apparent from (15) that $\omega$ should be scaled by $\epsilon$ in the equations for $\hat{A},\hat{B}$. It is also convenient to remove the rotation of $\hat{A},\hat{B}$ indicated by (13). We thus define

$$\begin{aligned} &\omega' = \epsilon\omega && \\ &z = z'/\epsilon^2 && L = L'/\epsilon^2 \\ &A^\epsilon = e^{-i\omega z}\hat{A} && B^\epsilon = e^{i\omega z}\hat{B} \quad . \end{aligned} \tag{16}$$

From equations (15), (16) we then have, after dropping primes,

$$g_\tau^\epsilon(\sigma) = \frac{1}{2\pi\sqrt{\epsilon}} \int_{\infty}^{\infty} e^{-i\omega\tau/\epsilon - i\omega\sigma}\, \hat{f}(\omega)\, B^\epsilon(\omega,0)\, d\omega \tag{17}$$

where $A^\epsilon$, $B^\epsilon$ satisfy

$$\begin{aligned} A_z^\epsilon &= (i\omega/\epsilon)(m(z/\epsilon^2)\, A^\epsilon + n(z/\epsilon^2) e^{-2i\omega z/\epsilon}\, B^\epsilon) \ , && A^\epsilon(\omega,0)=1 \\ B_z^\epsilon &= (-i\omega/\epsilon)(n(z/\epsilon^2) e^{2i\omega z/\epsilon}\, A^\epsilon + m(z/\epsilon^2) B^\epsilon) \ , && B^\epsilon(\omega,L)=0 \quad . \end{aligned} \tag{18}$$

Equations (17) and (18) now give a complete formulation of the problem with our scaling. To compute the covariance of $g_\tau^\epsilon(\sigma)$, we have from (17)

$$E[g_\tau^\epsilon(\sigma)\ g_\tau^\epsilon(0)] = \frac{1}{4\pi^2\epsilon}\int_\infty^\infty d\omega_1 \int_\infty^\infty d\omega_2 \exp(-i\omega_1\sigma)\ \hat{f}(\omega_1)\ \hat{f}(\omega_2)$$
$$\bullet\ \exp(\ i(\omega_1-\omega_2\ )\tau/\epsilon)\ E[B^\epsilon(\omega_1,0)\ B^{\epsilon *}(\omega_2,0)] \quad . \tag{19}$$

We change variables in (19), $\omega=(\omega_2+\omega_1)/2$, $h=(\omega_2-\omega_1)/\epsilon$. Let

$$v^\epsilon(L,h,\omega)=E[B^\epsilon(\omega-\epsilon h/2,0)\ B^{\epsilon *}(\omega+\epsilon h/2)] \quad . \tag{20}$$

It will be shown that $v^\epsilon$ has a limit v as $\epsilon\to 0$. Then

$$E[g_\tau^\epsilon(\sigma)\ g_\tau^\epsilon(0)] = \frac{1}{4\pi^2}\int_\infty^\infty d\omega\ e^{-i\omega\sigma}\int_\infty^\infty dh\ \hat{f}(\omega-\epsilon h/2)\ \hat{f}^*(\omega+\epsilon h/2)$$
$$\bullet e^{i\omega h/2}\ e^{ih\tau} v^\epsilon(L,h,\omega) \quad .$$
$$\xrightarrow[\epsilon\to 0]{} \frac{1}{4\pi^2}\int_\infty^\infty d\omega\ e^{-i\omega\sigma}\ |\hat{f}(\omega)|^2 \int_\infty^\infty dh\ e^{-ih\tau} v(L,h,\omega) \tag{21}$$

Next, v has a limit as $L\to\infty$. From scaling properties of this limit it can be shown that

$$\int_\infty^\infty dh\ e^{ih\tau} v(L,h,\omega) \xrightarrow[L\to\infty]{} \frac{1}{\tau}\ \mu(\sqrt{\alpha\tau}\ \omega) \tag{22}$$

where $\mu$ is a universal function determined below. Equations (21) and (22) yield the power spectral density $S_\tau(\omega)$ for $g_\tau^\epsilon(\sigma)$, which is stated in equation (1). Similar calculations for the higher order moments of $g_\tau^\epsilon$ indicate that the process is Gaussian.

In section 4 we derive the limit law for $v^\epsilon$, and verify the scaling properties that imply equation (22).

## 3. Propagator Matrices

To solve the linear boundary value problem (18), we introduce the fundamental solution, or propagator matrix $Y^\epsilon(z)$, which satisfies

$$\frac{d}{dz}\ Y^\epsilon = \frac{i\omega}{\epsilon}\begin{bmatrix} m(z/\epsilon^2) & e^{-2i\omega z/\epsilon} n(z/\epsilon^2) \\ -e^{2i\omega z/\epsilon} n(z/\epsilon^2) & -m(z/\epsilon^2) \end{bmatrix} Y^\epsilon \tag{23}$$
$$Y^\epsilon(0) = \begin{bmatrix} 1 & 0 \\ 0 & 1 \end{bmatrix} \quad .$$

It is clear from symmetries in (23) that if $(a,b^*)^T$ is a vector solution of (23), then so is $(b,a^*)^T$. Furthermore, the determinant of $Y^\epsilon$ is one, since the trace of the matrix multiplying it in (23) is zero. Therefore

$$Y^\epsilon = \begin{bmatrix} a & b \\ b^* & a^* \end{bmatrix} \quad , \quad \text{with } |a|^2 - |b|^2 = 1 \quad . \tag{24}$$

In terms of a,b, we can write the solution $A^\epsilon, B^\epsilon$ of the boundary value problem (18) from the relation

$$\begin{bmatrix} A^\epsilon(\omega,L) \\ 0 \end{bmatrix} = \begin{bmatrix} a(\omega,L) & b(\omega,L) \\ b^*(\omega,L) & a^*(\omega,L) \end{bmatrix} \begin{bmatrix} 1 \\ B^\epsilon(\omega,0) \end{bmatrix} \quad . \tag{25}$$

In particular, the reflection coefficient is given by

$$B^\epsilon(\omega,0) = -\frac{b^*(\omega,L)}{a^*(\omega,L)} \quad . \tag{26}$$

We next introduce "polar coordinates" $\theta,\phi,\psi$ for complex a,b satisfying $|a|^2 - |b|^2 = 1$.

$$\begin{aligned} a &= e^{i(\phi+\psi+\pi)/2} \cosh(\theta/2) \\ b &= e^{i(\phi-\psi-\pi)/2} \sinh(\theta/2) \\ 0 \le \theta < \infty, &\quad 0 \le \phi < 4\pi, \quad 0 \le \psi < 2\pi \quad . \end{aligned} \tag{27}$$

Substitution of (27) and (24) into (23) then gives equations for $\theta,\phi,\psi$

$$\begin{aligned} \frac{d\theta}{dz} &= \frac{2\omega}{\epsilon} n(z/\epsilon^2) \sin(\phi + 2\omega z/\epsilon) \\ \frac{d\phi}{dz} &= \frac{2\omega}{\epsilon} [m(z/\epsilon^2) + n(z/\epsilon^2) \coth(\theta) \cos(\phi + 2\omega z/\epsilon)] \\ \frac{d\psi}{dz} &= -\frac{2\omega}{\epsilon} n(z/\epsilon^2) \operatorname{csch}(\theta) \cos(\phi + 2\omega z/\epsilon) \quad . \end{aligned} \tag{28}$$

From (26), (27) the reflection coeficient $B^\epsilon$ can be written as

$$B^\epsilon(\omega,0) = e^{i\psi} \tanh(\theta/2) \quad . \tag{29}$$

In view of equation (20), we need to calculate the correlation of reflection coefficients at two near-by frequencies

$$\begin{aligned} \omega_1 &= \omega - \epsilon h \\ \omega_2 &= \omega + \epsilon h \end{aligned} \quad . \tag{30}$$

Let us consider $\theta_1, \phi_1, \psi_1$ corresponding to $\omega_1$, and $\theta_2, \phi_2, \psi_2$ corresponding to $\omega_2$. We define the vector $x \epsilon \mathbb{R}^6$ by $x=(\theta_1, \phi_1, \psi_1, \theta_2, \phi_2, \psi_2)^T$. Then substitution of (30) into (28) yields an equation of the form

$$\frac{dx}{dz} = \frac{1}{\epsilon} F(z, z/\epsilon, q, x, \omega, h) + G(z, z/\epsilon, q, x, \omega, h) \tag{31}$$

where $q=q(z/\epsilon^2)$ is a vector-valued stochastic process including as two of its components the random processes $m(z/\epsilon^2)$ and $n(z/\epsilon^2)$, and the functions F,G are defined by

$$F(z,\xi,q,x,\omega,h)= \begin{bmatrix} n \sin(\phi_1 + 2\omega\xi - hz) \\ m + n \coth(\theta_1) \cos(\phi_1 + 2\omega\xi - hz) \\ -n \operatorname{csch}(\theta_1) \cos(\phi_1 + 2\omega\xi - hz) \\ n \sin(\phi_2 + 2\omega\xi + hz) \\ m + n \coth(\theta_2) \cos(\phi_2 + 2\omega\xi + hz) \\ -n \operatorname{csch}(\theta_2) \cos(\phi_2 + 2\omega\xi + hz) \end{bmatrix} \tag{32}$$

$$G(z,\xi,q,x,\omega,h) = \begin{bmatrix} -n \sin(\phi_1 + 2\omega\xi - hz) \\ -(m + n \coth(\theta_1) \cos(\phi_1 + 2\omega\xi - hz)) \\ n \operatorname{csch}(\theta_1) \cos(\phi_1 + 2\omega\xi - hz) \\ n \sin(\phi_2 + 2\omega\xi + hz) \\ m + n \coth(\theta_2) \cos(\phi_2 + 2\omega\xi + hz) \\ -n \operatorname{csch}(\theta_2) \cos(\phi_2 + 2\omega\xi + hz) \end{bmatrix} .$$

In the next section, we derive a limit for solutions of equations (31), (32) as $\epsilon \to 0$. For comparison, we may consider the single frequency $\bar{x}=(\theta,\phi,\psi) \epsilon \mathbb{R}^3$. Then from (28), $\bar{x}$ satisfies an equation of the same type as (31), but with the corresponding $\bar{G}=0$.

## 4. A Stochastic Limit

The limit for x will be derived under the hypothesis that m,n are components of a d-dimensional stochastic process q, which is ergodic and Markovian. Since the state space $\mathbb{R}^d$ is of arbitrarily high dimension, a very large class of processes can be constructed in this way, *e.g.*, any stationary Gaussian process with power spectrum that is a rational function of frequency. This hypothesis is introduced for convenience. A large literature exists on theorems for equations only slightly different than our equation (31), for which this hypothesis is unnecessary[7]. What is needed is that the processes are mixing, *i.e.*, that values of the process at two times separated by a large time lag are asymptotically independent, so that the processes are essentially unpredictable over an infinite time interval.

Let Q be the infinitesimal generator of q, and define the conditional expectation of the product of reflection coefficients at $\omega_1$, $\omega_2$ by

$$u=u^{\epsilon}(L,z,q,x,\omega,h)=E\left[e^{i(\psi_1-\psi_2)}\tanh(\theta_1/2)\ \tanh(\theta_2/2)\ \middle|\ \begin{matrix} x(z)=x \\ q(z/\epsilon^2)=q \end{matrix}\right] \tag{33}$$

for z<L. Then from equation (31), (q,x) are jointly Markovian on the state space $\mathbb{R}^{d+6}$, so that u satisfies the Kolmogorov backward equation

$$\begin{aligned} &u_z + \frac{1}{\epsilon^2} Qu + \frac{1}{\epsilon} F.\nabla_x u + G.\nabla_x u = 0 \qquad \text{for } z<L \\ &u\big|_{z=L} = e^{i(\psi_1-\psi_2)}\tanh(\theta_1/2)\ \tanh(\theta_2/2) \quad . \end{aligned} \tag{34}$$

We will solve equation (34) as $\epsilon\to 0$ by the method of multiple scales. Let

$$u = \sum_{k=0}^{\infty} \epsilon^k\, u^{(k)}(L,z,\xi,q,x,\omega,h)\big|_{\xi=z/\epsilon} \quad . \tag{35}$$

We replace $\partial_z$ in equation (34) by $\partial_z+(1/\epsilon)\partial_\xi$, substitute from equation (35), and collect powers of $\epsilon$ to obtain the following three equations.

$$Q\, u^0 = 0 \quad . \tag{36}$$

$$Q\, u^1 + (F.\nabla_x + \partial_\xi)\, u^0 = 0 \tag{37}$$

$$Q\, u^2 + (F.\nabla_x + \partial_\xi)\, u^1 + (G.\nabla_x + \partial_z)\, u^0 = 0. \tag{38}$$

Equation (36) and the assumed ergodicity of q imply that $u^0$ does not depend on q. Also because of the ergodicity of q, there exists a unique stationary distribution $\bar{P}(dq)$ which satisfies the time-invariant forward Kolmogorov equation

$$Q^*\bar{P} = 0. \tag{39}$$

where $Q^*$ is the adjoint of the operator Q. We next multiply equation (37) by $\bar{P}$ and integrate on q. The first term is zero because of (39).From (32) and the assumption that m,n have mean zero (with respect to the stationary distribution) the second term also vanishes. Thus $\partial_\xi u^0=0$, and hence $u^0$ does not depend on $\xi$.

Q is not invertible, since it has as null space all functions which do not depend on q. We can, however, define an inverse on the subspace of functions which have $\bar{P}$ mean zero, since, by equation (39), this amounts to the Fredholm condition for invertibility of Q. Accordingly, we define the inverse T of -Q to also have mean zero . It is given by

$$T = -Q^{-1} = \int_0^\infty ds\ e^{sQ} \tag{40}$$

That is, if w is a function of q

$$Tw(q_0) = \int_0^\infty ds\ E\left[w(q(s))\ \middle|\ q(0)=q_0\right] \tag{41}$$

In terms of T the solution of (37) may be written

$$u^1=T\ (F.\nabla_x + \partial_\xi\ )\ u^0 + u^{1,0} \tag{42}$$

where $u^{1,0}$ does not depend on q.

We define the averaging operators

$$< \bullet >_q = \int \bullet\ \bar{P}(dq) \tag{43}$$

$$< \bullet >_\xi = \lim_{\xi\to\infty} \frac{1}{\xi} \int_0^\xi \bullet\ d\xi \tag{44}$$

Then insertion of (42) into (38), and averaging of the resulting equations via (43) and (44) yields, after using that F and G have q-average zero, and that $u^0, u^{1,0}$ are bounded functions of $\xi$, that

$$u_z^0 + \mathcal{L}\ u^0 = 0 \tag{45}$$

where the operator $\mathcal{L}$ is given by

$$\mathcal{L} = \langle \langle F.\nabla_x \; T \; F.\nabla_x \rangle_q \rangle_\xi \qquad (46)$$

We suppress most of the arguments of F to write $F=F(q,\xi,x)$. Then from (41), (46) $\mathcal{L}$ may be written as

$$\mathcal{L} = \int_0^\infty ds \, \langle E[\, F(q(0),\xi,x).\nabla_x F(q(s),\xi,x).\nabla_x] \rangle_\xi \qquad (47)$$

where q(0) is assumed to have its stationary distribution, $\bar{P}$. A straightforward calculation now yields a lengthy expression for $\mathcal{L}$. The expression can be simplified somewhat by changing to sum and difference variables

$$\begin{aligned} \phi &= \phi_2 - \phi_1 & \hat{\phi} &= (\phi_2 + \phi_1)/2 \\ \psi &= \psi_2 - \psi_1 & \hat{\psi} &= (\psi_2 + \psi_1)/2 \end{aligned} \qquad (48)$$

It can then be seen that the solution of equation (45) does not depend on $\hat{\phi},\hat{\psi}$. The equation can be further simplified by defining

$$z' = L - z \qquad \phi' = \phi + 2hz \qquad (49)$$

Let

$$\alpha = 2 \int_0^\infty E[n(0)\; n(s)]\, ds \qquad (50)$$

Then after dropping primes in (49) we obtain

$$u^0_z = 2hu^0_\phi + \alpha\omega^2\{\coth(\theta_1)\, u^0_{\theta_1} + \coth(\theta_2)\, u^0_{\theta_2}\}$$

$$+\alpha\omega^2\{u^0_{\theta_1\theta_1} + 2\cos(\phi)\, u^0_{\theta_1\theta_2} + u^0_{\theta_2\theta_2}$$

$$+[\coth^2(\theta_1) - 2\coth(\theta_1)\coth(\theta_2)\cos(\phi) + \coth^2(\theta_2)]\, u^0_{\phi\phi}$$

$$+[\mathrm{csch}^2(\theta_1) - 2\,\mathrm{csch}(\theta_1)\,\mathrm{csch}(\theta_2)\cos(\phi) + \mathrm{csch}^2(\theta_2)]\, u^0_{\psi\psi} \tag{51}$$

$$-2\sin(\phi)\coth(\theta_1)\, u^0_{\theta_2\phi} - 2\sin(\phi)\coth(\theta_2)\, u^0_{\theta_1\phi}$$

$$+2\sin(\phi)\,\mathrm{csch}(\theta_1)\, u^0_{\theta_2\psi} + 2\sin(\phi)\,\mathrm{csch}(\theta_2)\, u^0_{\theta_1\psi}$$

$$+2[-\coth(\theta_1)\,\mathrm{csch}(\theta_1) - \coth(\theta_2)\,\mathrm{csch}(\theta_2)$$

$$+\coth(\theta_1)\,\mathrm{csch}(\theta_2)\cos(\phi) + \coth(\theta_2)\,\mathrm{csch}(\theta_1)\cos(\phi)]\, u^0_{\phi\psi}\}$$

$$u^0\Big|_{z=0} = e^{i\psi}\tanh(\theta_1/2)\tanh(\theta_2/2)$$

We need the limit of $u^0$ for large z

$$\tilde{u}(\theta_1,\theta_2,\phi,\psi,h,\omega) = \lim_{z\to\infty} u^0 \tag{52}$$

In particular, $\tilde{u}(0,0,2hL,0,h,\omega)$ has the same limit as $L\to\infty$ as $v(L,h,\omega)$, which appears in equation (21). Using periodicity of equation (52) in $\phi$, we alternatively consider the limit of $\tilde{u}(0,0,0,0,h,\omega)$ by letting $L\to\infty$ in such a way that $2Lh=0$ mod $2\pi$. We next derive the scaling law claimed in equation(22).

It is readily apparent that with the change of variables $z'=\alpha\omega^2 z, h'=h/(\alpha\omega^2)$, equation (51) is transformed into itself, but with $\alpha\omega^2=1$. Therefore, after letting $z\to\infty$ in (51), we must have $\tilde{u}(0,0,0,0,h,\omega)$ a function of $h/(\alpha\omega^2)$ alone, and this function is universal, since the only dependence on m,n is through $\alpha$, and this factor is taken care of by the scaling. Thus the Fourier transform indicated in equation (22) is given as $L\to\infty$ by

$$\int_{-\infty}^{\infty} dh\, e^{ih\tau}\, \tilde{u}(0,0,0,0,h,\omega) = \int_{-\infty}^{\infty} dh\, e^{ih\tau}\, \bar{u}(h/(\alpha\omega^2))$$

$$= \frac{1}{\tau}\, \alpha\omega^2\tau \int_{-\infty}^{\infty} dh\, e^{i\alpha\omega^2\tau\, h}\, \bar{u}(h) \tag{53}$$

$$= \frac{1}{\tau}\, \mu(\sqrt{\alpha\tau}\,\omega) \quad .$$

where $\mu$ is a universal function.

The method used to derive the limit equation (51) may be applied to quantities depending only on a single frequency, $\bar{x}=(\theta,\phi,\psi)\epsilon\mathbb{R}^3$, as mentioned at the end of section 3. The perturbation scheme is identical in the abstract to that given here, and leads to an expression of the form (45), (47), but with F replaced by $\bar{F}\epsilon\mathbb{R}^3$ given in the right hand side of equation (28). In particular, the mean power reflected by a time harmonic wave of frequency $\omega$ from a slab of (scaled) length z is $E[\,|\,B^{\epsilon}(\omega,0)\,|^2] = E[\tanh^2(\theta/2)]$, and can be obtained from the following differential equation

$$\begin{aligned} \hat{u}_z &= \omega^2\alpha\, (\,\hat{u}_{\theta\theta} + \coth(\theta)\,\hat{u}_\theta + \operatorname{csch}^2(\theta)\,\hat{u}_{\psi\psi}) \\ \hat{u}\big|_{z=0} &= \tanh^2(\theta/2) \end{aligned} \tag{54}$$

Equation (54) was first obtained in [4], in the study of monochromatic waves in the frequency domain, and for a somewhat different scaling.Its solution is given there as

$$\hat{u} = 1 - 2\pi \int_0^{\infty} e^{-\alpha\omega^2 z(t^2+1/4)}\, \frac{t\,\sinh(\pi t)}{\cosh^2(\pi t)}\, dt \quad . \tag{55}$$

For large z, $\hat{u}$ is asymptotically

$$\hat{u} \sim 1 - \frac{\pi^{3/2}}{\sqrt{\alpha z}\,\omega}\, e^{-\alpha\,\omega^2 z/4} \quad , \; z\to\infty \quad . \tag{56}$$

Thus the localization length, $l_{loc}$ defined by $\log(1-\hat{u}) \sim -z/l_{loc}$ is given by

$$l_{loc} = \frac{4}{\alpha\omega^2} \qquad (57)$$

Equation (57) agrees with the well-known result that for low frequencies the localization length is inversely proportional to $\omega^2$, and shows that our results correctly account for the effects of very many multiple scatterings.

## References

1. E. Abrahams, P.W. Anderson, D. C. Licciardello, and T. V. Ramakrishnan, *Phys. Rev. Lett.* 42, 673 (1979).

2. Richards, P. G., and Menke, W., The Apparent Attenuation of a Scattering Medium, *Bull. Seism. Soc. Am.*, 73, 1005 (1983).

3. Burridge, R., Papanicolaou, G., and White, B., Statistics for Pulse Reflection from a Randomly Layered Medium, *S.I.A.M. J. Appl. Math.*, to appear.

4. Sheng, P., Zhang, Z. Q., White, B., and Papanicolaou, G., Multiple Scattering Noise in One-D: Universality Through Localization Length Scaling, *Phys. Rev. Lett*, submitted.

5. Kohler, W., and Papanicolaou, G., Power Statistics for Wave Propagation in One Dimension and Comparison with Radiative Transport Theory, *J. Math. Phys.*, 14, 1733 (1973); and 15 2186 (1974).

6. Sheng, p., Zhang, Z. Q., White, B., and Papanicolaou, G., unpublished.

7. Papanicolaou, G., Asymptotic Analysis of Stochastic Equations, In *M. A. A. Studies in Mathematics, vol. 18, Studies in Probability*, M. Rosenblatt ed., Math. Assoc. Am., 111 (1978).

EXERCISES FOR A BOOK ON RANDOM POTENTIALS

René Carmona*

Department of Mathematics
University of California at Irvine
Irvine, California 92717

**Abstract**

We first review known results on the spectral properties of one dimensional random Schrödinger operators and of their perturbations by deterministic potentials. We then give some new results in the form of exercises and we then apply them to the case of multidimensional spherically symmetric random potentials.

## I. Introduction

The purpose of these notes is to investigate the spectral properties of the self-adjoints operators:

$$H(\omega) = -\Delta + V(x,\omega) \tag{I.1}$$

on the Hilbert space $L^2(R^n,dx)$ when

$$V(x,\omega) = v(|x|,\omega) \tag{I.2}$$

where $\{v(r,\omega);r > 0,\omega \in \Omega\}$ is a real valued stationary ergodic stochastic process on some complete probability space $(\Omega,\mathcal{F},P)$.

Let us first recall how one can reduce the study of these Schrödinger operators with spherically symmetric potentials to the study of operators on the half-line $R_+ = [0,\infty)$. See for example the Appendix of Section X.1 of [13]. This procedure is deterministic and is used for $\omega \in \Omega$ fixed. So, we skip $\omega$ from our notations for a short while.

Let us recall that

* Partially supported by NSF DMS 850-3695

$$L^2(R^n,dx) = L^2(R_+,r^{n-1}dr) \otimes L^2(S^{n-1},d\sigma)$$

$$= \bigoplus_{k=0}^{\infty} L^2(R_+,r^{n-1}dr) \otimes L_k \qquad (I.3)$$

where $S^{n-1}$ denotes the unit sphere in $R^n$ and $d\sigma$ its normalized surface area measure, and where $L_k$ denotes the k-th eigenspace of the Laplace-Beltrami operators $\Delta_\sigma$ on $L^2(S^{n-1},d\sigma)$. We will denote by $-\kappa_k$ the corresponding eigenvalue. If $f(x) = \phi(r)\psi(\sigma)$ with $r = |x|$ and $\sigma = |x|^{-1}x$, we have:

$$[Hf](x) = [-\frac{d^2}{dr^2} - \frac{n-1}{r}\frac{d}{dr} + v(r)]\phi(r)\psi(\sigma) - \frac{1}{r^2}\phi(r)\Delta_\sigma\psi(\sigma)$$

and

$$= [-\frac{d^2}{dr^2} - \frac{n-1}{r}\frac{d}{dr} + v(r) + \frac{\kappa_k}{r^2}]\phi(r)\psi(\sigma)$$

if $\psi \in L_k$. Hence, we want to investigate the spectral properties of the formally self-adjoint operators:

$$H_k = -\frac{d^2}{dr^2} - \frac{n-1}{r}\frac{d}{dr} + v(r) + \frac{\kappa_k}{r^2} \qquad (I.4)$$

on $L^2(R_+,r^{n-1}dr)$. These operators are defined on the domain $C_c^\infty((0,\infty))$ of $C^\infty$-functions with compact supports contained in $(0,\infty)$. We will discuss their essential self-adjointness and the possible need for boundary conditions in Section V below.

$$[U\phi](r) = r^{(n-1)/2}\phi(r) \qquad (1.5)$$

defines a unitary operator from $L^2(R_+,r^{n-1}dr)$ onto $L^2(R_+,dr)$ which transforms the operator $H_k$ into the operator $\tilde{H}_k$ given by:

$$\tilde{H}_k = UH_kU^{-1} = -\frac{d^2}{dr^2} + [v(r) + (\frac{(n-1)(n-3)}{4} + \kappa_k)\frac{1}{r^2}]. \qquad (I.6)$$

Our original problem appears now as a particular case of the following more general one: what are the spectral properties of the random self-adjoint operators

$$\tilde{H}(\omega) = -\frac{d^2}{dt^2} + q(t,\omega)$$

where

$$q(t,\omega) = q_1(t,\omega) + p(t)$$

for some stationary ergodic stochastic process $\{q_1(t,\omega);\ t > 0, \omega \in \Omega\}$ and some (non-random) perturbation $p(t)$.

We switched to the variable $t$ instead of $r$ to follow the usual notations of the theory of stochastic processes and of the theory of ordinary differential equations.

We now review what is known on this subject. Most of the results we quote deal with the case of the real line $R$ instead of the half-line $R_+$.

In the lattice case, when the second derivative is replaced by the usual second order difference operator and when $\{q(t);t \in Z\}$ is an i.i.d. sequence with a smooth bounded density, it is shown in [7] that the spectrum is pure point with exponentially decaying eigenfunctions with probability one, irrespective of the deterministic perturbation $p(t)$.

This stability does not hold in the continuous case. Indeed, if $q_1(t,\omega)$ is bounded and smooth and if $p(t)$ behaves like $-t^\alpha$ for some $\alpha \in [1,2]$ when $t \to +\infty$, then the spectrum of $\tilde{H}(\omega)$ is the whole real line and it is purely absolutely continuous. See [1] or [4]. This result is deterministic in the sense that it is proven with $\omega \in \Omega$ fixed.

On the other hand, periodic perturbations, like in the lattice case, preserve the exponential localization (i.e., pure point spectrum with exponentially decaying eigenfunctions). See [2]. We will show that the same result is true if the perturbation $p(t)$ tends to zero when $t$ tends to infinity. See Section IV below. We believe that the same result should be true for all <u>bounded nonrandom perturbations</u> $p(t)$ but we do not know how to prove such a result.

Sections II and III contain the notations and preparatory technical estimates in the deterministic and random situations respectively. We prove the perturbation result in section IV and we come back to the spherically symmetric multidimensional case in Section V.

## II. Deterministic Preliminaries

Whenever $q(t)$ is a (random in the following sections) locally integrable function we will be concerned with solving the eigenvalue equation:

$$-y''(t) + q(t)y(t) = \lambda y(t) \tag{II.1}$$

which we will rewrite as a first order differential system:

$$Y'(t) = A_\lambda(t)Y(t) \tag{II.2}$$

with

$$A_\lambda(t) = \begin{bmatrix} 0 & 1 \\ q(t)-\lambda & 0 \end{bmatrix} .$$

The solutions of (II.2) are constructed with the propagator $\{U_\lambda(t,s); t,s \in R\}$ which is the flow of $2\times2$ unimodular matrices satisfying:

$$U_\lambda(t,s)Y(s) = Y(t).$$

Setting

$$Y(t) = \begin{bmatrix} y(t) \\ y'(t) \end{bmatrix}$$

and

$$y(t) = r(t)\sin\theta(t)$$
$$y'(t) = r(t)\cos\theta(t)$$

with $r(t) = |Y(t)| = [|y(t)|^2 + |y'(t)|^2]^{1/2}$, the function $y(t)$ is a solution of (II.1) if and only if the functions $r(t)$ and $\theta(t)$ are solutions of the first order nonlinear system:

$$\theta'(t) = \cos^2\theta(t) + [\lambda - q(t)]\sin^2\theta(t) \qquad \text{(II.3)}$$

$$r'(t) = -\tfrac{1}{2}\, r(t)[\lambda - q(t) - 1]\sin 2\theta(t) \qquad \text{(II.4)}$$

They are called the phase and the amplitude of the solution $y(t)$. Note that equation (II.4) can be integrated in the form

$$r(t) = r(s)\exp[-\tfrac{1}{2}\int_s^t [\lambda - q(u) - 1]\sin 2\theta(u)du] \qquad \text{(II.5)}$$

provided the phase function $\theta(t)$ is known. So the crucial problem is to solve (II.3). We will denote by $\theta(t,a,\alpha,\lambda)$ the value of the (unique) solution of (II.3) which is equal to $\alpha$ when $t = a$.

Finally, we will use the notation $A\alpha$ for $A\begin{bmatrix}\sin\alpha\\ \cos\alpha\end{bmatrix}$ whenever $A$ is a $2\times 2$ matrix and $\alpha$ a real number. We are now ready to tackle our first exercise.

**Problem II.1**:

Let $q(t) = F(t) + p(t)$ where $F(t)$ is a bounded measurable function and $p(t)$ a locally integrable function satisfying

$$\lim_{t\to\infty} p(t) = 0 \qquad \text{(II.6)}$$

If $\theta(t,a,\alpha,\lambda)$ is as above and if $\theta^{(0)}(t,a,\alpha,\lambda)$ is defined similarly with $q(t) = F(t)$ (i.e., $p(t) \equiv 0$), show that for each bounded interval $\Lambda$ and each $s' > 0$ we have:

$$\lim_{s_0\to\infty}\ \sup_{s\geq s_0}\ \sup_{t\in[s,s+s'],\alpha\in[0,\pi],\lambda\in\Lambda} |\theta(t,s,\alpha,\lambda) - \theta^{(0)}(t,s,\alpha,\lambda\varepsilon)| = 0 \qquad \text{(II.7)}$$

**Solution**:

Equation (II.3) is of the form $\theta'(t) = \Phi_\lambda(t,\theta)$ and $\Phi_\lambda$ satisfies a Lipchitz condition in $\theta$ uniformly in $t$ (at least if $t$ is large enough). Consequently the solutions $\theta(t,a,\alpha,\lambda)$ and $\theta^{(0)}(t,a,\alpha,\lambda)$ are given by the Picard's iterative procedures (see [6]):

$$\theta_0(t) = \alpha$$

$$\theta_{j+1}(t) = \alpha + \int_s^t \Phi_\lambda(u,\theta_j(u))du \qquad j \geq 1,$$

and:

$$\theta^{(0)}(t) = \alpha$$

$$\theta_{j+1}^{(0)} = \alpha + \int_s^t \Phi_\lambda^{(0)}(u,\theta_j^{(0)}(u))du \qquad j \geq 1.$$

where $\Phi_\lambda^{(0)}(t,\theta)$ denotes the right hand side of (II.3) with $q(t) = F(t)$. Now:

$$|\theta_{j+1}(t) - \theta_{j+1}^{(0)}(t)| \leq \int_s^t |\Phi_\lambda(u,\theta_j(u)) - \Phi_\lambda^{(0)}(u,\theta_j^{(0)}(u))|du$$

$$\leq c(s)|t - s| + \int_s^t |\Phi_\lambda^{(0)}(u,\theta_j(u)) - \Phi_\lambda^{(0)}(u,\theta_j^{(0)}(u))|du$$

with

$$c(s) = \sup |p(u)| \tag{II.8}$$

$$\leq c(s)|t - s| + k|\int_s^t|\theta_j(u) - \theta_j^{(0)}(u)|du$$

where $k$ is the Lipschitz constant of $\Phi_\lambda^{(0)}$.

This last relation implies:

$$|\theta_{j+1}(t) - \theta_{j+1}^{(0)}(t)| \leq \frac{c(s)}{k}[e^{k(t-s)}-1]$$

and this gives the desired result because:

$$|\theta(t,s,\alpha,\lambda) - \theta^{(0)}(t,s,\alpha,\lambda)| = \lim_{j\to\infty} |\theta_{j+1}(t) - \theta_{j+1}^{(0)}(t)|,$$

the definition (II.8) of $c(s)$, and our assumption (II.6) on the function $p(t)$.

$\square$

Problem II.2.

Let $q(t)$ be an integrable function on a closed bounded interval $I$ containing 0. For each $\alpha \varepsilon [0,\pi)$ and $\lambda$ in a closed bounded interval $\Lambda$ show that the function:

$$t \to \theta(0,-t,\alpha,\lambda)$$

is differentiable and show that, for each $\varepsilon > 0$ there is a finite constant $c(\varepsilon) > 0$ such that:

$$\int_a^b \Psi(\theta(0,-t,\alpha,\lambda))dt \leq c(\varepsilon) \int \Psi(\theta)d\theta \tag{II.9}$$

for all $a$ and $b$ in $I$ such that $\lambda > q(-t) + \varepsilon$ for all $t \varepsilon [a,b]$ and for each nonnegative continuous function $\psi$ with compact support.

Solution:

The matrix valued function $W(t) = U_\lambda(0,-t)$ is the unique solution of the matrix valued ordinary differential equation

$$\begin{cases} \dfrac{dW(t)}{dt} = W(t)A_\lambda(-t) \\ \\ W(0) = I \end{cases}$$

Consequently, using the notations $\bar{e}_1$ and $\bar{e}_2$ for the vectors $\begin{bmatrix}1\\0\end{bmatrix}$ and $\begin{bmatrix}0\\1\end{bmatrix}$ of $R^2$, we have:

$$\begin{aligned} \frac{d}{dt}\theta(0,-t,\alpha,\lambda) &= \frac{d}{dt}\tan^{-1}\frac{<U_\lambda(0,-t)\alpha,e_1>}{<U_\lambda(0,t)\alpha,e_2>} \\ &= \frac{\det\{A_\lambda(-t)\alpha,\alpha\}}{|U_\lambda(0,-t)\alpha|^2} \\ &= (\cos^2\alpha + [\lambda - q(-t)]\sin^2\alpha)|U_\lambda(0,-t)\alpha|^2 \end{aligned} \tag{II.10}$$

Note that $|U_\lambda(0,-t)\alpha|^{-2}$ can be bounded above and below away from zero uniformly in $(t,\alpha,\lambda)$ because $|U_\lambda(0,-t)\alpha|^2$ is jointly continuous and never vanishes. Consequently, if $[a,b]$ is any subinterval of $I$ on which $\lambda - q(-t) > \varepsilon$, we have

$$\frac{d}{dt}\,\theta(0,-t,\alpha,\varepsilon)) \geq (\cos^2\alpha + \varepsilon \sin^2\alpha)|U_\lambda(0,-t)\alpha|^{-2} \geq c'(\varepsilon)$$

for some constant $c'(\varepsilon) > 0$ independent of $\alpha, t$ and $\lambda$. We can perform a change of variables in the left hand side of (II.9) and we obtain (II.9) with $c(\varepsilon)$ equal to $\frac{1}{c'(\varepsilon)}$ times the number of time $\theta$ wraps around the torus when $t$ runs through the interval $[a,b]$. This number is bounded independently of $t$, $\alpha$ and $\lambda$ since $\theta(0,I,S^1,\Lambda)$ is bounded. □

## III. Probabilistic Background

We first introduce the notations which we will use throughout this section:

- $p(t)$ will be a non random integrable function which tends to $0$ as $t \to \infty$
- $\{x_t;\ t \geq 0\}$ is the stationary process of Brownian motion of a compact connected Riemannian manifold $X$ and $F\colon X \to \mathbf{R}$ is a Morse function such that $\inf F = 0$ and $\sup F = 1$.

For each $\omega$ in the probability space $\Omega$ on which the Brownian motion is defined we will consider the differential equation (II.3) with $q(t,\omega) = F(x_t(\omega)) + p(t)$ and we will denote its solution by $\theta_\lambda(t)$ or $\theta_\lambda(t,s,\alpha,\lambda)$ depending on whether or not we want to emphasize the dependence on the initial condition.

For each $\lambda$, $\{(x_t,\theta_\lambda(t));\ t \geq 0\}$ is a Markov process which is not homogeneous if the function $p(t)$ is not identically $0$.

We will use the notation $dx$ for the normalized Riemannian volume element of $X$, the notations $P_{s,x}$ and $E_{s,x}$ for the probability and expectation for the Brownian motion conditioned to be at $x$ at time $s$, and we will restrict the notations $P$ and $E$ to the case of the stationary Brownian motion (i.e.,

$$P = \int_X P_{0,x}(dx).$$

We will use the following notation for the transition operators of the Markov processes $(x_t, \theta_\lambda(t))$

$$[P^{(\lambda)} f](x,\theta) = E_{\lambda,s,(x,\theta)}\{f(x_t, \theta(t))\}$$
$$= E_{s,x}\{f(x_t, \theta(t,s,\theta,\lambda\varepsilon))\}$$

for every function $f$ on $X \times S^1$ for which the last expectation makes sense.

In the particular case $p(t) \equiv 0$, the transition operators actually depend on $t - s$, they form a semigroup, and the infinitesimal generator is easily seen to coincide with

$$\tfrac{1}{2}\Delta_x + [\cos^2\theta - (F(x) - \lambda)\sin^2\theta]\frac{\partial}{\partial\theta} \tag{III.1}$$

on smooth functions. This second order partial differential operator is degenerate but, it is not difficult to see that it is hypoelliptic and Hormander's theorem (see [9] or [10]) implies the existence of a smooth kernel $p_\lambda^{(0)}(t,(x,\theta),x',\theta'))$ for the operators $P_{t-s}^{(\lambda)}$.

For $\lambda > 0$, the process $(x_t, \theta_\lambda^{(0)}(t))$ is ergodic (see [8]) and possesses a unique invariant measure $\pi_\lambda^{(0)}(dx,d\theta) = \pi_\lambda^{(0)}(x,\theta)dxd\theta$. The compactness of the state space and standard arguments give:

$$\lim_{t\to\infty} \sup_{(x,\theta)\in X\times S'} |p_\lambda^{(0)}(t,(x,\theta),(x',\theta')) - \pi_\lambda^{(0)}(x',\theta')| = 0. \tag{III.2}$$

Finally we notice that all these properties are actually uniform in $\lambda$ restricted to compact intervals of $(0,\infty)$.

**Problem III.1:**

1) Show the existence of conditional probabilities $P_{\lambda,s,(x,\theta)}^{t,(x',\theta)}$ that the process $(x_t, \theta_\lambda(t))$ is in the state $(x',\theta')$ at time $t$ knowing that is was in the state $(x,\theta)$ at time $s < t$.

2) <u>Show that, for each fixed</u> $\lambda$ <u>and</u> $s < t$, <u>the map</u>:

$$((x,\theta),(x',\theta')) \to P^{t,(x',\theta')}_{\lambda,s,(x,\theta)}$$

<u>is continuous for the weak topology on the space of probability measures on path space</u>.

3) <u>If</u> $\lambda$ <u>and</u> $s < t$ <u>are fixed, show that for each continuous function</u> $V: [s,t] \times X \times S^1 \to \mathbf{R}$ <u>the function</u>

$$((x,\theta),(x',\theta')) \to \mathbf{E}^{t,(x',\theta')}\{e^{-\int_s^t V(u,X_u,\theta(u))du}\}$$

<u>is jointly continuous</u>.

<u>Solution</u>:

We first consider the case $p(t) \equiv 0$. In this case we notice that the probability measures

$$p_\lambda^{(0)}(t_1 - s,(x,\theta),(x_1,\theta_1))p_\lambda^{(0)}(t_2 - t_1,(x_1,\theta_1),(x_2,\theta_2))$$

$$\dots p_\lambda^{(0)}(t - t_n,(x_n\theta_n),(x',\theta'))dx_1d\theta_1dx_2d\theta_2 - dx_nd\theta_n \tag{III.3}$$

form a consistent family and give the desired conditional probabilities $P^{t(x',\theta)}_{\lambda,\lambda,(x_1\theta)}$ on path space via the usual Kolmegorov theorem. The continuity property 2) is equivalent to the continuity of

$$((x,\theta),(x',\theta')) \to E^{t,(x',\theta)}_{\lambda,s,(x,\theta)}\{\Phi\} \tag{III.4}$$

for each bounded continuous function on path space. In order to check this we need only to consider cylindrical function $\Phi$ of the form

$$\Phi = \phi_1(X_{t_1},\theta(t_1)) \dots \phi_n(X_{t_n},\theta(t_n))$$

and the continuity of (III.4) is then an immediate consequence of the smoothness of the transition densities $p_\lambda^{(0)}(t,(x,\theta),(x',\theta'))$ and the construction of

$p^{t,(x',\theta')}_{\lambda,s,(x_1\theta)}$ from (III.3). For $s,t$ and $V$ fixed, the function

$$w \to e^{-\int_s^t V(u,X_u,\theta(u))du}$$

is a bounded continuous function on path space and 3) follows trivially 2). To handle the general case when $p(t)$ is not identically $0$ we notice simply that it introduces a perturbation $p(t)\frac{\partial}{\partial\theta}$ for the infinitesimal generator (III.1) and this can easily be handled with a Cameron Martin formula type argument (see [14]). □

The propagators $U_\lambda(t,s)$ introduced in Section II are now random. For each real numbers $\lambda$ and $\delta$ and for each $s$ and $t$ such that $0 < s < t$ we set:

$$[P^{(\lambda,\delta)}f](x,\theta) = E_{\lambda,s,(x,\theta)}\{f(x_t,\theta(t))|U_\lambda(t,s)\theta|^\delta\}$$

$$= E_{s,x}\{f(x_t,\theta(t,s,\theta,\lambda))|U_\lambda(t,s)\theta|^\delta\} \qquad \text{(III.5)}$$

for each function $f$ on $X \times S^1$. As before, we used the notation $E_{s,x}$ **for the expectation over the Brownian path starting from $x$ at time $s$.**

<u>Problem II.3</u>:

<u>Each of the above operators has a bounded jointly continuous kernel. In particular, they map boundedly</u> $L^1(dx,d\theta)$ <u>into</u> $L^\infty(dx,d\theta)$, <u>they are compact on the Banach space</u> $C(X\times S^1)$ <u>of continuous functions on the compact space</u> $X \times S^1$, <u>and they are Hilbert-Schmidt on</u> $L^2(dx,d\theta)$.

<u>Solution</u>:

Taking conditional expectations in (III.5) gives the kernel:

$$E^{t,(x',\theta')}_{\lambda,s,(x,\theta)}\{|U_\lambda(t,s)\theta|^\delta\}.$$

According to the discussion in Section II (recall formula (II.5)) we have:

$$|U_\lambda(t,s)\theta|^\delta = e^{-\frac{\delta}{2}\int_s [\lambda-q(u)-1]\sin 2\theta(u,s,\theta,\lambda)du}$$

so that we can apply the result 3) of Problem III.1 to the function $\delta V_\lambda(u,x,\theta)$ where:

$$V_\lambda(u,x,\theta) = \tfrac{1}{2}[\lambda - F(x) - p(u) - 1]\sin 2\theta.$$

The rest is plain.

Problem III.3:

Under the above conditions (actually, $p(t)$ locally integrable is enough) show that for each compact $\Lambda$ in $\mathbb{R}$,

$$c = \sup_{x\in X,\ t\geq 1,\ \lambda\in\Lambda,\ \theta\in S} E_{0,x}\{|U_\lambda(t,0)\theta|^{-2}\}$$

is a finite constant depending only on $\sup\Lambda$ and $\int_0^1 |p(t)|dt$.

Solution:

Using the Markov property one can rewrite:

$$E_{0,x}\{|U_\lambda(t,0)\theta|^{-2}\} = P_{0,t}^{(\lambda,-2)}1](x,\theta)$$

$$\leq |P_{0,1}^{(\lambda,-2)}|_{L^1,L^\infty} |P_{t,1}^{(\lambda,-2)}1|_{L^1}$$

$$= |P_{0,1}^{(\lambda,-2)}|_{L^1,L^\infty} \int_{X\times S^1} E_{1,x}\{|U_\lambda(t,1)\theta|^{-2}\}dx d\theta$$

$$\leq c_{p,\Lambda} \sup_{\substack{(x,\theta)\in X\times S^1\\ |f|_{L^1}\leq 1}} |E_{0,(x,\theta)}\{f(x_t,\theta(t))|U_\lambda(1,0)\theta|^2\}|$$

where $c_{p,\Lambda}$ is a positive upper bound (depending only on $\int |p(t)|dt$ and $\sup\Lambda$) for $|U_\lambda(1,0)\theta|^{-2}$

$$\leq c_{p,\Lambda} \sup_{\lambda\in\Lambda,(x,\theta),(x',\theta')\in X\times S^1} p_\lambda(1,(x,\theta),(x',\theta')). \qquad \square$$

When $p(t) \equiv 0$, we denote by $U_\lambda^{(0)}(t,s)$ the random propagator. Since in this case the Markov process $(X_t, \theta_\lambda^{(0)}(t))$ is time homogeneous, we have the model introduced by the Russian school and the number:

$$\alpha(\lambda) = \lim \frac{1}{t} \operatorname{Log} |U_\lambda^{(0)}(t,0)|$$

exists and defines a strictly positive continuous function of $\lambda$. See [8] or [12]. $\alpha(\lambda)$ can also be obtained by applying the transfer matrices to non-random unit vectors. This gives:

$$\alpha(\lambda) = \lim_{t\to\infty} \frac{1}{t} \operatorname{Log} |U_\lambda^{(0)}(t,0)\theta|$$

P almost surely, for every $\theta$. Using (II.5) and the ergodic theorem for the process $(x_\lambda^t, \theta_\lambda^{(0)}(t))$ one gets:

$$\alpha(\lambda) = -1/2 \int_{X\times S'} [\lambda - F(x) - 1]\sin 2\theta \; \pi_\lambda^{(0)}(x,\theta)dxd\theta. \tag{III.6}$$

We set:

$$\alpha(\lambda) = \inf_{\lambda\varepsilon\Lambda} \alpha(\lambda)$$

whenever $\Lambda$ is a compact interval contained in $(0,\infty)$. Obviously $\alpha(\Lambda) > 0$.

The following estimate is a crucial step in all the proofs of localization in one dimension.

**Problem III.4:**

Show that there exist positive numbers $s_0, t_0$ and $\delta_0$ such that:

$$\sup_{\lambda\varepsilon\Lambda,\; s\geq s_0,\; \delta\varepsilon(0,\delta_0)} |p(\lambda,-\delta)| = \sup_{(x,\theta)\varepsilon X\times S^1} E_{s,x}\{e^{\delta/2\int_s^{\delta+t}[\lambda-q(u)-1]\sin 2\theta(u,s,\theta,\lambda)du}\}.$$

Using the inequality $e^y \leq 1 + y + y^2 e^{|y|}$ one obtains

$$e^{\delta/s\int_s^{s+t}[\lambda - q(u) - 1]\sin 2\theta(u,s,\theta,\lambda)du}$$

$$\leq 1 + \delta/2 \int_s^t [\lambda - q(u) - 1]\sin 2\theta(u,s,\theta,\lambda)du + t^2\delta^2 c(s)^2 e^{t\delta c(s)}$$

where

$$c(s) = \sup_{\lambda\varepsilon\Lambda,\ u\geq s} 1/2|\lambda - q(u) - 1| =$$

$$= 1 + \delta/2 \int [-F(x_u) - 1]\sin 2\theta(u,s,\theta,\lambda)du$$

$$+ \delta/2 \int_s^{s+t} [\lambda - F(X_u) - 1][\sin 2\theta(u,,\theta,\lambda) - \sin 2\theta^{(0)}(u,s,\theta,\lambda)]du$$

$$+ \delta/2 \int_s^{\delta+t} p(u)\sin 2\theta(u,s,\theta,\lambda)du$$

$$+ t^2\delta^2 c(s)^2 e^{t\delta c(s)} \tag{III.7}$$

$$= 1 + (i) + (ii) + (iii) + (iv).$$

(recall Problem II.1 and the beginning of this section for the definition of $\theta^{(0)}$). First we note that:

$$E_{s,x}\{(i)\} = \delta/2 \int_s^{s+t} E_{s,(x,\theta)}\{[\lambda - F(X_u) - 1]\sin 2\theta^{(0)}(u)\}du$$

$$= \delta/2 \int_0^t \int_{X\times S^1} [\lambda - F(x) - 1]\sin 2\theta' p^{(0)}(u,(x,\theta),(x',\theta'))dx'd\theta'$$

$$= \delta/2t \int_{X\times S} [\lambda - F(x') - 1]\sin 2\theta' \pi_\lambda^{(0)}(x',\theta')dx'd\theta'$$

$$+ \delta/2 \int_{X\times S^1} [\lambda - F(x') - 1]\sin 2\theta' \int_0^t (p^{(0)}(u,(x,\theta),(x',\theta')) -$$

$$- \pi^{(0)}(x',\theta')]dx'd\theta'$$

$$\leq -t\delta\alpha(\lambda) + \delta/2(\sup_{\lambda\varepsilon\Lambda(x,\theta)\varepsilon X\times S^1}[\lambda - F(x) - 1]\sin 2\theta)$$

$$(2 + \int_1^\infty \sup_{(x,\theta)\varepsilon X\times S^1}|p^{(0)}(u,(x,\theta),(x',\theta')) -$$

$$- \pi^{(0)}(x',\theta')|dx'd\theta')$$

if one recall formula (III.6) for the Lyapunov exponent, and

$$\leq -t\,\delta\alpha(\lambda) + k\delta \tag{III.8}$$

for some positive constant $k$ which depends only on $\Lambda$ and $F$ but not on $t$, if one uses the exponential convergence of the transition densities in the homogeneous case.

Second, we remark that:

$$E_{s,x}\{(iii)\} \leq \frac{\delta t}{2} \sup_{u \geq s} |p(u)| \tag{III.9}$$

and finally, putting (III.7) through (III.9) together we obtain:

$$|P^{(\lambda,-\delta)}_{s,s+t}| \leq 1 - t\delta[\alpha(\Lambda) - k/t - 1/2 \sup_{u \geq s} |p(u)| - c(s)^2 \delta t\, e^{c(s)\delta t}$$

$$- \sup_{\lambda\varepsilon\Lambda,\ x\varepsilon X} (\lambda - F(x) - 1) \sup_{\lambda\varepsilon\Lambda,\ \theta\varepsilon S^1, u\varepsilon[s,s+t]} |\theta(u,s,\theta,\lambda) -$$

$$- \theta^{(0)}(u,s,\theta,\lambda)|].$$

Now if $\varepsilon' > 0$ is given, we pick $t > 0$ large enough so that $\frac{k}{t} \leq -\frac{\varepsilon'}{4}\alpha(\Lambda)$. Then, once such a $t > 0$ is fixed, we pick $s_0$ large enough so that:

$$\sup_{\lambda\varepsilon\Lambda, x\varepsilon X} |\lambda - F(x) - 1| \sup_{\lambda\varepsilon\Lambda, \theta\varepsilon S', u\varepsilon[s,s+t], s \geq s_0} |\theta(u,s,\theta,\lambda) - \theta^{(0)} < \frac{\varepsilon'}{4}\alpha(\Lambda)$$

(recall Problem II.1) and $\sup_{u \geq s_0} |p(u)| < \frac{\varepsilon'}{2}\alpha(\Lambda)$. Finally we choose $\delta_0 > 0$ small enough so that $c(s_0)^2 \delta t e^{c(s_0)\delta t} < \frac{\varepsilon'}{4}\alpha(\Lambda)$ whenever $0 < \delta < \delta_0$. Under these conditions we have:

$$|P^{(\lambda,-\delta)}_{s,s+t}| \leq 1 - \delta t\alpha(\Lambda)(1 - \varepsilon'). \quad \square \tag{III.10}$$

## IV. Localization Under Perturbation

Let

$$q(t) = F(x_t) + p(t)$$

where $\{F(x_t);\ t \geq 0\}$ is a random potential of the Brownian type as described in the previous section and $\{p(t);\ t \geq 0\}$ is a nonrandom locally integrable function which tends to zero as $t \to \infty$.

For each realization $\omega$ of the Brownian motion process and for each $\alpha \in S^1$ we denote by $H_\alpha(\omega)$ the self-adjoint operator on $L^2([0,\infty),dt)$ which coincides on $C^\infty$ functions with compact supports in $(0,\infty)$ with the formal differential operator:

$$-\frac{d^2}{dt^2} + q(t)$$

and with boundary condition at $0$ given by:

$$y(0)\cos\alpha - y'(0)\sin\alpha = 0 \tag{IV.1}$$

**Problem IV.1**

Show that for each $\alpha \in S^1$ we have, with probability one:

i) $[0,\infty)$ $\Sigma(H_\alpha)$

ii) $E^{(H_\alpha)}_{|[0,\infty)} = E^{(H_\alpha)}_{pp|[0,\infty)}$

iii) The eigenfunctions corresponding to the (dense) pure point spectrum in $[0,\infty)$ decay exponentially, the rate of exponential fall off being given by the Lyapunov exponent of the unperturbed problem as defined in (III.6).

Solution

As explained in [3] and in Section VII.2 of [5] it is enough to show that, for each open interval $\Lambda$ whose closure is contained in $(0,\infty)$ one can find positive real numbers $\gamma = \gamma(\Lambda)$, $\gamma' = \gamma(\Lambda)$ and $K = K(\Lambda)$ such that the following holds for all $L > 0$:

$$E\{\int_\Lambda \int_0^L |U_\lambda(t,0)\alpha|^{\gamma'} e^{\gamma t}\, dt\ \sigma_\alpha(d\lambda)\} < K \tag{IV.2}$$

where $\sigma_\alpha(d\lambda)$ denotes the (random) spectral measure of the (random) operator $H_\alpha$. See below for the definition. We will assume from now on that $L > 0$ is fixed.

Let us assume momentarily that the realization $\omega$ of the Brownian motion is fixed. $H_\alpha$ is in the limit point case at $+\infty$ (see [6] for example). Consequently, one has:

$$\sigma_\alpha = \lim_{b\to\infty} \frac{1}{\pi} \int_0^\pi \sigma_{0,\alpha,b,\beta} d\beta$$

in the sense of the vague convergence of measures where $\sigma_{0,\alpha,b,\beta}$ is the spectral measure of the operator $H_\alpha$ restricted to the interval $[0,b]$ with boundary condition at $b$ given by $y(b)\cos\beta - y'(b)\sin\beta = 0$. According to Lemma II.2 of [4] one has:

$$\frac{1}{\pi}\int_0^\pi \sigma_{0,\alpha,b,\beta}(d\lambda)d\beta = \frac{1}{\pi}|U_\lambda(b,0)\alpha|^{-2}d\lambda.$$

Moreover, the function

$$\lambda \to \int_0^L |U_\lambda(t,0)\alpha|^{\gamma'}\ e^{\gamma t}dt$$

is continuous and consequently (we will always take $b > L + 1$) we must have the following upper bound for the left hand side of (IV.2):

$$\underline{\lim}_{b\to\infty} \frac{1}{\pi} \int_\Lambda \int_0^L E\{|U_\lambda(t,0)\alpha|^\gamma |U_\lambda(b,0)\alpha|^{-2}\} e^{\gamma t} dt d\lambda$$

$$= \underline{\lim}_{b\to\infty} \frac{1}{\pi} \int_\Lambda \int_0^L \{|U_\lambda(t,0)\alpha|^{\gamma'-2} |U_\lambda(b,t) \frac{U_\lambda(t,0)\alpha}{|U_\lambda(t,0)\alpha|}|^{-2}\} e^{\gamma t} dt d\lambda$$

$$= \underline{\lim}_{b\to\infty} \frac{1}{\pi} \int_\Lambda \int_0^L E\ \{|U_\lambda(y,0)\alpha|^{\gamma'-2}$$

$$E\{|U_\lambda(b,t)\theta|^{-2} | [0,t]\}|\theta = \frac{U_\lambda(t,0)\alpha}{|U_\lambda(t,0)\alpha|} \} e^{\gamma t} dt d\lambda \qquad (IV.3)$$

Note that, P almost surely we have:

$$E\{|U_\lambda(b,t)\theta|^{-2} | \ [0,t]\} = E_{0,x_t}\{|\tilde{U}_\lambda^t(b-t,0)\theta|^{-2}\} \qquad (IV.4)$$

where $\tilde{U}_\lambda^t(\cdot,\cdot)$ denotes the propagator when the function $s \to p(s)$ is replaced by the function $s \to p(t+s)$. Because of our assumptions we have:

$$\sup_{t\leq 0} \int_0^1 |p(t+s)| ds < +\infty,$$

and we can use Problem III.3 to bound uniformly (IV.4) from above by a constant $c$ depending only on $\Lambda$ and $F$. Consequently, coming back to (IV.3) we obtain the following upper bound for the left hand side of (IV.2):

$$\frac{1}{\pi} \int_\Lambda \int_0^L E\{|U_\lambda(t,0)\alpha|^{\gamma'-2}\} e^{\gamma t} dt d\lambda$$

$$\leq \frac{c}{\pi} \int_\Lambda \int_0^L \left(\int_X [P_{0,t}^{(\lambda,\gamma'-2)} 1](x,\alpha) dx\right) d^{\gamma t} dt d\lambda$$

$$\leq \frac{c}{\pi} \int_\Lambda \int_0^\infty e^{-\gamma'' t} e^{\gamma t} dt d\lambda$$

if we pick $\gamma' \in (0,2)$ such that $2 - \gamma' < \delta_0$ and if we use Problem IV.3. This gives the desired result since $\gamma''$ can be taken independent of $\lambda \in \Lambda$ and provided $\gamma > 0$ small enough so that the above integral converges. This shows the so-called "<u>exponential localization phenomenon</u>" in $[0,\infty)$. The point (iii) in the rate of the exponential decay of the eigenfunctions can be argued in the following way: first one remarks that $|P_{0,t}^{(\lambda,\gamma'-2)}| = |P_{0,t}^{(\lambda,\gamma')}|$ if $0 < \gamma' < 2$ so that, by picking $\gamma'$ small enough one will be able to choose $\gamma$ as close to $\alpha(\Lambda)$ as we like according to the estimate (III.10). □

Let us emphasize that the spectrum of $H_\alpha$ could have a negative part if the function $p(t)$ takes negative values. Typically we expect this part of the spectrum to be made of isolated eigenvalues with exponentially decaying corresponding eigenfunctions. This can be proven under some restrictions on $p(t)$ by using the Weyl's criteria for essential spectra.

It is elegantly shown in [11] that for any stationary non-deterministic bounded stochastic process $\{q(t);\ t \geq 0\}$ the operator $H_\alpha$ has dense pure point spectrum with exponentially localized eigenfunctions with probability one, for almost every (with respect to lebesque's measure $d\alpha$) choice of the boundary condition $\alpha$. Moreover some sufficient condition is given to get rid of the "almost every boundary condition" assumption. Unfortunately it seems very difficult to perturb such a general potential and still be able to control quantities like the Lyapunov exponent and show that the localization phenomenon is preserved.

## V. Spherically Symmetric Random Potentials:

We now come back to the case of potential functions $V(x,\omega)$ of the form $v(|x|,\omega)$ where $\{v(t,\omega);\ t \geq 0,\ \omega \in \Omega\}$ is a random potential of the Brownian type as described in Section III. For each integer $k \geq 1$ the operaor $\tilde{H}_k$ defined in (I.6) can be written in the form:

$$\tilde{H}_k = -\frac{d^2}{dt^2} + F(X_t) + p_k(t) \tag{V.1}$$

with $p_k(t) = [\frac{(n-1)(n-3)}{4} + \kappa_k]\frac{1}{t^2}$. So, except for the singularity of $p(t)$ at $t = 0$ we are in the situation investigated in the previous section. Unfortunately this singularity forces the study of the operator $\tilde{H}_k$ on intervals of the form $[a,b]$ with $a > 0$, with boundary conditions at both ends $a$ and $b$ and then, the limits $a \to 0$ and $b \to \infty$ have to be controlled. The limit $b \to \infty$ was handled (for $a$ fixed) in the previous section. The limit $a \to 0$ is more subtle.

If the dimension $n$ is larger than 3 (i.e., $n \geq 4$) then

$$\frac{(n-1)(n-3)}{4} + \kappa_k \geq \frac{3}{4} \qquad (V.2)$$

and $\tilde{H}_k$ is always in the limit point case at 0. The limit $a \to 0$ is handled in the same way the limit $a \to -\infty$ is controlled in the usual case of the whole line (see for example [3]) and since the generalized eigenfunctions have to vanish at 0 one can conclude that the spectrum of $\tilde{H}_k$ is pure point with eigenfunctions exponentially decaying at $+\infty$, and since this is true with probability one for each $k \geq 1$ the conclusion remains true for the operator $H$ defined by (I.1) on $L^2(R^n,dx)$.

If the dmension is 2 or 3, (V.2) is satisifed only for the large values of k (in which case the conclusion on the almost sure spectral properties of $\tilde{H}_k$ remains valid) while for the smaller values of $k$, the operator $\tilde{H}_k$ given by (V.1) is not in the limit point case at 0 and so, it is not essentially self adjoint on the space of smooth functions with compact supports contained in $(0,\infty)$. Even though the right hand side of (V.1) needs to be supplemented with a boundary condition at 0 to define a self-adjoint operator on $L^2([0,\infty),dt)$, the self-adjoint operator $\tilde{H}_k$ obtained from the direct sum decomposition of the operator $H$ on $L^2(R^n,dx)$ corresponds to only one of these boundary conditions. A simple computation shows that the generalized eigenfunctions of $H_k$ have to vanish at zero so that the Dirichlet boundary condition (corresponding to $\alpha = 0$ in the notation of the previous section) is the one. One can then control the limit $a \to 0$ as before and obtain the same conclusion. The details are left as an extra exercise for the reader.

**References**

[1] F. Bentosela, R. Carmona, P. Duclos, B. Simon, B. Souillard and R. Weder: Schrödinger Operators with an Electric Field and Random or Deterministic Potentials. Comm. Math. Phys. 88 (1983) 387-397.

[2] J. Brossard: Perturbations Aleatoires de Potentiels Periodiques (preprint) Sept. 1982.

[3] R. Carmona: Exponential Localization in One-Dimensional Disordered Systems. Duke Math. J. 49 (1982) 191-213.

[4] R. Carmona: One dimensional Schrödinger Operators with Random or Deterministic Potentials: New Spectral Types, J. Functional Anal. 51 (1983) 229-258.

[5] R. Carmona: Random Schrödinger Operators. Ecole d'Eté de Probabilités. Saint Flour (1984) to appear in Lect. Notes in Math. Springer Verlag.

[6] E.A. Coddington and N. Levinson: Theory of Ordinary Differential Equations, McGraw Hill, New York (1955).

[7] F. Deylon, H. Kunz and B. Souillard: One dimensional wave equations in random media, J. Phys. A16 (1983), 25-42.

[8] I. Ya. Goldsheid, S.A. Molcanov and L.A. Pastur: A pure point spectrum of the stochastic one-dimensional Schrödinger operator. Funct. Analysis Appl. 11 (1977), 1-10.

[9] L. Hörmander: Hypoelliptic differential equations of second order, Acta Math. 119 (1967) 147-171.

[10] K. Ichihara and H. Kunita: A Classification of the Second Order Degenerate Elliptic Operators and its Probability Characterization. Z. Wahrscheinl. verw. Geb. 30 91974), 235-254.

[11] S. Kotani: Lyapunov exponents and spectra for one-dimensional random Schrödinger operators. Proc. A.M.S. Conf. on "Random Matrices and their Applications", June 1984.

[12] S.A. Molcanov: The structure of eigenfunctions of one dimensional unordered structures, Math. U.S.S.R. Izvestija 12-1 (1978), 69-101.

[13] M. Reed and B. Simon: Methods of Modern Mathematical Physics II, Fourier Analysis-Self-Adjointness, Academic Press (1975) New York.

[14] D. Stroock and S.R.S. Varadhan: Multidimensional Diffusion Processes, Springer Verlag, (1979).

# TRAVEL-TIME PROBLEMS FOR WAVES IN RANDOM MEDIA*

P.L. Chow**

Department of Mathematics
Wayne State University
Detroit, Michigan 48202

**Abstract:**

By method of characteristics, the travel time was introduced as a probabilistic hitting time by an ideal particle moving along a characteristic curve. For a simple wave model, the moments of travel time are investigated and are found to satisfy appropriate boundary-value problems in the diffusion approximation. Two problems, corresponding to plane and cylindrical waves, are solved explicitly. For the random wave equation, the mean and variance of the travel time are obtained in a simple wave approximation, and some numerical results are provided. The general problem is discussed briefly to indicate some technical difficulties.

## 1. Introduction

For wave propagation in random media, the problem of pulse propagation has been investigated by many authors (for references, see [1]), due to its important applications to atmospheric and underwater sound problems. In practical problems, one of the most important wave parameters to be observed is the pulse arrival time. Others include the reflection and transmission coefficients, the mean intensity etc.

In the engineering literature, there are numerous articles concerning the statistical problem of pulse arrival time in random media, (see, e.g., [2]). The popular method of treating such problem is the so-called "temporal moment" method. This is, in our viewpoint, an "effective media" approach. To be specific, let $u(x,t,\omega)$ be the pulse-wave function. By examining the temporal moment method more closely, (which was done in time-harmonic case), this approach amounts to

---

* This work was supported in parts by the ARO Contract DAAG 29-83-K-0014 and by the NASA Grant NAG-1-460.

** Visiting the Institute for Mathematics and its Applications, University of Minnesota, Minneapolis, Minnesota.

introducing a probability density function

$$p(x,t) = \frac{E|u(x,t)|^2}{\int_0^\infty E|u(x,s)|^2 ds},$$

where, as $t = 0$, the pulse shape $u(x,0) = f(x)$ is assumed to have a narrow support in a neighborhood of the origin. Then one regards the pulse arrival time $\tau$ at $x$ from the origin at $t = 0$ as a random variable with respect to the probability density $p(x,t)$ given above. This is, of course, only an effective pulse arrival time, which does not correspond to the actual arrival time. This obvious discrepency has led us to the current investigation.

The paper is divided into four parts, corresponding to Sections 2-5. In Section2, we formulate the travel time problems for a simple random wave equation and the standard random wave equation. The travel time is defined probabilistically. A summary of the main results is given. Some general results for travel time of simple waves are presented in Section 3. The first two moments of travel time for plane and cylindrical waves are solved for in Section 4. The last section deals with the travel time problem for the random wave equation.

We remark that the method of characteristics combined with the diffusion approximation forms the basic mathematical tool for this study. This approach was used in an earlier work of ours [3] to study progressing waves in random media. Also we wish to mention that the reflection and transmission of a pulse through a random medium was reported recently in [4].

## 2. Preliminaries and Main Results

A simple model for wave propagation in random media is provided by the following first-order random partial differential equation:

(2.1) $$\partial_t u + v(x,t,\omega) \cdot \partial_x u = b(x,t)u, \quad t > 0,\ x \in R^d,$$

where $\partial_t = \frac{\partial}{\partial t}$, $\partial_x = \nabla$ the gradient operator, $v(x,t,\omega)$ is the local wave velocity which is a random function of $x$ and $t$, and $b(x,t)$ is a given scalar function. It is elementary in partial differential equations [5] that, by the method of characteristics, the characteristic equation for (2.1) is given by

$$(2.2)\qquad \begin{cases} \dfrac{dx_t}{dt} = v(x_t, t, \omega), & t > 0, \\ x_0 = x \in R^d \end{cases}$$

in which we have set $x_t = x(t,\omega)$.

Next we consider the standard wave equation in a random medium

$$(2.3)\qquad \partial_t^2 u - c^2(x,t,\omega)\Delta u = 0, \quad t > 0, \ x \in R^d,$$

where the local wave speed $c(x,t,\omega)$ is a random function of $x$ and $t$, and $\Delta$ denotes the Laplacian as usual. According to the geometric theory of waves [5], the motion of wave-front (or the characteristic surface) is governed by the random Hamilton-Jacobi equation

$$(2.4)\qquad (\partial_t \phi)^2 - c^2(x,t,\omega)|\partial_x \phi|^2 = 0.$$

Again, via the method of characteristics, the associated characteristic equations are just the random Hamilton equation

$$(2.5)\qquad \begin{cases} \dfrac{dx_t}{dt} = c(x_t, t, \omega)\hat{p}_t, & x_0 = x, \\ \dfrac{dp_t}{dt} = -\partial_x c(x_t, t, \omega)|p_t|, & p_0 = p, \end{cases}$$

where, for a vector $p$, $\hat{p}$ denotes the unit vector along $p$. In contrast with (2.2), the system (2.5) is sometimes called the bi-characteristic equations. Physically the set of characteristic curves

$$(2.6)\qquad \Gamma_x : y = x_t, \quad 0 \leq t \leq T,$$

is known as the rays through the point $x$ in the physical space. Here, for physical reasons, we allow nonunique solutions for (2.2) or (2.5). As is well known, for a wave equation, the disturbance propagates along the rays. If a disturbance starts from the point $x$, then it will propagate into the range $R_x(\omega)$ of influence by $x$ [5], which depends on $\omega$. Define

(2.7) $$R_x = \overline{\bigcup_{\omega} R_x(\omega)}$$

to be the maximal range of influence for $x$. Let $B$ be a region or a subset of $R^d$. Clearly, when a ray through $x$ first enters the set $B$, the disturbance from $x$ has reached the region $B$. It seems natural to define the (wave) travel time as the least time for a ray through $x$ to reach $B$, which is, of course, a random variable. In probabilistic language, regarding $\{x_t, t > 0\}$ as a stochastic process, what is called the travel time is actually a hitting time [6]. This observation led us to define the travel time as follows.

Let $\Gamma_x = \{x_t: x_0 = x, t > 0\}$ denote the set of random rays through $x$, and let $B$ be a space region in $R^d$. The (wave) travel time from $x$ to $B$ is defined as

(2.8) $$\tau_x(B) = \inf_{x_\cdot \in \Gamma_x} \inf \{t > 0: x_t \in B\}.$$

and $\tau_x(B) = \infty$ if $x_t \notin B$ for all $x_t$ and $t > 0$.

According to this definition, if $B \cap R_x = \phi$, the travel time $\tau_x(B) = \infty$. We remark that, physically, the random fields $v(\cdot,t,\omega)$ and $c(\cdot,t,\omega)$ are causal. This means probabilistically they are nonanticipating (w.r.t. an underlying increasing $\sigma$-field $\mathcal{F}_t$). Consequently, since $x_t$ is a solution of the stochastic equation (2.2) or (2.5), it is also adapted to $\mathcal{F}_t$. Therefore the travel time (2.8) is also a stopping time [6].

As mentioned previously, the study of travel time was motivated by pulse propagation problems. Once the travel time $\tau_x(B)$ was defined, it is reasonable to define a version of pulse-arrival time as the corresponding travel time $\tau_x(B)$ weighted by the initial pulse-shape. To be precise, let $u(x,0) = f(x)$ be the initial pulse shape. Define

(2.9) $$p(x) = \frac{|f(x)|}{\int |f(y)|dy},$$

where $f$ is assumed to be integrable. Therefore, from now on we will only be concerned with the travel time alone. As it stands, the problem is rather difficult. To make it tractable, we will assume that the random fluctuations are weak. Then an asymptotic analysis of the problem can be carried out to yield some results of interest.

To be specific, in the simple wave problem, we assume in (2.1) that $v$ depends on a small parameter $\varepsilon > 0$, denoted by $v^{\varepsilon}$, such that

$$v^{\varepsilon}(x,t,\omega) = \hat{v} + \varepsilon\tilde{v}(x,t,\omega). \tag{2.10}$$

Here $\hat{v}$ is a constant velocity, and $\tilde{v}$ is a centered random field with the correlations

$$R_{ij}(x,y;t,s) = E\tilde{v}_i(x,t)\tilde{v}_j(y,s). \tag{2.11}$$

For the wave equation (2.3), writing $c = c^{\varepsilon}$, the dependence of $c^{\varepsilon}$ on $\varepsilon$ is of the form

$$c^{\varepsilon}(x,t,\omega) = c_0 + \varepsilon\xi(x,t,\omega), \tag{2.12}$$

where $c_0$ is a positive constant, the unperturbed wave speed, and $\xi$ is a centered random function with the correlation

$$R(x,y;t,s) = E\xi(x,t)\xi(y,s). \tag{2.13}$$

Under the weak fluctuations and strong mixing assumptions for the random medium, we are able to apply a diffusion approximation to replace the random characteristics in (2.2) or (2.5) by a diffusion process. Thereby the asymptotic statistics of the travel time can be computed accordingly. The results are summarized in the following.

The principal results pertain to the simple waves described in §§3-4. First we show that, in the diffusion limit, all moments $T_m(x,B) = E\tau_x^m(B)$ can be computed by solving the respective boundary-value problems (3.19) and (3.20) associated with the generator of the diffusion. As far as we know, our method of deriving such equations are new. Then we calculate the moments $T_m$ of travel time in two special cases. For a plane wave, we constructed the Green's fucntion $g(x,y)$ for travel time. The mean travel time $T(\ell)$ and its second moment $T_2(\ell)$ are computed for the wave starting at $x_1 = \ell$ to reach $x_1 = \ell_1$ or $\ell_2$, where

$\ell_1 < \ell < \ell_2$. These results are given by (4.7), (4.10) and (4.12), respectively. In the case of a cylindrical wave, the associated Green's fucntion $g$ and the mean travel time are obtained for the same boundaries, by means of a Fourier transform. They are shown by (4.23) and (4.26), respectively. In both cases, it is shown that the mean travel time is approximately equal to the effective travel time, i.e. the distance divided by the effective wave speed $\hat{c}_\varepsilon$, (see (4.14) and (4.28)). The correction terms are exponentially small.

For the random wave equation the main results are concerned with a simple wave approximation. As discussed in the sub-section 5.1, if the small random fluctuation in the direction of wave propagation is neglected, it reduces to a simple wave problem discussed above. A closer examination of the results for plane wave shows that, in physically realistic case, the mean travel time decreases due to random scattering. Also the standard deviation of travel time is proportional to standard deviation of the random medium and $(\hat{c}_\varepsilon)^{-3/2}$, and it growns like $x^{1/2}$, where $x$ is the travel distance. Some numerical results are obtained to show the dependence of the mean travel time and its standard deviation on $x$, $\varepsilon$ and the parameter $\alpha_1$, appearing in special form of correlation function (5.7). Some results are shown in Figs. 1-4. Finally, in the sub-section 5.2, we make some observation and remarks about the general problem. The coupling of characteristic equations gives rise to a travel time problem in the higher-dimensional phase space. In the diffusion approximation, the associated boundary-value problem is degenerate. The analytical questions, such as well-posedness and computational problem remain to be resolved.

## 3. Travel Time Problems for Simple Waves

Let us call the solutions of the model wave equation (2.1) simple waves. In view of the assumptions (2.10), the associated characteristic equation (2.2) becomes

$$\begin{cases} \dfrac{dx_t^\varepsilon}{dt} = \hat{v} + \varepsilon\tilde{v}(x_t, t, \omega), \quad t > 0, \\ x_0^\varepsilon = x. \end{cases} \tag{3.1}$$

Let us rename $x_t$ the unperturbed solution

$$x_t = x + t\hat{v}. \tag{3.2}$$

Define

$$y_t^\varepsilon = x_t^\varepsilon - x_t. \tag{3.3}$$

Then, in view of (3.1), (3.2), $y^\varepsilon$ satisfies

$$\begin{cases} \dfrac{dy_t^\varepsilon}{ds} = \varepsilon\eta(y_t^\varepsilon, x, t, \omega), \\ \\ y_0^\varepsilon = 0, \end{cases} \tag{3.4}$$

where

$$\eta(y,x,t,\omega) = \tilde{v}(x_t + y, t, \omega). \tag{3.5}$$

In order to obtain a diffusion approximation for $y^\varepsilon$, we assume that the random field $v(x,t,\omega)$ satisfies a strong mixing condition in space and time so that $v(x,t,\cdot)$ and $v(y,s,\cdot)$ become asymptotically independent when either $|t - s|$ or $|x - y|$ gets large. The former condition is due to Khasminskii [7], later generalized by Kohler and Papanicolaou [8] to prove limit theorems for stochastic equations of the form (3.4). The latter, the spatial mixing condition was introduced by Kesten and Papanicolaou [9,10] to treat some limit theorems in turbulent diffusion and stochastic acceleration. To include the degenerate cases, where $v$ is independent of $x$ or $t$, we assume that both mixing conditions are satisfied. They will be called the Khasminskii-Kesten-Papanicolaou conditions or K.K.P. conditions, in short. In the diffusion limit, $\varepsilon \downarrow 0$, $t \to \infty$ with $r = \varepsilon^2 t$ held fixed, we invoke either the Khasminskii Theorem [7] or the Kesten-Papanicolaou Theorem in [9] to assert that the process $y_t^\varepsilon$ converges weakly to $y(r)$ or

$$y_t^\varepsilon = y^\varepsilon(r/\varepsilon^2) \Rightarrow y(r), \text{ as } \varepsilon \downarrow 0, \tag{3.6}$$

where $y(r)$, $r > 0$, is a diffusion process with known drift $a(x)$ and the diffusion matrix $b(x)$. Furthermore they can be computed as follows:

$$(3.7)\qquad \begin{aligned} b_{ij}(x) &= \lim_{T\to\infty} \frac{1}{T} \int_0^T\int_0^T R_{ij}(x + \hat{v}t,\ x + \hat{v}s;\ t,s)dtds,\\ a_i(x) &= \lim_{T\to\infty} \frac{1}{T} \int_0^T\int_0^T R_i(x + \hat{v}t, x + \hat{v}s; t,s)dtds \end{aligned}$$

where $R_i(x,y;t,s) = \partial_{x_j}(R_{ij}(x,y;t,s)$ and the summation over repeated indices is in effect. of course, here it was tacitly assumed that the limits in (3.7) are finite. Now we set

$$(3.8)\qquad z_t^\varepsilon = x_t + y(r) = x_t + y(\varepsilon^2 t).$$

Then we see that $x_t^\varepsilon \sim z_t^\varepsilon$ in the diffusion limit and $z_t^\varepsilon$ satisfies the Itô equation

$$(3.9)\qquad \begin{cases} dz_t^\varepsilon = [\hat{v} + \varepsilon^2 a(z_t^\varepsilon)]dt + \varepsilon\sigma(z_t^\varepsilon)dw_t, \\ z_0^\varepsilon = x. \end{cases}$$

where $\sigma$ is the root of $a$ or $\sigma\sigma^* = b$, and $w_t$ is the standard Brownian motion in $R^d$. Since $z_t^\varepsilon$ is a diffusion process, the travel time problem may be treated asymptotically by the method of differential equations.

To this end we define the travel times

$$(3.10)\qquad \begin{aligned} \tau_x^\varepsilon(B) &= \inf\{t > 0:\ x_t^\varepsilon \in B\} \\ \tau_x(B) &= \inf\{t > 0:\ z_t^\varepsilon \in B\} \end{aligned}$$

Here we assumed that the characteristic curve through each point is unique. Otherwise they should be modified as in (2.8). Since $\tau_x^\varepsilon \sim \tau_x$ in the diffusion limit, we shall be concerned with the statistical properties of $\tau_x$ in this approximation. Consider the probability

$$(3.11)\qquad q_t(x,B) = p\{\tau_x(B) > t\}$$

Then it is known that $q_t$ satisfies the parabolic equation [11].

$$(3.12)\qquad \begin{cases} \partial_t q_t = Lq_t, & x \in D = B^c,\ t > 0 \\ q_0(x,D) = 1, \\ q_t|_{\partial D} = 0, \end{cases}$$

where

$$(3.13)\qquad Lu = \frac{1}{2}\,\varepsilon^2 b_{ij}(x)\,\frac{\partial^2 u}{\partial x_i\,\partial x_j} + [v_i + \varepsilon^2 a_i(x)]\,\frac{\partial u}{\partial x_i},$$

and the boundary $\partial D = \partial B$ is smooth. We let

$$(3.14)\qquad v_\lambda(x) = E\{e^{-\lambda\tau_x(B)}\},\ \text{with } \lambda \text{ being a parameter.}$$

Then, noting (3.11), we have

$$(3.15)\qquad v_\lambda(x) = -\int_0^\infty e^{-\lambda t}\,dq_t(x,B),$$

or $(-v_\lambda)$ is the Laplace transform of $q_t$. Suppose the coefficients $a,b$ of $L$ in (3.13) are smooth so that $q_t$ has a smooth density, say, $C^1$ in $t$. Then a Laplace transform of (3.12) yields the boundary-value problem

$$(3.16)\qquad \begin{cases} Lv_\lambda - \lambda v_\lambda = 0, & v \in D, \\ v_\lambda|_{\partial D} = 1. \end{cases}$$

In view of (3.14), $v_\lambda$ is a moment-generating function for the travel time $\tau_x$. In fact we have

$$(3.17)\qquad T_m(x,B) = E[\tau_x(B)]^m = (-1)^m\,\frac{\partial}{\partial\lambda}\,v_\lambda(x)\Big|_{\lambda=0},\quad m = 1,2,\ldots\ .$$

Here we have assumed a $C^m$-dependence of $v_\lambda$ on $\lambda$ at $\lambda = 0$. We expect this to be true if the domain $D$ is bounded and convex. If such a $\lambda$-dependence holds also for $Lv_\lambda$, uniformly in $x$, then by a repeated differentiation of the equation (3.16) with respect to $\lambda$, we get

$$(3.18)\qquad \begin{cases} Lv_\lambda^{(m)} - \lambda v_\lambda^{(m)} - mv_\lambda^{(m-1)} + \lambda v_\lambda^{(m)} = 0, & x \in D,\ m = 1,2,\ldots, \\ v_\lambda^{(m)}|_{\partial D} = 0, \end{cases}$$

where, by definition, $v_\lambda^{(0)} = 1$.

Now, by virture of (3.17), we set $\lambda = 0$ in (3.18) to obtain the moment equations:

$$(3.19)\qquad \begin{cases} LT(x,B) = -1, \quad x \in D = B^c, \\ T(x,B)|_{x \in \partial D} = 0; \end{cases}$$

$$(3.20)\qquad \begin{cases} LT_m(x,B) = -mT_{m-1}(x,D), \quad x \in D,\ m > 2, \\ T_m(x,B)|_{x \in \partial D} = 0, \end{cases}$$

where $T(x,B) = T_1(x,B)$ being the mean travel time is of particular interest. Thus, if $g(x,y)$ is the Green's function associated with the Dirichlet problem for $(-L)$, the moments $T_m$ of $\tau_x$, given by the solutions of (2.19), or (2.20), can be written simply as

$$(3.21)\qquad T_m(x,B) = m \int_D g(x,y)\, T_{m-1}(y,B)dy, \quad m = 1,2,\dots ,$$

where we recall $T_0 = 1$. In what follows we shall consider two special cases.

## 4. Plane and Cylindrical Waves

For concrete results, it is instructive to consider two special cases, corresponding to the travel times for plane and cylindrical waves, respectively.

### 4.1 Plane Wave Problems

Let us assume that

$$(4.1)\qquad \begin{aligned} \hat{v}_i &= v_0 \delta_{1i}, \\ \tilde{v}_i &= \rho(x,t,\omega)\delta_{1i}, \quad i = 1,2,3, \end{aligned}$$

where $v_0$ is the wave speed in the direction $e_1 = (1,0,0)$, and $\rho$ is a scalar random function, which is stationary homogeneous and isotropic so that

$$(4.2)\qquad \begin{aligned} &E\rho = 0, \\ &E\rho(x,t)\rho(y,t) = \gamma(|x - y|,t - s). \end{aligned}$$

Then the problem admits a plane wave solution propagating in the direction $e_1$.

Upon substituting (4.1) and (4.2) into (3.7), we get

$$(4.3)\qquad \begin{aligned} b_{ij} &= \beta\delta_{1i}\delta_{1j}, \\ a_i &= \alpha\delta_{1i}, \end{aligned}$$

where

$$(4.4)\qquad \begin{aligned} \beta &= \int_0^\infty \gamma(v_0 s, s)\,ds, \\ \alpha &= \int_0^\infty \gamma'(v_0 s, s)\,ds, \end{aligned}$$

are constant with $\gamma'(r,s) = \frac{\partial}{\partial r}\gamma(r,s)$.

If we take $D$ to be the slab $\{x \in R^d: \ell_1 < x_1 < \ell_2\}$, then it is easy to check that there exist solutions of the systems (3.19) and (3.20) which satisfy

$$(4.5)\qquad \begin{cases} L_1 T_m(x_1,B) = -mT_{m-1}(x_1,B), & \ell_1 < x < \ell_2, \\ T_m(\ell_1,B) = T_m(\ell_2,B) = 0, \end{cases}$$

where

$$(4.6)\qquad \begin{aligned} L_1 u(x_1) &= \frac{1}{2}\,\sigma_\varepsilon^2\, u''(x_1) + \hat{c}_\varepsilon u'(x_1), \\ \sigma_\varepsilon^2 &= \beta\varepsilon^2, \\ \hat{c}_\varepsilon &= (v_0 + \alpha\varepsilon^2). \end{aligned}$$

Now the Green's function $g(x_1,y_1)$ for the system (3.26) can be constructed by an elementary method [5]. It is found to be

$$(4.7)\qquad g(x_1,y_1) = \frac{e^{\nu_\varepsilon y_1}}{\hat{c}_\varepsilon(e_2 - e_1)} \begin{cases} (1 - e_1 e^{-\nu_\varepsilon x_1})(1 - e_2 e^{\nu_\varepsilon y_1}), & x_1 < y_1, \\ (1 - e_2 e^{-\nu_\varepsilon x_1})(1 - e_1 e^{-\nu_\varepsilon y_1}), & x_1 > y_1, \end{cases}$$

where

$$(4.8)\qquad \begin{aligned} \nu_\varepsilon &= 2\hat{c}_\varepsilon/\sigma_\varepsilon^2 = 2(v_0 + \alpha\varepsilon^2/\beta\varepsilon^2), \\ e_i &= e^{\nu_\varepsilon \ell_i}, \quad i = 1,2. \end{aligned}$$

In terms of the Green's function (4.7), the moments of $\tau_x$ can be computed recursively by

(4.9) $$T_m(x_1,B) = m \int_{\ell_1}^{\ell_2} g(x_1,y_1)T_{m-1}(y_1,B)dy_1,$$

where $B = \{x: x_1 > \ell_2 \text{ or } x_1 < \ell_1\}$.

In particular the mean travel time $T(\ell)$ for the plane wvae to propagate from the plane $x_1 = \ell$ to $x_1 = \ell_1$ or $\ell_2$ is given by

(4.10) $$T(\ell) = d_1(1 - e_2 e^{-\nu_\varepsilon \ell}) + \frac{(\ell_2 - \ell)}{\hat{c}_\varepsilon},$$

where $\ell_1 < \ell < \ell_2$, and

(4.11) $$d_1 = e_1(\ell_2 - \ell_1)/\hat{c}_\varepsilon(e_2 - e_1).$$

Similarly for $m = 2$, we get

(4.12) $$\begin{aligned} T_2(\ell) &= d_2(1 - e_2 e^{-\nu_\varepsilon \ell}) + 2\frac{d_1}{\hat{c}_\varepsilon}(\ell_2 - \ell)(1 + e_2 e^{-\nu_\varepsilon \ell}) \\ &+ 4(\frac{\sigma_\varepsilon}{\hat{c}_\varepsilon})^2(\ell_2 - \ell) + \frac{4}{\hat{c}_\varepsilon}(\ell_2 - \ell)^2, \end{aligned}$$

where

(4.13) $$d_2 = \frac{(\ell_2 - \ell_1)}{\hat{c}_\varepsilon(e_2 - e_1)}\{2(\ell_2 - \ell_1) + 2\frac{\sigma_\varepsilon^2}{\hat{c}_\varepsilon} + d_1(e_1 + e_2)\}.$$

The result (4.10) has a simple physical interpretation. The second term on the right-hand side is the effective travel time, the distance divided by the effective travel speed $\hat{c}_\varepsilon$. The first term, which is negative, yields a decrease of mean travel time due to the presence of the boundary $x_1 = \ell_1$. In fact, as $\ell_1 \to -\infty$, we get $d_1 \to 0$ so that

(4.14) $$T(\ell) = (\ell_2 - \ell)/\hat{c}_\varepsilon.$$

That is, the mean travel time coincides with the effective travel time in the diffusion limit.

## 4.2 Cylinderical Wave Problem

Consider the travel time for a cylindrical wave. As before, let

$$\hat{v}_i = v_0 \delta_{1i}. \tag{4.15}$$

But the fluctuation velocity $\hat{v}(x,t,\omega)$ has independent components $\tilde{v}_i$ so that the correlation $R_{ij}$ defined by (2.11) are nonzero for $i = j$, zero otherwise. Also $\tilde{v}$ is assumed to be stationary, homogeneous and isotropic such that

$$R_{ij}(x,y;t,s) = \gamma(|x - y|,|t - s|)\delta_{ij}. \tag{4.16}$$

The limits in (3.7) can be computed as follows:

$$\begin{aligned} b_{ij} &= \beta\delta_{ij}, \\ a_i &= \alpha\delta_{1i}, \end{aligned} \tag{4.17}$$

where $\alpha$, $\beta$ are defined as before in (3.25). The corresponding operator $L$ defined by (3.13) becomes

$$Lu = \frac{1}{2}\sigma_\varepsilon^2 \Delta u + \hat{c}_\varepsilon \frac{\partial u}{\partial x_1}, \tag{4.18}$$

where $\Delta$ denotes the Laplacian in $R^d$; $\sigma_\varepsilon$ and $\hat{c}_\varepsilon$ are given in (4.6). Let us construct the Green function $g(x,y)$ in a slab region $D = \{\ell_1 < x_1 < \ell_2\}$ in three dimensions $(d = 3)$. It satisfies the following:

$$\begin{cases} Lg(x,y) = -\delta(x - y), \\ g|_{x_1 = \ell_1} = g|_{x_1 = \ell_2} = 0 \\ g, \dfrac{\partial g}{\partial x_i} \to 0 \quad \text{sufficiently fast as } |x| \to \infty. \end{cases} \tag{4.19}$$

The above problem can be reduced to a one-dimensional problem by a Fourier transform in the transverse variable $x_\perp$, where $x = (x_1, x_\perp)$. Define the Fourier transform of $f(x_\perp)$ by $\hat{f}(\kappa)$ as follows

$$\hat{f}(\kappa) = \int_{R^2} f(x_\perp)e^{i\kappa\cdot x_\perp}dx_\perp. \tag{4.20}$$

After a Fourier transform, the resulting one-dimensional problem can be readily solved to get the transformed Green's function $\hat{g}(x_1,\kappa;y)$ given by

$$\hat{g}(x_1,\kappa;y) = \frac{2}{\sigma^2\theta \operatorname{sh}\theta(\ell_2-\ell_1)} \exp\{i\kappa\cdot y_\perp - \tfrac{1}{2}\nu_\varepsilon(x_1-y_1)\} \tag{4.21}$$

$$\times \begin{cases} \operatorname{sh}\theta(x_1-\ell_1)\operatorname{sh}\theta(y_1-\ell_2), & \ell_1 < x_1 < y_1, \\ \operatorname{sh}\theta(x_1-\ell_2)\operatorname{sh}\theta(y_1-\ell_1), & y_1 < x_1 < \ell_2, \end{cases}$$

where $\sigma_\varepsilon$, $\nu_\varepsilon$ are as before, sh(•) is the hyperbolic sine, $\theta = \theta(k)$ is given by

$$\theta(k) = \sqrt{\hat{c}_\varepsilon^2 + k^2\sigma_\varepsilon^4/\sigma^2}, \quad \text{with } k = |\kappa|. \tag{4.22}$$

Thus an inverse Fourier transform of (4.21) yields the Green's function

$$g(x,y) = \frac{1}{(2\pi)^2}\int_{R^2} \hat{g}(x_1,\kappa;y)e^{-i\kappa\cdot x_\perp}\,d\kappa. \tag{4.23}$$

Using the same transform techniques, one can solve the boundary-value problem (3.19) for the mean travel time T. However, since its right-hand side (-1) has no Fourier transform as a regular function, we first set

$$\tilde{T} = T - \frac{(\ell_2 - x)}{\hat{c}_\varepsilon}. \tag{4.24}$$

Then $\tilde{T}$ satisfies a homogeneous equation with an inhomogeneous boundary condition:

$$\begin{cases} L\tilde{T} = 0, & \ell_1 < x_1 < \ell_2, \\ \tilde{T}|_{x_1=\ell_1} = -(\ell_2-\ell_1)/\hat{c}_\varepsilon, \\ \tilde{T}|_{x_1=\ell_2} = 0. \end{cases} \tag{4.25}$$

Now we can apply the Fourier transform method to (4.25). This solution, after simplification, is substituted into (4.24) to give

$$T(x;\ell_1,\ell_2) = \frac{(\ell_2-x_1)}{\hat{c}_\varepsilon}\{1 - \frac{1}{2\pi}\int_0^\infty kJ_0(k|x_\perp|) \times \frac{\operatorname{sh}[(\ell_2-x_1)\theta(k)]}{\operatorname{sh}[(\ell_2-\ell_1)\theta(k)]}\,dk\}, \tag{4.26}$$

where $J_0(r)$ is the Bessel function of order zero. To simplify the expression (3.47), we notice the fact that, from (4.22) $\theta(k) > \hat{c}_\varepsilon/\sigma_\varepsilon^2$ for $k > 0$ and

(4.27) $\quad \mathrm{sh}[\ell\Theta(k)] \sim \frac{1}{2} e^{\ell\{1/2\nu_\varepsilon + k^2/\nu_\varepsilon\}}$ for $\ell > 0$ and $(k\varepsilon^2) \ll 1$,

Making use of (4.27) in (4.26), we obtain the asymptotic expansion for T:

$$T(x;\ell_1,\ell_2) \sim \frac{(\ell_2 - x_1)}{\hat{c}_\varepsilon}\{1 - \frac{1}{2\pi} e^{-(1/2\nu_\varepsilon)(x_1 - \ell_1)} \times \tag{4.28}$$

$$\times \int kJ_0(k|x_\perp|)e^{-(k^2/\nu_\varepsilon)(x_1 - \ell_1)}dk\}, \quad \text{as} \quad \varepsilon \downarrow 0.$$

Here again, we see clearly the presence of the plane boundary $x_1 = \ell_1$ introduces an exponentially small correction term, which vanishes as $\ell_1 \to -\infty$. In this limit, similar to the case of plane wave, the mean travel time equals to the effective travel time with a modified wave speed $\hat{c}_\varepsilon$, since $\alpha$'s were defined differently.

## 5. Travel Time for Random Wave Equation

For the random wave equation (2.3), under the weak fluctuation assumption (2.12), we are interested in the asymptotic properties of the travel time in the diffusion limit. Upon substituting (2.12) into the characteristic equations (2.5), we get

$$\begin{cases} \dfrac{dx_t^\varepsilon}{dt} = c^\varepsilon(x_t^\varepsilon,t,\omega)\hat{p}_t^\varepsilon, & x_0^\varepsilon = x, \\[2ex] \dfrac{dp_t^\varepsilon}{dt} = -\varepsilon\partial_x\xi(x_t^\varepsilon,t,\omega)|p_t^\varepsilon|, & p_0^\varepsilon = p. \end{cases} \tag{5.1}$$

where, referring to (2.12), $c^\varepsilon = c_0 + \varepsilon\xi$. To rewrite it in a form of stochastic acceleration, we set

$$v^\varepsilon = c^\varepsilon\hat{p}_t^\varepsilon. \tag{5.2}$$

Then it is simple to check that (5.1) and (5.2) imply that

$$\begin{cases} \dfrac{dx_t^\varepsilon}{dt} = v^\varepsilon(x_t^\varepsilon,t,\omega), & x_0^\varepsilon = x, \\[2ex] \dfrac{dv_t^\varepsilon}{dt} = \varepsilon F^\varepsilon(x_t^\varepsilon,v_t^\varepsilon,t,\omega), & v_0^\varepsilon = v, \end{cases} \tag{5.2}$$

where

$$F^{\varepsilon}(x,v,t,\omega) = [c^{\varepsilon}(x,t,\omega)]^{-1}[\partial_t \xi(x,t,\omega) + 2\hat{v} \cdot \partial_x \xi(x,t,\omega)]v - c^{\varepsilon}(x,t,\omega)\partial_x \xi(x,t,\omega). \tag{5.3}$$

In contrast with the simple wave problem, the present problem is more complex. From (5.1) or (5.21) one sees that, in order to study the travel time, we have to go to the $x$ - $p$ (phase) space. But it is instructive to consider a special case before we discuss the full problem.

## 5.1 A Simple Wave Approximation

In view of (5.1), the first equation says the wave velocity equal to the product of the wave speed $c^{\varepsilon}$ and the direction $\hat{p}_t^{\varepsilon}$ of propagation, both fluctuating randomly. The direction $\hat{p}_t^{\varepsilon}$ is governed by the second equation which has a solution $p_t = p + O(\varepsilon)$ as $\varepsilon \downarrow 0$. If we approximate $\hat{p}_t^{\varepsilon}$ in the first equation by $\hat{p}$, it reduces to a simple wave problem

$$\begin{cases} \dfrac{dx_t^{\varepsilon}}{dt} = [c_0 + \varepsilon\xi(x_t^{\varepsilon},t,\omega)]e, \\ \\ x_0^{\varepsilon} = x, \end{cases} \tag{5.4}$$

where $e = \hat{p}$. Without the loss of generality, let us set $e = (1,0,0)$ or

$$e_i = \delta_{1i}.$$

Then the problem becomes identical with case 4.1 of Section 4, where $v_0$ and $\rho$ in (4.1) are replaced by $c_0$ and $\xi$, respectively. Again consider the travel time $\tau_x(D)$ for a plane wave from $x_1 = \ell$ to the planes $x_1 = \ell_1$ or $\ell_2$ in the $x_1$-direction. Then, assuming that $\xi$ has the same statistical properties of $\rho$, the first two moments of $\tau_x$ are given by (4.10) and (4.12), respectively. As $\ell_1 \to -\infty$, we get the moments of $\tau_\ell(\ell_2)$ from $x_1 = \ell$ to $x_1 = \ell_2$

$$T(\ell) = (\ell_2 - \ell)/\hat{c}_\varepsilon$$

(5.5)
$$T_2(\ell) = \frac{(\ell_2 - \ell)^2}{(\hat{c}_\varepsilon)^2} + \frac{\sigma_\varepsilon^2}{2(\hat{c}_\varepsilon)^3}(\ell_2 - \ell),$$

or the standard deviation of $\tau_\ell(\ell_2)$ is

(5.6)
$$\sigma_\ell(\ell_2) = \sigma_\varepsilon(\ell_2 - \ell)^{1/2}/\sqrt{2}(\hat{c}_\varepsilon)^{3/2}.$$

A closer look at the expressions (5.5) and (5.6) reveal the following. The mean travel time $T(\ell)$, in comparison with the unperturbed value $T(\ell) = (\ell_2 - \ell)/c_0$, increases or decreases depending on $\hat{c}_\varepsilon$. Since $\hat{c}_\varepsilon < c_0$ if $\alpha < 0$ $(\alpha > 0)$, the mean travel time increases (decreases). But, noting (3.25),

$$\alpha = \int_0^\infty \gamma'(c_0 s, s)ds$$

where $\gamma(|x|,t) = E\xi(x,t)\ \xi(0,0), \gamma'(r,t) = \partial_r\gamma(r,t)$. From physical consideration, the correlation function $\gamma(r,t)$ falls off as the distance $r$ increases. Thus $\gamma' < 0$, or $\alpha < 0$. As expected, the mean travel time $T$ increases due to random scattering. The standard deviation from the mean $\sigma_\ell(\ell_2)$ given by (5.6) is proportional to $\sigma_\varepsilon = \varepsilon\sqrt{\beta}$, inversely to $(\hat{c}_\varepsilon)^{3/2}$. Also it grows like the square-root of the travel distance $\Delta\ell = (\ell_2 - \ell)$. Therefore, the mean value $T(\ell)$ is a good measure of the travel time only when $(\varepsilon^2\Delta\ell) \ll 1$.

For numerical examples, we take $c_0 = 1$,

(5.7)
$$\gamma(r,t) = \left(\frac{e^{-\alpha_1 t}}{1 + r^2}\right), \quad \alpha_1 > 0.$$

Some numerical results are displayed in four graphs, Figs. 1-4 at the end of the paper. In Fig. 1, the mean travel time $T$ is plotted against the distance $x = \Delta\ell$, with $\alpha_1 = 0.7$, $\varepsilon = 0$, 0.2 and 0.5, respectively, while $\alpha_1$ is changed to 1.0 in Fig. 2. They clearly showed the increase in mean travel time with $\varepsilon$. In Figs. 3 and 4, the three lines (or curves) correspond to the mean travel time $T$ and its deviations $T \pm \sigma_\ell(\ell_1)$ v.s. $x = \Delta\ell$, with $\varepsilon = 0.5$ fixed and $\alpha_1 = 0.7$ and 1.0, respectively. They show the increase in deviations with the distance and the deviations decrease as the decay rate $\alpha_1$ increases. Even though they are approximate results, the qualitative feature may hold in general.

## 5.2 General Problem

If we take the random change of direction $\hat{p}_t^\varepsilon$ into acount, the coupling of the equations in (5.1) or (5.2) must be considered. For simplicity we first suppose the medium, in addition to the previous assumptions, is time-independent. Let us rewrite (5.3) as

$$F^\varepsilon = F + \varepsilon G + O(\varepsilon^2), \tag{5.8}$$

where

$$\begin{aligned} F &= \frac{1}{c_0}\,[2(\hat{v}\cdot\partial_x\xi)v - c_0^2\partial_x\xi] \\ G &= -2c_0^{-2}\xi(\hat{v}\cdot\partial_x\xi)v - \xi\partial_x\xi \end{aligned} \tag{5.9}$$

Here we define, as in [10]

$$\begin{aligned} b_{ij}(v) &= \int_{-\infty}^{\infty} EF_i(0,v)F_j(tv,v)dt, \\ a_i(v) &= g_i(v) + \int_0^\infty EF_j(0,v)\,\frac{d}{dv_j}\,F_i(tv,v)dt, \end{aligned} \tag{5.10}$$

where

$$\begin{aligned} &\frac{d}{dv_j}\,F_i(tv,v) = \{t\partial_{x_j}F_i(x,v) + \partial_{v_j}F_i(x,v)\}|_{x=tv}, \\ &\text{and} \\ &g_i(v) = EG_i(x,v). \end{aligned} \tag{5.11}$$

In view of (5.9), the coefficients $a_i$ and $b_{ij}$ in (5.10) can be expressed in terms of the correlation function $\gamma$ for $\xi$ and its derivatives. However they will not be written down. Now, in the diffusion limit, we invoke a theorem of Kesten and Papanicolaou [10] to infer that

$$(x_t^\varepsilon, v_t^\varepsilon) \Rightarrow (x_t, v_t), \quad \text{as} \quad \varepsilon \downarrow 0, \quad t \uparrow \infty, \ (\varepsilon^2 t) \ \text{fixed},$$

which is jointly a diffusion process with the generator

$$\begin{aligned} Lu &= \frac{1}{2}\,\varepsilon^2 b_{ij}(v)\partial^2_{v_i v_j}u + \varepsilon^2 a_j(v)\partial_{v_j}u + v_j\partial_{x_j}u \\ &= L_1 u + v_j\partial_{x_j}u. \end{aligned} \tag{5.12}$$

In order to compute the statistical properties of the travel time $\tau_x(D)$ in physical space, we must first find the invariance distribution of $v_t$. If there exists a smooth density $q(v)$, it satisfies, see e.g. [12]

$$(5.13)\qquad \begin{cases} L_1^* q = 0, \\ q > 0;\ \int q(v)dv = 1 \end{cases}$$

where $L_1^*$ is the formal adjoint of $L_1$. Next we define the travel time in the phase-space

$$(5.14)\qquad \tau_{x,v}(B_1,B_2) = \inf\{t > 0: x_t \in B_1 \text{ and } v_t \in B_2\}$$

where $B_1$, $B_2$ are some regions in $R^d$. It seems plausible that the moments of $\tau_{x,v}$, if exist, may be calculated by solving the boundary value problems similar to (3.20). In particular the mean travel time $\tilde{T}$ would satisfy the boundary-value problem

$$(5.15)\qquad \begin{cases} L\tilde{T}(x,v) = -1 \text{ in } (B_1 \times B_2)^c, \\ \tilde{T} = 0 \text{ on } \partial(B_1 \times B_2). \end{cases}$$

However, since $L$ is a degenerate elliptic operator, the boundary-value problem (5.15) may not be well-posed for a certain region $B_1$. This basic question has to be carefully investigated. Secondly, even if the problem (5.15) has a unique solution, since $L$ has variable coefficients, analytic solutions would be difficult without further approximations. We note that once $\tilde{T}$ is found, the physical mean travel time is obtained by an integration

$$T(x) = \int \tilde{T}(x,v)q(v)dv.$$

As seen from the above discussion, the physicsally interesting general problem is still an open problem. This is a subject of our further investigation.

**Acknowledgements**

The author wishes to thank Professor George Papanicolaou for inviting him to participate in the Workshop on Random Media at IMA. To Professor Joe Keller the author is grateful for his interest shown in this work and his helpful suggestion of a needed numerical computation which resulted in the graphs displayed in Figs. 1-4. Also he is thankful to the IMA, in particular to Professors Hans Weinberger and George Sell, for their hospitality and the support that made it possible to cmplete this work.

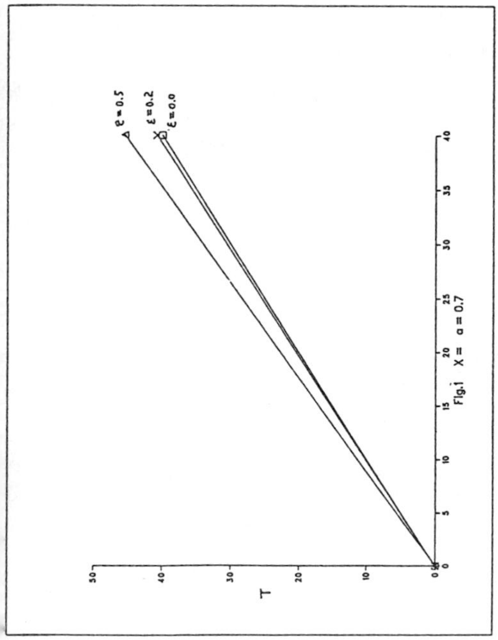


Fig.1 X = a = 0.7

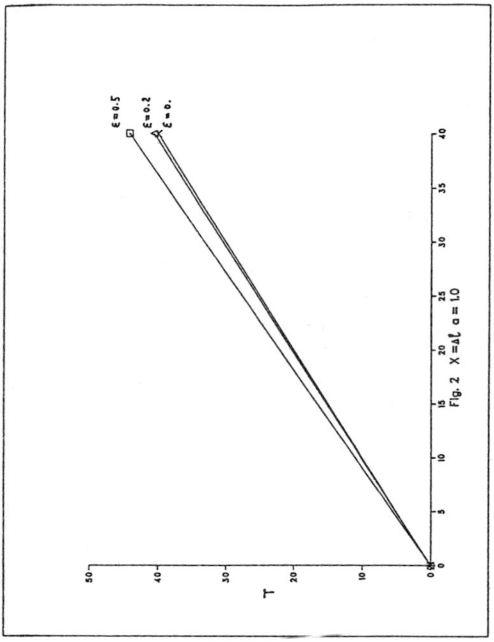


Fig. 2 X = Δℓ a = 1.0

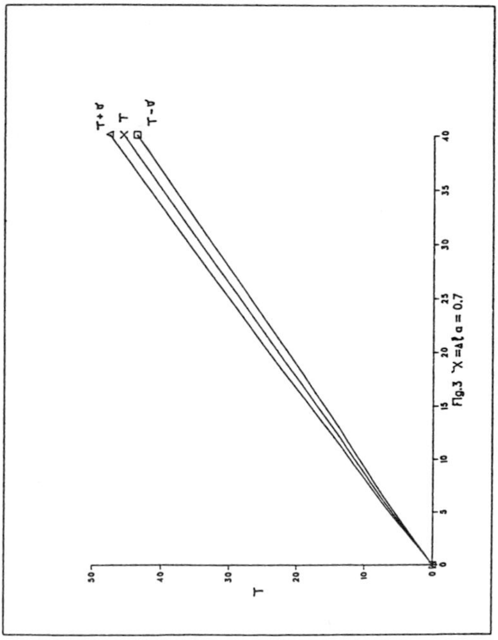


Fig.3 X = Δℓ a = 0.7

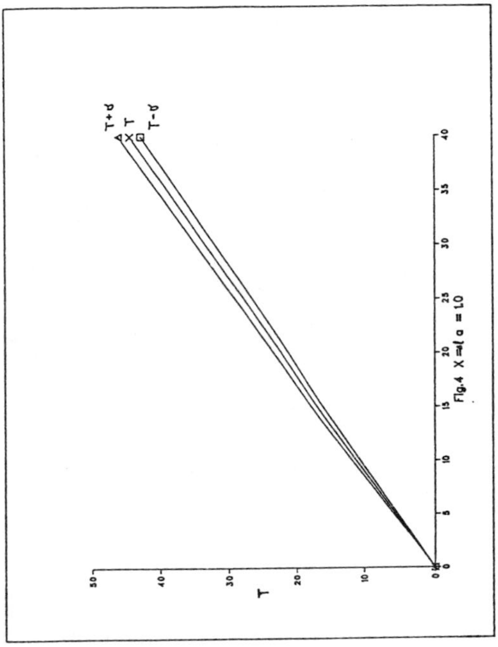


Fig.4 X = ℓ a = 1.0

## References

1. A. Ishimaru, Wave Propagations and Scattering in Random Media, vols. I, II, Acad. Press, N.Y., 1982.

2. C.H. Liu and K.C. Yeh, Opt. Soc. Amer., 70, (1980), p. 168.

3. P.L. Chow, Method of progressing in waves in random media, Proc. Symp. on Multiple Scattering in Random Media, to appear.

4. R. Burridge, G. Papanicolaou and B. White, Statistics for pulse reflection from a randomly layered medium, this volume.

5. R. Courant and D. Hilbert, Methods of Mathematical Physics, Vols. I, II, Wiley-Interscience, N.Y., 1962.

6. S.C. Port and C.J. Stone, Brownian Motion and Classical Potential Theory, Acad. Press, N.Y., 1978.

7. R.Z. Khasiminskii, Theory Prob. Appl. 11, (1966), p. 211.

8. G.C. Papanicolaou and W. Kohler, Comm. Pure and Appl. Math., 27 (1974), p. 641.

9. H. Kesten and G. Papanicolaou, Comm. Math. Phys. 65, (1979), p. 97.

10. ______________________________, Comm. Math. Phys. 78, (1980), p. 19.

11. M. Freidlin, Functional Integration and Partial Differential Equations, Princeton University Press, 1985.

12. S.R.S. Varadhan, Diffusion Problems and Partial Differential Equations, T.I.F.R. Lectures, Springer-Verlag, N.Y., 1980.

# ELEMENTS OF THE THEORY OF AMORPHOUS SEMICONDUCTORS*

Morrel H. Cohen

Exxon Research and Engineering Company
Annandale, NJ 08801 USA

**Abstract**

The main elements of the theory of amorphous semiconductors are reviewed. The kinds of disorder occuring in these materials are classified. The consequences of the various kinds of disorder for electronic states and energies at and near the band edges and in the gaps are discussed. The energy spectrum is divided into ranges according to the principal characteristics of the states and the spectrum, localized vs. nonuniversal, smooth vs. fractal, etc. The effects of interactions are discussed. The general theory of transport is reviewed. The consequences of the above for the optical and the transport properties are discussed briefly.

**Introduction**

The investigation of amorphous semiconductors constitutes a large, highly developed field of science. There is a broad range of materials studied. Diverse methods are used to prepare them. Many different physical properties are measured. A wide and flexible set of experimental techniques is used for the measurements. As a consequence vast amounts of data exist.

Nonetheless, much of that information is poorly understood. There are conflicting interpretations found within the literature. Inapplicable concepts are used, sometimes carried over incorrectly from crystalline semiconductor physics and sometimes generated de novo. There is, in my opinion, much confusion in the field.

Why is this so? Again in my opinion, the intrinsic complexity of the materials has inhibited the development of the theory. As a consequence, theory

---

* Substantially the same material will appear in "Hydrogen in Amorphous Solids", Nato ASI, Series B, Plenum Press (in press).

has had little impact on the development of the subject, in contrast to the case of crystalline semiconductors, where there has been a healthy symbiosis between theory and experiment.

It is the presence of disorder which leads to the great complexity of amorphous semiconductors; that is, it is the loss of long-range order which makes the theory of amorphous semiconductors so much more difficult than that of crystalline semiconductors. The materials have compositional disorder; that is, they are rarely pure materials, and the constituents are not regularly arranged (e.g., a-SiHx, a-si$F_x$, a-$H_xF_y$). The materials also contain structural disorder, both geometric and topological in nature. Because the materials are covalently bonded, the geometric and topological disorder have certain characteristic features. The geometric disorder is comprised of random bond-length variation, bond-angle variation, and dihedral-angle variation. The topological disorder derives from the presence of coordination defects, odd rings, and other, subtler features. There may be inhomogeneity in composition, in the range of order (e.g., microcrystallites embedded in an amorphous matrix), in the distribution of defects (e.g., dangling bonds may cluster on internal surfaces, odd rings are threaded by line defects). Another complication is the multiplicity of atomic orbitals required to describe the electronic structure of the material. At the very least, there is needed a set of one s- and three p-orbitals for each host element and possibly an excited s-orbital and 5 d-orbitals for a more accurate treatment of the conduction band. Beyond that, one needs orbitals for any hydrogen and fluorine present. This multiplicity is a major complication; most exact or accurate theoretical results have been obtained only for models containing a single s-orbital. The final complication is the presence of interactions. Electron-phonon interactions are always important, and electron-electron interactions are important in certain specific circumstances.

Despite this complexity some progress has been made recently towards a definitive theory. The purpose of the present paper is to provide a selective review of the present state of the theory, concentrating on the contributions of my colleagues and/or myself.

The paper is organized as follows. Section II contains a discussion of band bounds in disordered materials, distinguishing between band edges and band limits. Section III introduces the mobility-edge concept. Section IV introduces the current version of the simplest band model of an amorphous semiconductor. Section V contains a discussion of the band-edge features in the electronic structure of a disordered material and classifies them according to the degree to which they can be represented as universal. In Section VI the effects of electron-phonon interaction are discussed. In Section VII there is a brief discussion of fast processes such as the optical absorption and in Section VIII of slow processes such as the dc transport properties. We conclude in Section VIII with an overall summary of the present status of the theory.

**Band Bounds, Limits, and Edges**

From the existence of an optical-absorption edge and an activation energy in the d.c. conductivity of amorphous semiconductors, it has been inferred that the materials possess energy gaps in the presence of disorder. At the simplest level, this has been explained in terms of the covalent character of the material. Most atoms have their valence requirements locally satisfied so that it costs energy to break a bond, hence an energy gap. However, it is possible to go beyond such primitive arguments and prove rigorously that for certain simple models of disordered systems, the energy bands can have sharply defined bonds. This was shown first in 1964 by Lifshitz, and Weaire and Thorpe showed in 1971 that gaps can exist.

Consider the example of a tightly bound s-band for which the states are linear combinations of s-orbitals centered on the sites of a regular lattice. The matrix elements of the Hamiltonian are

$$H_{\ell m} = \varepsilon_\ell \delta_{\ell m} + V_{\ell m}(1 - \delta_{\ell m}) \qquad (1)$$

where $\ell$, m indicate the s-orbitals on sites $\ell$ and m of the lattice. Randomness in the diagonal (off-diagonal) elements is termed diagonal (off-diagonal) disorder.

There exist two kinds of bounds. To illustrate the first kind, suppose there is diagonal disorder only, with the $\varepsilon_\ell$ independent random variables each having an average value $\varepsilon$, the same as for the perfect crystal, and with the $V_{\ell m}$ as in the crystal. Let $p(\varepsilon_\ell)$ be the probability distribution of the individual $\varepsilon_\ell$. If $p(\varepsilon_\ell)$ is bounded, that is

$$p(\varepsilon_\ell) = 0, \quad \varepsilon_\ell < \varepsilon_L \quad \text{or} \quad \varepsilon_U, \tag{2}$$

then the energy band has a lower bound $\varepsilon - \varepsilon_L$ below the bottom of the unperturbed energy band and an upper bound $\varepsilon_U - \varepsilon$ above its top. The states just inside the bounds are confined to large regions within which the $\varepsilon_\ell$ are all nearly equal to $\varepsilon_L$ or $\varepsilon_U$, respectively. Such regions are highly improbable, and the density of states vanishes strongly as the band bounds are approached,

$$n(E) \propto e^{-A_1/|E-E_L|^{d/2}} \tag{3}$$

as $E \to E_L^+$, the lower bound, and similarly for $E_U$, the upper bound. Such bounds we call band limits or Lifshitz limits, after their discoverer.

If $p(\varepsilon_\ell)$ is unbounded, i.e. $\varepsilon_L \to -\infty$ and $\varepsilon_U \to \infty$, as is the case for, e.g., a Gaussian or a Lorentzian and if $|E - E_{L,U}| \gg B$, the unperturbed band width,

$$n(E) \sim p(E) \tag{4}$$

holds, and $n(E)$ has no bounds or limits.

The second kind of bound is a normal band edge $E_N$ at which

$$n(E) \propto |E - E_N|^{(d/2 - 1)} \tag{5}$$

inside the band and zero outside, where $d$ is the Euclidean dimension of the material. The behavior in (5) is the same as for a crystal and occurs when the disordered material has a hidden symmetry. Consider a tight-binding s-band for which all $\varepsilon_\ell = 0$, all $V_{\ell m} = V < 0$ for $\ell m$ nearest neighbors and zero otherwise, all sites have the fixed coordination number $z$, and otherwise the structure is arbitrary. Such a structure is an ideal continuous random net, without coordination defects, and the Hamiltonian contains topological disorder only, no quantitative disorder. It is easy to prove that the Hamiltonian (1) has a spectrum bounded by $\pm zV$. The wave function at the bound $zV$ is of bonding type and is translationally invariant. As a consequence of this hidden symmetry, the bonding bound is a normal band edge with $n(E)$ given by (5). The wave function at the other bound, $-zV$, is antibonding and of equal amplitude everywhere but of opposite sign for all nearest neighbor pairs. Such a state can be realized only in a net having no odd rings, a bichromatic net. The antibonding bound is thus a normal band edge only in the presence of another hidden symmetry, evenness of all rings in the structure. By rings we mean a closed graph of lines connecting nearest neighbors, an even ring having an even number of lines. In the presence of odd rings, the antibonding bound becomes a Lifshitz limit.

To summarize, if $p(\varepsilon_\ell)$ is unbounded the energy band is unbounded. If the band is bounded and there is sufficient symmetry remaining, the bound is a normal band edge. Otherwise, it is a Lifshitz limit.

Let us turn now to a more complex model, one considered by Singh for the representation of a-Si in a 4-coordinated CRN structure. It is a tight-binding model with a basis set consisting of an $s$, $p_x$, $p_y$, and $p_z$ orbital on each atomic site. If the structure has odd rings, but no bond-length, bond-angle, or dihedral-angle disorder, the top of the valence band is a normal band edge corresponding to a 3-fold degenerate bonding $p$ state. As soon as dihedral-angle disorder is introduced, the top of the valence band turns into a Lifshitz limit. When bond-length and bond-angle disorder are added, the Lifshitz limit is smeared out. Similarly, in the presence of dihedral-angle disorder but no odd rings or bond-length or bond-angle disorder, the bottom of the conduction band is a normal band edge corresponding to a doubly-degenerate antibonding $\sigma$ state (for values of the matrix elements which would put the bottom of the conduction band at the zone boundary for the crystal). Adding odd rings turns the normal band edge into a Lifshitz limit, and including bond-length and bond-angle disorder smears out the Lifshitz limit.

From this simplified but still realistic model, we are led to two conclusions. First, band bounds are smeared out in general, and therefore we must determine $n(E)$ and the features of the wave functions explicitly for each class of materials or models. Second, in real materials a fairly large number of matrix elements is required for a tight-binding model which accurately represents the electronic structure. The various kinds of disorder present affect these matrix elements and the states near the different band edges in quite different ways. These two circumstances lead to the approximate validity of two simplifying assumptions which we shall explore in the following:

1) We can use a tight-binding model without orbital multiplicity to represent the conduction or valence band of real materials, but we must suppose the disorder potential to have a Gaussian probability distribution. That is, the complexity of the real problem allows us to invoke the central limit theorem.
2) The valence and conduction bands can be taken as statistically independent because individual elements of the disorder affect the valence and conduction band edges quite differently.

**The Mobility Edge Concept**

We now suppose that either there is no electron-phonon interaction or $T = 0$. In a crystal, there are extended states within the energy bands and no states within the gaps. The extended states are Bloch states which have equal amplitudes in every unit cell and complete phase coherence. If an isolated defect or impurity is introduced into the material, localized states can appear within the energy gaps. By localized states, we mean states which decay exponentially in amplitude asymptotically far away from the center of the region of localization, in this case the impurity or defect responsible for this existence of the state. Even though the band edge is smeared out in a disordered material, localized defects or impurity states can still occur deeper in the gap. These will no longer be sharp in energy, as in a crystal, but will be broadened by the disorder. These, however, are not the only localized states. Even in the absence of such defect or impurities, as in an ideal CRN, there is an energy $E_c$, called the mobility edge for reasons which will emerge below, which divides localized states in the tail of the energy band from extended states within the band. These localized states also decay exponentially, but the decay length, the so-called localization length, diverges as the mobility edge is approached. The extended states are also quite different from those in a crystal. The phase becomes incoherent over distances greater than the mean-free-path $\ell$, and violent amplitude fluctuations set in on distance scales between $\ell$ and the amplitude coherence length $\xi$, which also diverges at the mobility edge. In fact, the wave functions are fractal between those limits.

In the absence of the electron-phonon interaction or when the important phonon energies are substantially less than $k_B T$, we can write the Kubo-Greenwood formulas for the dc conductivity in the following form,

$$\sigma = \int dE\ \sigma(E)\left(-\frac{df(E)}{dE}\right) \tag{6}$$

$$\sigma = -\int dE n(E) e\mu(E) f(E) \tag{7}$$

$$\frac{d\sigma(E)}{dE} = -n(E)e\mu(E) \tag{8}$$

In eq. (6), $\sigma(E)$ is the microscopic conductivity, and $f(E)$ is the electron occupation number for a state of energy $E$. In Eq. (7), $\mu(E)$ is the microscopic mobility; Eq. (8) relates the two. It has been shown that in the absence of the electron-phonon interaction,

$$\sigma(E) \propto (E - E_c)^s, \quad E \to E_c^+ \tag{9a}$$

$$= 0 \qquad , \quad E < E_c$$

Eq. (9b) follows from the fact that electrons in localized states $(E < E_c)$ are immobile, whereas electrons in extended states $(E > E_c)$ can move macroscopic distances. Eq. (8) implies that $s \geq 1$, and the best current theories give $s = 1$. In that case Eq. (8) and Eq. (9a,b) imply that there is a step in the mobility from zero to a finite value at $E_c$. Thus $E_c$ is termed a mobility edge.

Eq. (9a) holds only near the mobility edge, where amplitude fluctuations suppress the conductivity. A major task of the theory is to attach an energy scale to the variation of $\sigma(E)$ and other quantities with energy. However, supposing that the characteristic energy scale is larger than $k_BT$ leads immediately to the observed activated temperature dependence of the conductivity. Thus with the two concepts of somewhat smeared band edges but sharp mobility edges in disordered bands, we can understand the most characteristic properties of amorphous semiconductors, the apparent optical absorption edge and the activated dc conductivity.

**The Simplest Band Model Of An Amorphous Semiconductor**

I shall use as an updated version of the original Mott-CFO model of the energy bands of an amorphous semiconductor a sketch of the density of states in a-$SiH_x$ as it is now emerging from a wide variety of experiments. In Fig. 1a we show $n(E)$ vs $E$. There is a valence band and within it a mobility edge $E_v$, and a conduction band and within it a mobility edge $E_c$. Below $E_v$ and above $E_c$ the states are extended. Between $E_v$ and $E_c$ the states are localized. Both bands have an exponential tail, the width being about 500K for the valence band and 300K for the conduction band.

Near the center of the gap there is a bump associated with neutral 3-fold coordinated Si atoms, the dangling bond state $D^\circ$, and above that a bump corresponding to the same state doubly occupied, $D^-$, the shift being due to the repulsion between the two electrons of opposite spin occupying the dangling bond state in $D^-$. The region between the two mobility edges is called the mobility gap $E_c - E_v$ because the mobility vanishes there even though the density of states remains finite (at $T = 0$ or neglecting the electron-phonon interaction).

In Fig. (1b) are shown the corresponding microscopic conductivities and mobilities. However, to complete the correspondence between Figs. (1a) and (1b) it is necessary to establish the energy scale for Fig. (1b), that for Fig. (1a) being known in most respects. Moreover, the effects of the electro-phonon interaction must be included as well. Thus, Figs. (1a) and (1b) enable us to understand the gross features of the optical absorption and the dc conductivity, but they are incomplete.

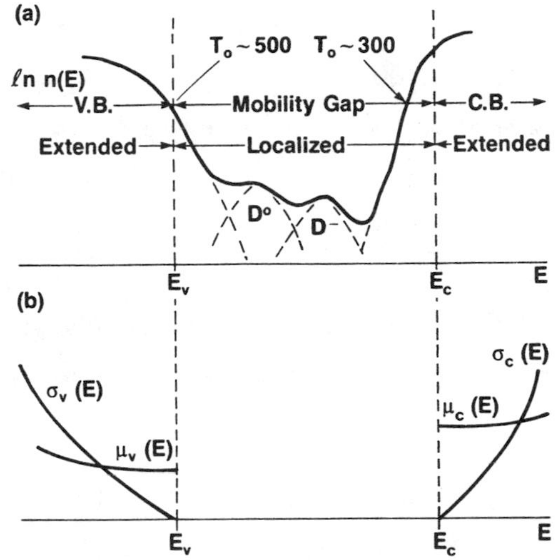


**FIGURE 1**

Fig. 1. (a) The density of states of a-$SiH_x$ after Fritzsche's synthesis of the experimental data. (b) A sketch of the microscopic conductivities and mobilities for the conduction and valence bands.

## Universality, Quasiuniversality, and Pseudouniversality in the Electronic Structure at Band Edges

Electronic properties arising from states near band edges appear to be quite similar in very different amorphous semiconductors. For example, the optical absorption $\alpha$ always displays a Tauc region in which

$$\{\alpha/\hbar\omega\}^{1/2} = B[\hbar\omega - E_{Go}] \tag{10}$$

and at lower photon energies $h\omega$ an Urbach region in which

$$\alpha = \alpha_0 e^{\hbar\omega/E_{00}}, \tag{11}$$

where B and $\alpha_0$ are constants, $E_{G0}$ is the apparent optical or Tauc gap, and $E_{00}$ is the width of the Urbach tail. Another example is provided by the dc transport in which similar patterns of behavior of $\sigma$ and the thermpower S are found in dissimilar materials. These features of the data lead one to infer that there are universal or nearly universal features in the electronic structure near band edges, independent of structure, composition and the details of disorder, or nearly so.

It is a prime task for the theorist to establish whether and when such strict or approximate universality can exist. At present this task has been accomplished only for a specific class of models under the following conditions:

1. There exists a reference model from which the disorder can be measured.

2. The reference model need not be ordered but must have a normal band edge.

3. There is no orbital multiplicity in a tight-binding representation of the Hamiltonian.

4. The disorder is measured quantitatively via the variance $w^2$ of the random potential.

5. $w$ must be substantially less than the unperturbed band width B.

6. Only states of energies E deviating from the unperturbed band edge by much less than the band width are considered.

7. Only distance scales much greater than $\lambda$, the larger of the interatomic separation $a$ or the correlation length $\ell_c$ of the random potential are considered.

8. The continuum limit of the tight-binding model most have no ultraviolet catastrophe, that is, no sensitivity to the short-wavelength cutoff $\lambda$ or the high-energy cutoff B. The continuum limit is taken in such a way that $w \to \infty$, $B \to \infty$, $a \to 0$, $\lambda \to 0$ so that

$$Va^2 = \frac{\hbar^2}{2m^*} = \text{const}, \quad w^2\lambda^d = \gamma = \text{const}. \tag{12}$$

In Eq. (12) $V$ is an effective nearest neighbor electron-transfer matrix element ($V \to \infty$ as $B \to \infty$), $m^*$ is the effective mass of the reference model, and $d$ is the dimension of the model. No ultraviolet catastrophe occurs for $d < 2$, but it does occur for $d \geq 2$ leading to a shift of the unperturbed band edge to $E_{Bd}$. When energies are measured relative to $E_{Bd}$, the possibility of universality is restored.

The proof of universality follows from the observation that the model contains only 2 parameters, $\hbar^2/2m^*$ and $V$. A natural unit of energy $E_{od}$ and one of length $L_{od}$ can be defined from these two parameters. In the simpler case that $\lambda = a$, $E_{od}$ and $L_{od}$ have the particularly simple forms

$$E_{od} = W^{\frac{4}{4-d}} V^{-\frac{d}{4-d}} \tag{12}$$

$$L_{od} = a(V/w)^{\frac{2}{4-d}}, \tag{13}$$

$d \neq 4$. When all energies are measured relative to $E_{Bd}$ in units of $E_{od}$, lengths in units of $L_{od}$, and conditions (1.) to (7.) are met, then all physical properties can be expressed either as universal functions or universal numbers. As an example of the former, consider the density of states,

$$n(E) = L_{od}^{-d} E_{od}^{-1} f_d((E - E_{Bd})/E_{od}), \tag{14}$$

where $f_d(x)$ is a universal function of its argument $x$ for each $d$. As an example of the latter consider the position of the mobility edge $E_c$,

$$E_c - E_{bd} = C_{1d} E_{od}, \quad 2 < d, \tag{15}$$

where $C_{13} = 4.5 \times 10^{-3}$ and $C_{1d} = 0$, $d > 4$. There are two limitations to universality deriving from the energy dependence of conditions (6.) and (7.) even when all the conditions are met. Condition (6.) breaks down both in the continuum and in the gap when $|E - E_{Bd}|$ approaches $V$. Condition (7.) breaks down in the gap when the scale $L$ of the potential fluctuations important for localization becomes smaller than $\lambda$.

We have obtained a broad range of detailed quantitative results relating to universality in the absence of the electron-phonon interaction by various theoretical techniques. The methods include perturbation theory, the coherent potential approximation (CPA), field theory, path integral methods, numerical calculations, and the potential well analogy. The results include the density of states, the nature of the wave functions, the mean free path, the energy dependent conductivity and mobility, and the frequency-dependent dielectric function. We know how to add the electron-phonon interaction for fast processes, and some progress has been made for slow processes.

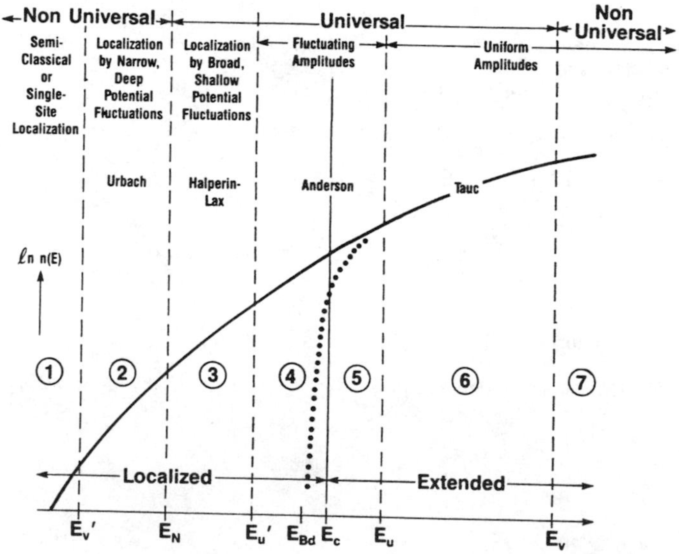


**FIGURE 2**

Fig. 2. The seven distinguishable energy regions near the band edge of a simple model of a disordered material (no multiplicity of orbitals).

In Fig. 2, we show the distinct energy regions into which the spectrum divides near the band edge and summarize the principal characteristics of the density of states and the corresponding features of the wave functions. There is of course a sharp transition from localized to extended states at the mobility edge $E_c$. However, there is far more structure in the spectrum than a simple division into localized and extended states. There are in fact seven distinguishable regions, four within the localized domain and three within the extended domain. Only the transition from localized to extended states at $E_c$ is sharp. The remaining five transitions are in fact smooth crossovers.

In addition to the localized/extended classification there are two other classifications used in Fig. 2: universal/nonuniversal and fluctuating/smooth. Universal behavior is found in regions 4, 5, and 6 of Fig. 2 (between $E_N$ and $E_V$) including both localized states in 4, and extended states in 5 and 6. Nonuniversality occurs in region 7 because the condition $|E - E_{Bd}| << V$ is violated at $E_V$, and the electronic properties can no longer be represented by universal forms such as in Eqs. (14) and (15). Nonuniversality occurs also in regions (1.) and (2.) because the length scale $L$ of the relevant potential fluctuations becomes smaller than $\lambda$ at $E_N$. Amplitudes are smooth below $E_u$, and above $E_{u'}$ in regions (1.)-(3.) and (6.), (7.), respectively. Within regions (4.) and (5.) there are strong amplitude fluctuations on length scales lying between the mean free path $\ell$ (roughly) and the localization length $L_{loc}$ in (4.) and the amplitude coherence length $\xi$ in (5). This amplitude fluctuation occurs because of resonances in potential fluctuations not quite strong enough for localization. The crossover energy $E_N$ occurs when $\xi$ falls below $\ell$, and the other one, $E_{u'}$, occurs when $L_{loc}$ approaches within a factor of 2 of $\ell$. We think it appropriate to term regions (4.) and (5.) the Anderson regime. In more quantitative terms, n(E) is accurately given by the CPA in regions (6.) and (7.). In region (6.), which we term the Tauc region, the density of states has the normal form, Eq. (5), with $E_N$ replaced by $E_{Bd}$ and the coefficient fixed by universality. Transitions from region (6.) in the valence band to region (6.) in the conduction band are responsible for the Tauc region of the optical absorption ($d = 3$), Eq. (10), and give

$$E_{Go} = E_{B3c} - E_{B3v} \tag{16}$$

Because $C_{13}$ is so small, cf. Eq. (15), $E_{Go}$ is nearly equal to the mobility gap $E_c - E_v$. These two statements must be understood with reference to optical absorption being a fast process, as discussed below.

In region (3.), the states are localized by broad, shallow potential fluctuations. This is the Halperin-Lax region, where the density of states is given by

$$\left.\begin{aligned} \ln n(E) &\propto -[|E - E_c|/E_{13}]^{1/2} \\ E_{13} &= 7 \times 10^{-4}\, E_{03} \end{aligned}\right\} \tag{17}$$

The interesting range between $E_N$ and $E_u$ comprising regions (3.), (4.), and (5.) is, if one chooses parameters for $w$ and $V$ appropriate to real materials for which $w/V \sim 1$, extremely narrow, and $E_{13}$ is clearly very small. Accordingly, we have used a highly nonlinear energy scale in Fig. 2, expanding the range $E_N$ to $E_u$ greatly. The range itself and its characteristic phenomena, amplitude fluctuations and the Halperin-Lax tail, should be very hard to observe. Inelastic effects at finite temperature should wipe out or smear out these characteristic phenomena, which should clearly be observable only at low temperatures.

In region (2.), the states are localized by narrow, deep potential fluctuations, and the density of states becomes a simple exponential,

$$\left.\begin{aligned} \ln n(E) &\propto - |E|/E_{23} \\ E_{23} &\cong \frac{1}{8} w^2/V \end{aligned}\right\} \tag{18}$$

but only for a Gaussian probability distribution, the one which should be used according to §2. We term this the Urbach region. The numerical coefficient in $E_{23}$ is nonuniversal, and, in fact, is probably much smaller for real materials than in the simple models. We therefore, term the Urbach region pseudouniversal.

We note that thermal disorder is also Gaussian and adds in the square to $E_{23}$ in Eq. (18), giving $E_{23}$ the observed T-dependence.

In new work by S. John, we have obtained an explicit analytic form for the density of states below $E_{u'}$, including regions (1.), (2.), and (3.), which accurately fits our exact results.

## Effects of Electron-Phonon Interactions

In amorphous semiconductors, electron-electron interactions are important only when two electrons of opposite spin occupy the same localized state or for Mott hopping at the Fermi level at very low temperature. Electron-phonon interactions are always present and more important in disordered than ordered semiconductors. The electron-phonon (e-p) interaction has three kinds of effects: coherent, elastic scattering, and inelastic scattering. The coherent effects favor small polaron formation at $T = 0$, and the eleastic scattering leads to thermal disorder at finite $T$. Both increase localization. On the other hand, the inelastic effects destroy localization. The polaron effects shift $E_c$ to higher energies and cause already localized states to be more tightly bound. Polaron resonances are formed in the range $E_c$ to $E_u$ with both energies shifted by the e-p interaction. As discussed above, this is a narrow range of energies, but within it we expect $\xi(E)$ to be substantially decreased by the presence of the small-polaron resonances and possibly the exponent $s$ in Eq. (9a) to be increased above unity.

Thermal disorder merely adds an additional Gaussian random potential whose variance adds linearly to $w^2$, as already discussed.

The inelastic processes lead to tunneling between localized states and between localized and extended and to scattering between extended states. This should substantially reduce the fractal behavior in regions (4.) and (5.) of Fig. 2, increasing the fractal dimension $D$ towards $d$, reducing $\xi$, and decreasing $E_u - E_{u'}$. In other words, inelastic scattering tends to suppress amplitude fluctuations. Moreover, there is a profound modification of the transport processes. In particular, $\sigma(E)$ is decreased above $E_u$ and increased everywhere below $E_u$, the increase growing rapidly with T.

The effect of the e-p interactions depends on the time scale of the physical process in question. In fast processes such as the optical absorption in which the frequency $\omega \gg \tau_{pol}^{-1}, \tau_{in}^{-1}$, where $\tau_{pol}$ is the time required for polaron effects to manifes themselves and $\tau_{in}$ is the inelastic scattering time, only thermal disorder plays a role, and it can be treated on the same basis as the static disorder previously described. On the other hand, in slow processes in which $\omega \ll \tau_{pol}^{-1}, \tau_{in}^{-1}$, all three effects play important roles, polaron effects, thermal disorder, and inelastic scattering.

**Fast processes: Optical Absorption**

In fast processes, the effect of the electron-phonon interaction is solely to increase $w^2$ through the additional contribution of thermal disorder. Thus, up to some energy above $E_V$, one can write the imaginary part of the dielectric constant as

$$\varepsilon_2(\hbar\omega) = (2\pi e)[d(\hbar\omega)]^2 V_a \int dE\, n_v(E) n_c(E - \hbar\omega), \tag{19}$$

where $V_a$ is the atomic volume and $[d(h\omega)]^2$ is the mean squared matrix element of the coordinate operator. Both experimentally and theoretically, the latter has been shown to be constant, independent of $\omega$, up to some value of $\hbar\omega$ above the upper limit of the Tauc regime, i.e. above $E_{cV} - E_v$. This constancy of the matrix element is easily understood in terms of statistical independence of the valence and conduction band wave functions and their phase incoherence beyond the mean free path. In a-si$H_x$, the observed value of d implies a mean free path on the atomic scale and incomplete randomness of phases. However, a detailed calculation of vertex corrections is still lacking because of the required multiplicity of the basis orbitals. The combined density of states, on the other hand, is understood quantitatively in simple models, and when multiplicity is included it is understood at the level of the CPA.

In summary, the optical absorption is the best understood of all physical properties.

## Slow Processes: DC Transport

Without electron-phonon interactions or when $h\omega_{ph} < k_B T$, the case we shall actually consider, one can write exact expressions for the dc conductivity and thermpower S

$$\sigma = \int dE\,\sigma(E)\left(-\frac{df(E)}{dE}\right) \tag{6.}$$

$$S = -\frac{k_B}{e}\frac{\beta}{\sigma}\int dE(E - E_p)\sigma(E)\left(-\frac{df}{dE}\right) \tag{20}$$

$$\int[n(E)f(E) - n_o(E)f_o(E)]dE = 0 \tag{21}$$

$$Q \equiv \ln \sigma/\sigma_o - (e/k_B)S \tag{22}$$

In Eq. (21) a subscript o indicates $T = 0$ values, and $\sigma_o$ is any convenient constant with dimensions of conductivity. Eq. (21) determines the temperature dependent Fermi level $E_F(T)$.

The experimental data fall into three categories: (1.) $\ln \sigma$, S, and therefore Q are linear functions of $1/k_B T$, and the thermopower and the conductivity relate in the way expected if the dominant carriers are in extended states above the mobility edge and inelastic scattering is not important; (2.) $\ln \sigma$, S and other dc transport properties are individually anomalous; (3.) $\ln \sigma$ and S show two distinct linear regions in $k_B T$ jointed by a kink with the lower temperature region behaving as (1.) and the higher temperature region as (2.), while Q remains linear in $1/k_B T$.

One can understand the anomalous behavior in cases (2.) and (3.) in the following way. As shown by Thomas and coworkers, inelastic e-p effects wipe out the mobility edge, delocalizing the electron states. The mobility step at $E_c$ for $T = 0$ is replaced by a mobility tail which falls off rapidly with $E_c - E$ and increases rapidly with T. The integral in Eq. (6.) or Eq. (20) therefore has a sharp maximum at $E^*$. Exponentiating the integrand and expanding the result to lowest order in $(E - E^*)$ leads to

$$\sigma = \sqrt{2\pi}\,\frac{\delta E}{k_B T}\,\sigma(E^*)e^{-\beta(E^*-E_F)} \tag{23}$$

$$S = \frac{-k_B}{e}\left[\frac{E^* - E_F}{k_B T} + \frac{a}{\sqrt{2\pi}}\,\frac{\delta E}{k_B T}\right] \tag{24}$$

$$Q = \ln\left[\sqrt{2\pi}\,\frac{\delta E}{K_B T}\,\frac{\sigma(E^*)}{\sigma_o}\right] + \frac{a\,\delta E}{\sqrt{2\pi}\,k_B T} \tag{25}$$

$$\delta E = \left\{\left[\frac{\Sigma'(E^*)}{\Sigma(E^*)}\right]^2 - \frac{\Sigma''(E^*)}{\Sigma(E^*)}\right\}^{-1/2} \tag{26}$$

$$a = (\delta E)^{-2} \int (E - E^*)e^{f(E,E^*)}dE \tag{27a}$$

$$f(E,E^*) = \ln\left[\Sigma(E)/\Sigma(E^*)\right] \approx -\beta(E - E^*) \tag{27b}$$

Note that the temperature dependent activation term $\beta(E^* - E_F)$ present in $\sigma$, Eq. (23), and S, Eq. (24), cancels out of Q, Eq. (25), as stressed by Overhof. If the complex and as yet uncertain temperature dependence of $\sigma(E)^*$ has the form

$$\sigma(E^*) = \sigma^* e^{-\beta\Delta E}, \tag{28}$$

then Q becomes

$$Q = \ln[\sqrt{2\pi}\ \beta\ \delta E\ \sigma^*/\sigma_o] - \beta[\Delta E - \frac{a}{\sqrt{2\pi}}\,\delta E], \tag{29}$$

consistent with the observations. The break in $\sigma$ and S separately must therefore be due to variations of $E^* - E_F$ with T. Overhof has proposed in fact that the transport anomalies arise from the T dependence of $E_F$. I find this less plausible than the variation of $E^*$ with T shown in Fig. 3. At low T, $E^*$ remains at $E_c$. Around a temperature $T_0$, it rapidly drops below the T = 0 mobility edge and continues to decrease for $T > T_0$ at a slower rate. This motion of $E^*$ and the associated transport via tail states, I propose, is responsible for the kinks in case (2.) and the anomalies in cases (2.) and (3.).

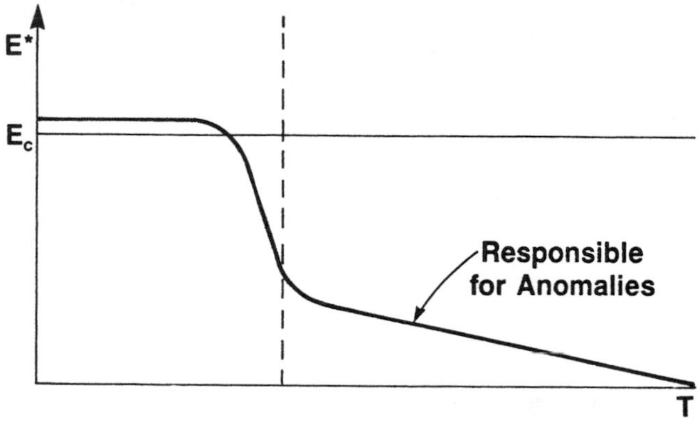


**FIGURE 3**

Fig. 3. Variation of $E^*$ with T proposed to account for the transport anomalies.

**Conclusions**

The density of states, the optical absorption, and related quantities are now becoming well understood for simple models, and substantial progress has been made for more complex models containing multiplicity and actually applicable to real materials.

The transport properties will be well understood when there is a more complete treatment of the e-p interaction. Our degree of understanding of polaron effects is only partial, that of thermal disorder is good, and that of inelastic effects is partial. However, each of these aspects has been treated independently. The most pressing need is an integrated treatment of all three aspects of the e-p interaction.

After the e-p interaction is fully integrated into the theory of the simple models, the then pressing need will be to introduce more features of real materials into the simple models. In particular, our understanding of off-diagonal disorder, topological disorder, and multiplicity of the basis set and their interplay with diagonal disorder must be further developed.

Clearly, the outlook is optimistic, but, even though the rate of progress is accelerating, the time scale will be long.

# MULTIDIMENSIONAL RWRE WITH SUBCLASSICAL LIMITING BEHAVIOR

Richard Durrett

Department of Mathematics
Cornell University
White Hall
Ithaca, New York 14853-7901

This paper is based on a talk given at the workshop on Random Media, September 17-24 at the IMA, and like the talk it describes two results proved in [4] and [5].

The first thing to do is to describe the model which is called random walk on a random hillside. Let $V: R^d \to R$ be a (random) function. If $x,y \in Z^d$ and $|x - y| = 1$ let

$$\alpha(x,y) = \exp(-V(\frac{x+y}{2}))$$

$$\alpha(x) = \sum_y \alpha(x,y)$$

$$p(x,y) = \alpha(x,y)/\alpha(x).$$

Since $\alpha(x,y) > 0$, we have $p(x,y) > 0$ and $\sum p(x,y) = 1$ i.e. $p$ is a transition probability. It is easy to see that the definition of $p$ is unchanged if we replace $\alpha(x,y)$ by

$$\tilde{\alpha}(x,y) = \exp(-(V(\frac{x+y}{2}) - V(x)))$$

so (a) the distribution of $p(x,y)$ only depends on that of the increments $V(\frac{x+\cdot}{2}) - V(x)$ and (b) we have put the minus sign in to get a process which likes to go downhill.

Example 1. When $d = 1$ a little arithmetic shows

$$p(x,x+1) = (1 + \exp(V(x+\frac{1}{2}) - V(x-\frac{1}{2})))^{-1}$$

we can generate any i.i.d. sequence $p(x,x+1)$ by taking $V(x+\frac{1}{2})x \in Z^d$ to be a random walk, and we can get any stationary sequence by choosing

$V(x + \frac{1}{2})x \in Z^d$ to have stationary increments.

The last observation shows our model contains the usual one dimensional nearest neighbor model as a special case. We will give other examples after we state

Theorem 1. Suppose $V(0) = 0$ and

(i) as $a \to \infty$ $V(az)/a^\alpha$ converges weakly to a limit $w(z)$

(ii) $z \to w(z)$ is continuous

(iii) with probability 1, $G(\varepsilon) \equiv$ the component of $\{z:w(z) < \varepsilon\}$ which contains 0 is bounded for some $\varepsilon > 0$ (which may depend on $w$).

Then with probability 1,

(a) $X_n$ is recurrent and

(b) $\sup_{m \leq n} |X_m| = O(\log^{1/\alpha} n)$

i.e. for any $\varepsilon > 0$ there is a $K < \infty$ so that

$$\limsup_{n \to \infty} P(\sup_{m \leq n} |X_m| > K \log^{1/\alpha} n) \leq \varepsilon.$$

The reasons for and content of these assumptions will become clear as we proceed. The first comment we would like to make is that this result makes precise a heuristic argument of Marinari, Parisi, Ruelle, and Windey (1983) (which learned about after we proved our theorem!) On page 2 of their paper they say (at various places up and down the page): "we assume that $V$ belongs to an ensemble which is invariant (at least for the large distance behavior) under the transformation $V \to V^*$ where

$$V^*(\lambda x) - V^*(0) = \lambda^\alpha (V(x) - V(0))"$$

"We choose the diffusion equation to be

$$(*) \qquad \frac{\partial c}{\partial t} = k\nabla \cdot J \quad J = \nabla c + c\nabla V$$

This is the continuous limit of a random walk on a lattice with nearest neighbor transition $i \to J$ proportional to $\exp 1/2[v_i - v_j]$."

"The diffusion (*) is chosen such that it has an equilibrium distribution $c \sim \exp(-\nu)$ in a bounded box. In the landscape created by the potential $V$ in $R^N$ one expects that $X(t)$ will occasionally go through a mountain pass and then rapidly relax to equilibrium in the intermediate valleys"

"Other things being similar, the flux through the mountain pass is proportional to the density $c = \exp(-v)$. When distances are multiplied by $\lambda$ the height of the mountain pass is multiplied by $\lambda^\alpha$ and the flux through the pass changes from $g(1)\exp(-v)$ to $g(\lambda)\exp(-\lambda^\alpha v)$ where $g(\lambda)$ is a polynomial. The time scale is correspondingly multiplied by $g(1)/g(\lambda)\exp((\lambda^\alpha - 1)V)$. Conversely multiplication of time by $\tau$ corresponds to multiplication of distances by a factor $\lambda(\tau) \sim (\log \tau)^{1/\alpha}$."

The key to our proof is the Dirichlet principle: Let $\Lambda \subset Z^d$ be a set with $0 \notin \Lambda$ and $Z^d - \Lambda$ finite. The function which minimizes the "energy" $\frac{1}{2} \sum_{xy} \alpha(x,y)(h(x) - h(y))^2$ among all the functions with $h(0) = 1$ and $h(x) = 0$ for $x \in \Lambda$ is $h(x) = P_x(T_0 < T_\Lambda)$ where $T_0 = \inf\{n > 0: X_n = 0\}$ and $T_\Lambda = \inf\{n > 0: X_n \in \Lambda\}$, and furthermore the minimum energy is $\alpha(0)P_0(T_\Lambda < T_0^+)$ where $T_0 = \inf\{n > 1: X_n = 0\}$. (see Griffeath and Liggett (1982), p. 885 or Liggett (1985) Section 2.6.)

The first part of this is the analogue of the classical Dirichlet principle i.e. the function which minimzes $\int_G |\nabla u|^2 dx$ among all the functions (in a suitable class) which are $=f$ on $\partial G$ is the harmonic function with those boundary values. The second part of the conclusion which gives the minimum energy is the more important for it implies that the minimum energy $\downarrow 0$ as $\Lambda \downarrow \phi$ if and only if the process is recurrent (i.e. $P_0(T_\Lambda < T_0^+) \to 0$).

Proof. Let $\Lambda_n^c = \{x \in Z^d: x \in G((1 + 3\varepsilon)\log n)\}$. Having assumed that $G(\varepsilon)$ is bounded for some $\varepsilon$ which may depend upon $w$ it follows from scaling that $G(\varepsilon)$ is bounded for all $\varepsilon < \infty$ (a little dirt that we swept under the rug in formulating our theorem) so $\Lambda_n$ is finite. Let $\partial\Lambda_n = \{(x,y): x \in \Lambda_n, y \in \Lambda_n^c\}$. The continuity of $W$ implies $W(x) = (1 + 3\varepsilon)\log n$ for all $x \in \partial G$ so if $n$ is large $V(\frac{x + y}{2}) > (1 + 2\varepsilon)\log n$ for all $(x,y) \in \partial\Lambda_n$.

To get an upper bound on the minimum energy and hence on $P_0(T_{\Lambda_n} < T_0^+)$ we plug in a function which at first looks like a stupid guess but turns out to be quite accurate: $\phi_n(x) = 1$ if $x \in \Lambda_n^c$ and $=0$ for $x \in \Lambda_n$. The energy of $\phi_n$ is $< |\partial\Lambda_n|\exp(-(1 + 2\varepsilon)\log n)$ and since $G(c \log n) \overset{d}{=} (\log n)^{1/\alpha}G(c)$ this is with high probability $< n^{-(1+\varepsilon)}$ if $n$ is large. From the last observation we see

$$P_0(T_{\Lambda_n} < T_0^+) < \frac{1}{\alpha(0)} n^{-(1+\varepsilon)}$$

which proves (a). To prove (b) we observe that starting from $0$ the probability of returning $< n$ times before hitting $\Lambda_n$ is $< cn^{-\varepsilon}$ and since each return takes at least 2 steps $P_0(T_{\Lambda_n} < 2n) \to 0$ as $n \to \infty$. The last result when combined with the scaling result quoted above completes the proof.

Our first application of Theorem 1 is to Example 1 with $p(x,x + 1)$ i.i.d. A little arithmetic shows

$$\log \frac{P(x,x + 1)}{P(x,x - 1)} = V(x - \frac{1}{2}) - V(x + \frac{1}{2}) \equiv (*).$$

so if $E(*) = 0$ and Variance$(*) \equiv \sigma^2 \in (0,\infty)$ then $V(at)/a^{1/2} \Rightarrow \sigma B_t$ where $B_t$ $t \in (-\infty,\infty)$ is a (two-sided) Brownian motion. This shows (i) holds. It is well known that Brownian motion has continuous paths so (ii) is true and it is easy to see that (iii) is satisfied so Theorem 1 can be applied. Conclusion (a) is (almost) half of the result of Solomon (1975) that in one dimension $E(*) = 0$ is a necessary and sufficient condition for recurrence. We will prove the other half below. Conclusion (b) strengthens a result of Sinai (1982) who showed that if $E(*) = 0$ and $\varepsilon \leq p(x,x+1) \leq 1-\varepsilon$ then as $n \to \infty$ $X_n/(\log n)^2$ converges to a nondegenerate limit. The last conclusion is stronger than the one proved above but follows from the proof of parts (c), (d), and (e) of the theorem given in the paper but not described above. The reader should note that our theorem does not require that the environments at different sites be independent but only that the central limit theorem holds for $V$.

To justify the title of this article we have to show that there are examples in any $d < \infty$ which satisfy the hypotheses of Theorem 1. This is done by

Example 2. Let $V$ be Gaussian have $V(0) = 0$, $EV(z) = 0$ and

$$EV(x)V(y) = |x|^\beta + |y|^\beta - |x-y|^\beta$$

where $0 < \beta < 2$. A theorem of Schoenberg shows that the expression above is positive definite and hence a legitimate covariance function. (see Gangolli (1965) or Pitt (1978) for this and other facts we will use below). It is easy to see that $V$ has stationary increments. When $\beta = 1$ $V$ is called Lévy's multiparameter Brownian motion because it is a Brownian motion along any line.

Brownian motion has the scaling property $B(ut) = a^{1/2}B(t)$. A moments thought generalizes this to the examples above as $V(ax) = a^{\beta/2}V(x)$ so (i) holds with $\alpha = \beta/2$ and $W = V$. Lévy (1948) has shown that (ii) holds so this brings us to the task of checking (iii). A simple argument using scaling and stationary increments shows that if (iii) fails there are no local maxima or minima (i.e. our hillside has no peaks or valleys!) The last state of affairs is clearly ridiculous and can be ruled out by using Lemma 7.1 of Pitt (1978).

Example 2 shows that there are RWRE in any $d < \infty$ in which

$$\vec{p}(x) \equiv (p(x,x+e_1), p(x,x-e_1), \ldots\ p(x,x-e_d))$$

(where $e_1, \ldots e_d$ are the standard basis vectors) is a stationary sequence but $x_n$ is (a) recurrent and (b) $o(\sqrt{n})$ or subdiffusive. The first conclusion contradicts a generalization of a conjecture of Kalikov that every RWRE in $d \geq 3$ is transient. He made this conjecture only for the case in which the $\vec{p}(x)$ are i.i.d. but... (b) casts some doubt on renormalization group (D. Fisher (1984)) and series expansion (Derrida and Luck (1983)) arguments that the "critical dimension is 2" i.e.

$$x_n \sim \begin{cases} (\log n)^2 & d = 1 \\ n^{1/2} & d \geq 2 . \end{cases}$$

Again these speculations are made only for the model in which the $\vec{p}(x)$ are independent but...

From the two but...'s you can see that we have not yet made a dent in the original problem: what happens when the $\vec{p}(x)$ are i.i.d.? It seems to us that the key to deciding the behavior in the i.i.d. case is to understand the behavior of stationary measures $\pi(x)$, i.e. solutions of

$$\sum_x \pi(x)p(x,y) = \pi(y)$$

so we will spend the rest of the paper arguing for the following

<u>Metatheorem</u>. The nature of the stationary measure controls the asymptotic behavior of the process.

Our first evidence for this is

<u>Theorem 2</u>. Let $P$, $Q$ be transition probabilities with stationary measures $u$, $v$. Suppose

(i) $c^{-1} \leq \frac{du}{dv} \leq c$,

(ii) $P \geq \varepsilon Q$, i.e. if $f \geq 0$ then $Pf \geq \varepsilon Qf$,

(iii) $Q$ is symmetric w.r.t. $v$, i.e.

$$\langle f,Qg\rangle_v = \langle Qf,g\rangle_v$$

where $\langle f,g\rangle_v = \int fg\, dv$,

then for all $\lambda < 1$, $f > 0$ we have

$$\sum_{n=0}^{\infty} \left( \frac{\lambda}{2 - \lambda} \right)^n \langle f, P^n f \rangle_u < \frac{2c^4}{\varepsilon} \sum_{n=0}^{\infty} \lambda^n \langle f, Q^n f \rangle_u .$$

In particular if $Q$ is transient then $P$ is.

Theorem 2 is essentially due to Varopoulos (1983) who considered the case $c = 1$ (where this result is true with $\lambda$ replacing $\lambda/(2 - \lambda)$ and $1$ instead of $2c^4$). The key to the proof in this case is a result due to Baldi, Lohoué and Peyriere (1977):

Lemma. Let $A$ and $B$ be two invertible operators on a real Hilbert space $H$ with

$$0 < (Ax,x) < (Bx,x)$$

To get started observe that $(y,Az)^2 < (y,Ay)(z,Az)$. The rest is easy but if you get stuck the answer can be found in Durrett (1985b).

A simple consequence of Theorem 2 is

Corollary 1. Let $X_t$ be a (nice) diffusion process in $R^d$ with a stationary measure $h(x)dx$ which has $c^{-1} < h(x) < c$. If $d > 3$ then $X$ is transient.

This corollary covers all the diffusions with random coefficients considered by Papanicolaou and coauthors ([6], [15], [16], [17], [18], [19]) but to be fair I need to admit that in some cases (e.g. [19]) the hard part is to show that the stationary measure has this property.

A second set of examples covered by Corollary 1 is diffusions of the form $\Delta + b \cdot \nabla$ where the $b(x)$ is nice, (or to be specific bounded and Lipschitz continuous) and has $b^i(x)$ independent of $x^i$. The last hypothesis implies $\sum_i \partial_i b^i(x) = 0$ which implies $dx$ is a stationary measure so in $d > 3$ these diffusions must be transient. Intuitively such drift cannot help you "find your way home" and it would be nice to have a proof which makes this intuition precise.

A second consequence of Theorem 2 is

Corollary 2. Let $X$ be a RWRE with a stationary ergodic environment in which $p(x,y) \geq \delta > 0$ when $|x - y| = 1$, then a necessary condition for recurrence is that $X$ has a stationary measure which is unique up to constant multiples and has

$$E \log \frac{\pi(x + e_i)}{\pi(x)} = 0 \quad i = 1,\ldots d.$$

Proof. The first step is to construct a reversible chain with the same stationary measure. To do this we use an idea we learned from T. Liggett. Let

$$q(x,y) = \frac{\pi(x)p(x,y) + \pi(y)p(y,x)}{2\pi(x)} .$$

From the definition it is clear that $\pi(x)q(x,y) = \pi(y)q(x,y)$, $q(x,y) \geq 0$ and

$$\sum_y q(x,y) = \frac{\pi(x)\cdot 1 + \pi(x)}{2\pi(x)} = 1$$

so $q$ is a transition probability with reversible measure $\pi$. To check that $p \geq \varepsilon q$ we write

$$p(x,y) - \varepsilon q(x,y) = p(x,y) - \varepsilon\left( \frac{p(x,y)}{2} + \frac{\pi(y)p(y,x)}{2\pi(x)} \right)$$

and

$$\pi(y)\ p(y,x) \leq \sum_z \pi(z)p(z,x) = \pi(x)$$

so

$$p(x,y) - \varepsilon q(x,y) \geq (1 - \frac{\varepsilon}{2})p(x,y) - \frac{\varepsilon}{2}$$

$$\geq (1 - \frac{\varepsilon}{2})\delta - \frac{\varepsilon}{2} \geq \frac{\delta}{4}$$

if $\varepsilon = \delta/2$. (recall $\delta \leq 1$)

At this point we have checked that the hypothesis of Theorem 2 are satisfied so it remains to show that $Q$ is transient. To do this we observe that the uniqueness of $\pi$ implies that $\log(\pi(x + e_i)/\pi(x))$ is stationary and ergodic so if (without loss of generality) $E \log(\pi(x + e_1)/\pi(x)) = c > 0$ then

$$\frac{1}{n} \log \pi(x + ne_1) \to c$$

as $n \to \infty$ and an easy argument using the Dirichlet principle (see [2] for details) shows that Q is transient.

Again since it is hard to compute stationary measures it is difficult to use Corollary 2 in concrete situations but there are some examples to which it can be applied. Let $p_i = (p_i(e_1), p_i(-e_1), \ldots p_i(-e_d))$ $i = 1, \ldots K$ be fixed environments and for each $\theta \in (0,\infty)^K$ and $z \in 2^d$ with $|z| = 1$ let

$$p_i^\theta(z) = p_i(z)\theta^z/c(\theta)$$

where

$$\theta^z = \prod_{j=1}^{d} \theta_i^{z_i}$$

and $c(\theta)$ is a constant which makes

$$\vec{p}_i^{\,\theta} = (p_i^\theta(e_1), \ldots, p_i^\theta(-e_d))$$

a probability distribution. We call $\vec{p}_i^{\,\theta}$ the exponential family generated by the $\vec{p}_i$.

Consider the RWRE in which sites are independently assigned $\vec{p}_i^{\,\theta}$ with probability $\alpha_i$. It is easy to see that if $\pi(x)$ is the stationary measure when $\theta = (1,\ldots,1)$ then $\pi(x)\theta^x$ is stationary for $\theta \neq (1,\ldots,1)$ so Corollary 2 tells us that at most one of these RWRE is recurrent. If we take a concrete case $\vec{p}_1 = (1/3, 1/6, 1/3, 1/6)$, $\vec{p}_2 = (1/6, 1/3, 1/6, 1/3)$, $\alpha_1 = \alpha_2 = 1/2$ then from symmetry we see that recurrence for $\theta$ implies recurrence for $(\theta_2, \theta_1)$ and for $(1/\theta_1, 1/\theta_2)$ and the last observation allows us to conclude that the RWRE must be transient for all $\theta \neq (1,1)$.

The last conclusion is very special but it is the best we can do at this point. Two other results which should be corollaries of Corollary 2 are.

A. Half of Erick Key's (1984) result which gives necessary and sufficient conditions for the recurrence of one dimensional finite range RWRE.

B. Kalikow's (1981) result that extremely biased two dimensional RWRE are transient.

In view of the difficulty of the proofs of the last two results it would be nice to obtain them from Corollary 2 and we invite the reader to try this.

**References**


1. B. Derrida and J.M. Luck (1983), Diffusion on a random lattice: weak disorder expansion in orbitrary dimension, Phys. Rev. B, 28, 7183-7190.

2. R. Durrett (1985a), Reversible Diffusion Processes to appear in Probability and Harmonic Analysis, edited by J. Chao, and W. Woyczynski, Publ. by Marcel Dekker, New York.

3. R. Durrett (1985b), Particle systems, random media, and large deviations. Conference proceedings, AMS Contemporary Math Series, vol. 41.

4. R. Durrett (1986a), Some multidmensional RWRE with subclassical limiting behavior. Comm. Math. Phys., 104, 87-102.

5. R. Durrett (1986b), Two comparison theorems for the recurrence of Markov chains. Am. Prob. submitted for publication.

6. R. Figari, E. Orlandi, and G. Papanicolaou (1983), Diffusive behavior of a random walk in random media, Taniguichi Symp. Katata, publ. by Kodansha, Japan.

7. D. Fisher (1984), Random walks in random environments, Bell Laboratories, preprint.

8. R. Gangoli (1965), Abstract harmonic Analysis and Lévy's Brownian motion of several parameters, 5th Berkeley Symp. Vol. II, 13-30.

9. D. Griffeath and T. Liggett (1982), Critical phenomena for Spitzer's reversible nearest particle systems, Ann. Prob. 10, 881-895.

10. S. Kalikow (1981), Generalized random walk in a random environment. Ann. Prob. 9, 753-768.

11. E.S. Key (1984), Recurrence and transience criteria for random walk in a random environment, Ann. Prob. 12, 529-560.

12. T. Liggett (1985), Interacting Particle Systems, Springer-Verlag, New York.

13. E. Marinari, G. Parisi, D. Ruelle, and P. Windey (1983a), Random walks in a random environment and 1/f noise, Phys. Rev. Letters 50, 1223-1225.

14. E. Marinari, G. Parisi, D. Ruelle and P. Windey (1983b), On the interpretation of 1/f noise, Comm. Math. Phys., 89, 1-12.

15. H. Osada (1983), Homogenization of diffusion processes with random coefficients, p. 507-517 in Proceedings of the 4th Japan USSR Probability Symposium, Springer LNM 1021.

16. G.C. Papanicolaou (1983), Diffusions and random walks in random media in The Mathematics and Physics of Disordered Media ed. by B.D. Hughes and B.W. Ninham, Springer LNM 1035.

17. G. Papanicolaou and O. Pironeau (1981), On the asymptotic behavior of motions in random flows, p. 36-41 in Stochastic Nonlinear Systems in Phsyics, Chemistry, and Biology, edited by L. Arnold and R. Lefever, Springer-Verlag.

18. G. Papanicolaou and S.R.S. Varadhan (1979), Boundary value problems with rapidly oscillating coefficients, p. 835-873 in Random Fields edited by J. Lebowitz and D. Szaxa, North Holland.

19. G. Papanicolaou and S.R.S. Varadhan (1981), Diffusions with random coefficients, p. 253-262 in Essays in Statistics and Probability edited by G. Kallianpur, P. Krishnaiah, and J. Ghosh, North Holland.

20. L.D. Pitt (1978), Local times for Gaussian vector fields, Indiana Math. J. 27, 309-330.

21. S. Schumacher (1984), Diffusions with random coefficients, Ph.D. Thesis, UCLA, A summary appears in Durrett (1985b).

22. Ya. Sinai (1982), Limit behavior of one-dimensional random walks in random environments, Theor. Probab. Appl. 27, 247-258.

23. F. Solomon (1975), Random walks in a random environment, Ann. Prob. 3, 1-31.

# LOCALIZATION
# FOR A
# RANDOM DISCRETE WAVE EQUATION

William G. Faris
Department of Mathematics
University of Arizona
Tucson, Arizona 85721

## Abstract

It is possible for a discrete wave equation with random local propagation speed to have only dense point spectrum in a certain range of frequency. This means that the only waves in this frequency are localized standing waves.

The argument depends on a principle that says that the expectation of the spectral measure is absolutely continuous with respect to Lebesgue measure, even if the spectral measure itself has only point masses. This principle gives one of the estimates that controls the effects of near resonance. It also is useful in deriving from the estimates that a frequency range contains only point spectrum.

## 1. The random discrete wave equation

Wave propagation within a random medium is extraordinarily complicated. The wave is scattered and the scattered waves are themselves rescattered, and so on. It would seem that one should expect a diffusive behaviour. However in some circumstances there is a particularly dramatic effect: localization. This is when the scattered waves conspire to produce only standing waves, with no propagation throughout the medium.

Rigorous understanding of this effect has been slow, expecially in more than one space dimension. The 1983 paper by J. Fröhlich and T. Spencer [5] was a key contribution. They derived decay estimates on the resolvent for a random discrete Schrödinger operator with large disorder, in any number of dimensions. These estimates are sufficient to imply localization.

One of the most transparent derivations of this implication is based on a principle that arose in work of S. Kotani [6]. This principle is found in work by F. Delyon, Y. Lévy, and B. Souillard [1], and by B. Simon and T. Wolff [2]. Another derivation of localization was given by J. Fröhlich, F. Martinelli, E. Scoppola, and T. Spencer [4]. This derivation gives more powerful results on the time dependence of the solutions.

This lecture is an application of the ideas of Fröhlich and Spencer and of the Kotani principle in the form used by Simon and Wolff to a random discrete wave equation. There are some new features of the estimates due to the fact that the randomness is off-diagonal. Also, the resolvents have a more complicated dependence on the frequency and coupling constant parameters than in the case of additive perturbations. The material is based on the results of [2] and [3]. Also thanks are due to Barry Simon for a useful discussion.

In the kind of application we have in mind, the Hilbert space is $\mathcal{H} = \ell^2(\mathbf{Z}^\nu)$, the space of functions $f$ defined on the $\nu$ dimensional integer lattice $\mathbf{Z}^\nu$ and satisfying

$$\sum_{\mathbf{x}\in\mathbf{Z}^\nu} |f(\mathbf{x})|^2 < \infty. \tag{1}$$

The Kronecker delta functions $\delta_{\mathbf{x}}, \mathbf{x} \in \mathbf{Z}^\nu$ form a basis for this space. The finite difference operator $-\Delta$ is a positive self-adjoint operator defined by

$$-\Delta f(x) = \sum_{|\mathbf{y}-\mathbf{x}|=1} f(\mathbf{x}) - f(\mathbf{y}). \tag{2}$$

Let $f$ have the Fourier expansion

$$f(\mathbf{x}) = \int_{\mathbf{T}^\nu} e^{i\mathbf{kx}} \hat{f}(\mathbf{k}) \frac{d^\nu \mathbf{k}}{(2\pi)^\nu}, \tag{3}$$

where the integral is over the $\nu$ dimensional torus with periods $2\pi$. Then

$$-\Delta f(\mathbf{x}) = \int_{\mathbf{T}^\nu} \omega(\mathbf{k}) e^{i\mathbf{kx}} \hat{f}(\mathbf{k}) \frac{d^\nu \mathbf{k}}{(2\pi)^\nu}, \tag{4}$$

where

$$\omega(\mathbf{k}) = \sum_{j=1}^{\nu} 4\sin^2(\frac{\mathbf{k}_j}{2}). \tag{5}$$

Since

$$0 \le \omega(\mathbf{k}) \le 4\nu, \tag{6}$$

the operator $-\Delta$ has absolutely continuous spectrum $[0, 4\nu]$.

Now let $c \ge 0$ be a function on $\mathbf{Z}^\nu$. Assume that there is a bound

$$c(\mathbf{x})^2 \le M, \qquad \mathbf{x} \in \mathbf{Z}^\nu. \tag{7}$$

Define

$$H = -c\Delta c. \tag{8}$$

Thus $H$ is the positive self-adjoint operator that first multiplies by the function $c$, then applies the operator $-\Delta$, and then multiplies again by $c$. The spectrum of $H$ is contained in the interval $[0, 4\nu M]$. The discrete wave equation is defined by

$$\frac{d^2 u}{dt^2} + Hu = 0. \tag{9}$$

If we make the substitution $w = cu$ this takes the familiar form

$$\frac{\partial^2 w}{\partial t^2} = c^2 \Delta w. \tag{10}$$

In order to make the operator random, we take the values $c^2(\mathbf{x}), \mathbf{x} \in \mathbf{Z}^\nu$ to be independent and identically distributed random variables. We assume that the common distribution of these random variables is absolutely continuous with density $\rho$.

It is easy to identify the spectrum of $H$. Let $a$ be in the essential range of the random variables $c(\mathbf{x})^2$. Then there will be arbitrarily large regions in the space $\mathbf{Z}^\nu$ on which all the $c(\mathbf{x})^2$ for $\mathbf{x}$ in the region are close to $a$. On such a region $H$ resembles the operator $-a\Delta$, which has spectrum $[0, 4\nu a]$. Furthermore $H$ will have approximate eigenfunctions localized in the interior of this region with eigenvalues close to any point in this interval. Thus the spectrum of $H$ includes the interval $[0, 4\nu a]$. This sort of argument may be made into a proof. The conclusion is that if the density of the $c(\mathbf{x})^2$ has even a small tail at high frequency, then there will be spectrum at high frequency.

**Theorem 1.** *For every $M$ with $0 < M < \infty$ and every $\beta$ with $0 < \beta < 1$, there is a $b > 0$ such that if the density $\rho$ is supported on $[0, M]$ and satisfies*

$$\rho(s) \leq \frac{b}{s}, \tag{11}$$

*and if $E$ is in the high frequency interval $[\beta 4\nu M, 4\nu M]$, then for every $\mathbf{x} \in \mathbf{Z}^\nu$,*

$$\langle \delta_{\mathbf{x}}, (H - E)^{-2} \delta_{\mathbf{x}} \rangle < \infty \tag{12}$$

*with probability one.*

The proof of Theorem 1 relies heavily on the techniques of Fröhlich and Spencer. Some of the ideas of the proof will be sketched later.

In order to understand what this theorem says, fix $M$ and $\beta$. This determines an interval $[\beta 4\nu M, 4\nu M]$ in the high frequency range. Then $b$ may be selected so that most of the probability distribution of the random variable $4\nu c(\mathbf{x})^2$ is below this interval. Then the conclusion is that for every frequency $E$ in this interval the decay estimate (12) is satisfied.

The meaning of the decay estimate (12) is seen by writing it as

$$\langle \delta_{\mathbf{x}}, (H - E)^{-2} \delta_{\mathbf{x}} \rangle = \|(H - E)^{-1} \delta_{\mathbf{x}}\|^2 = \sum_{\mathbf{y} \in \mathbf{Z}^\nu} |\langle \delta_{\mathbf{x}}, (H - E)^{-1} \delta_{\mathbf{y}} \rangle|^2 < \infty. \tag{13}$$

We see that it says that the matrix elements of the resolvent operator $(H - E)^{-1}$ must decay sufficiently rapidly. Actually the proof shows that they decay exponentially in the separation $|\mathbf{x} - \mathbf{y}|$.

It is somewhat surprising at first that for *every* $E$ in the interval the estimate holds with probability one . After all there should be a dense set of eigenvalues in the interval. The reason, of course, is that the probability that a given $E$ is an eigenvalue is zero.

On the other hand, it follows from the theorem that with probability one the estimate holds for Lebesgue almost every $E$ in the interval.

**Theorem 2.** *Under the same hypotheses, there is no absolutely continuous spectrum in the high frequency interval.*

Proof: Theorem 2 is an immediate consequence of Theorem 1. For $\epsilon > 0$ define the approximate Dirac delta function by

$$\delta_\epsilon(x) = \frac{1}{\pi} \frac{\epsilon}{x^2 + \epsilon^2}. \tag{14}$$

Then

$$\langle \delta_{\mathbf{x}}, \delta_\epsilon (H - E) \delta_{\mathbf{x}} \rangle \leq \frac{\epsilon}{\pi} \langle \delta_{\mathbf{x}}, (H - E)^{-2} \delta_{\mathbf{x}} \rangle, \tag{15}$$

so

$$\lim_{\epsilon \to 0} \langle \delta_{\mathbf{x}}, \delta_\epsilon (H - E) \delta_{\mathbf{x}} \rangle = 0 \tag{16}$$

for almost every $E$. However this limit is the absolutely continuous part of the matrix element of the spectral measure. ∎

**Theorem 3.** *Under the same hypotheses, there is no singular continuous spectrum in the high frequency interval.*

The proof that Theorem 3 is also a consequence of Theorem 1 will be given later. It follows from Theorems 2 and 3 that only point spectrum is found in the interval. The eigenfunctions are all square summable. Actually it is also possible to prove that the eigenfunctions have exponential decay.

Let $E_i$ be the eigenvalues and $\psi_i$ the corresponding eigenvectors. Then

$$\langle \delta_{\mathbf{x}}, (H-E)^{-2}\delta_{\mathbf{x}}\rangle = \|(H-E)^{-1}\delta_{\mathbf{x}}\|^2 \geq \sum_i |\langle \delta_{\mathbf{x}}, (H-E)^{-1}\psi_i\rangle|^2 = \sum_i \frac{1}{(E_i-E)^2}|\psi_i(\mathbf{x})|^2. \quad (17)$$

The fact that the $E_i$ cluster at $E$ is compensated by the fact that the $\psi_i$ are eventually each concentrated far from a given $\mathbf{x}$. Thus the decay estimate may hold even though $E$ is a limit of eigenvalues.

Outline of proof of Theorem 1:

The plan of the proof is to introduce a (random) singular region $S_0 \subset \mathbf{Z}^\nu$ on whose complement $\mathbf{Z}^\nu \setminus S_0$ the decay estimate is almost evident. The main problem is thus to estimate the resolvent on the singular region $S_0$. This is done by an inductive construction.

A sequence of distance scales $d_i$ and resonance scales $\gamma_i$ are defined for $i = 0,1,2,3,\ldots$ in such a way that $d_i \to \infty$ and $\gamma_i \to \infty$ as $i \to \infty$. The region $S_0$ is written as a disjoint union of a sequence of gentle singular regions $S_i^g \subset S_0$, for $i = 0,1,2,3,\ldots$. These regions are made of components that are no larger than $d_i$ and no more resonant than $\gamma_i$ and that are separated from each other by a distance $d_{i+1}$. Let the very singular region $S_i$ be the union of the $S_j^g$ for $j \geq i$. The induction step is to pass from an estimate in the region $\mathbf{Z}^\nu \setminus S_i$ to an estimate in the region $\mathbf{Z}^\nu \setminus S_{i+1}$, by controlling the contribution from the region $S_i^g$.

There are two deterministic estimates that are used in the induction argument:

(i) Matrix elements of the resolvent decay exponentially in the complement of $S_0$.

(ii) The interaction between neighboring points is not too strong.

Estimate (i) is the starting point. Then estimate (ii) is used as an ingredient in a perturbation argument that shows that the inductive step of introducing the region $S_i^g$ does not spoil the decay. The reason this works is that the region $S_i^g$ is made up of components separated by $d_{i+1}$, and the resonance of order $\gamma_i$ in the component of size $d_i$ is compensated by the decay in the region $\mathbf{Z}^\nu \setminus S_i$ in between.

There are then two probabilistic estimates that show that this procedure is effective:

(iii) The region $S_0$ has low density.

(iv) Strong resonances are unlikely in moderate sized regions.

The estimates (iii) and (iv) combined with statistical mechanical arguments then show that the density of the $S_i$ approach zero rapidly. Thus the inductive estimate is effective with high probability.

Discussion of the estimates:

The estimate (i) comes from a path expansion of the type discussed by Tom Spencer at this conference.

The second estimate (ii) depends on the fact that it is assumed that $c(\mathbf{x})^2 \leq M$. It should be possible to remove this assumption, but it would considerably complicate the argument.

The estimate (iii) is easy. Let $0 < \alpha < 1$. Define

$$S_0 = \{\mathbf{y} \in \mathbf{Z}^\nu : c(\mathbf{y})^2 > \alpha\beta M\}. \tag{18}$$

Let $\mathbf{x} \in \mathbf{Z}^\nu$. Then

$$P[\mathbf{x} \in S_0] = \int_{\alpha\beta M}^{M} \rho(s)\, ds \leq \int_{\alpha\beta M}^{M} \frac{b}{s}\, ds = b \log(\frac{1}{\alpha\beta}). \tag{19}$$

This may be made arbitrarily small by taking $b$ sufficiently small. Since it is not too difficult to show that the resolvent has exponential decay in the region $\mathbf{Z}^\nu \setminus S_0$, we see that the region where wave propagation could possibly be allowed already occupies only a small fraction of space.

The estimate (iv) will be discussed in Section 3.

## 2. The Kotani principle

Now we turn to some operator theory. Let $H \geq 0$ be a positive self-adjoint operator. Let $\phi$ be a unit vector in the Hilbert space. Then $\phi\langle\phi, \cdot\rangle$ is the projection operator onto the space spanned by $\phi$. For $\lambda > 0$ define $C_\lambda$ by

$$C_\lambda = \lambda\phi\langle\phi, \cdot\rangle + 1 - \phi\langle\phi, \cdot\rangle. \tag{20}$$

The operator $C_\lambda$ is to be thought of as a perturbation that differs from the identity in only one direction. The perturbed operator $H_\lambda$ is defined by

$$H_\lambda = C_\lambda H C_\lambda. \tag{21}$$

Write $1_S$ for the indicator function of $S$, so that $1_S(H)$ is the spectral projection of $H$ corresponding to the part of the spectrum in $S$.

The Kotani principle is that the averaged spectral projections are absolutely continuous. In our case it takes the form of the following theorem.

**Theorem 4.** *The spectral projections of the multiplicatively perturbed operators $H_\lambda$ satisfy*

$$\int_0^\infty \langle\phi, 1_S(H_\lambda)\phi\rangle \frac{d\lambda^2}{\lambda^2} \leq \int_S \frac{dE}{E}. \tag{22}$$

Proof: Let

$$\Gamma = -\frac{1}{\langle\phi, H(H - E - i\epsilon)^{-1}\phi\rangle} = \Gamma_1 + i\Gamma_2. \tag{23}$$

Then

$$\langle\phi, H\delta_\epsilon(H - E)\phi\rangle > 0 \Rightarrow \Gamma_2 > 0. \tag{24}$$

The reason for this definition is the perturbation identity

$$\langle\phi, H(H_\lambda - E - i\epsilon)^{-1}\phi\rangle = \frac{\lambda^2}{\lambda^2 - 1 - \Gamma}. \tag{25}$$

This identity is the key to the proof. By taking imaginary parts it follows that

$$\langle\phi, H\delta_\epsilon(H_\lambda)\phi\rangle = \delta_{\Gamma_2}(\lambda^2 - 1 - \Gamma_1)\lambda^2. \tag{26}$$

Integration gives

$$\int_0^\infty \langle \phi, H\delta_\epsilon(H_\lambda)\phi\rangle \frac{d\lambda^2}{\lambda^2} = \int_0^\infty \delta_{\Gamma_2}(\lambda^2 - 1 - \Gamma_1)\, d\lambda^2 \le 1. \tag{27}$$

By letting $\epsilon \to 0$ we see that the densities satisfy the required bound. ∎

Next we sketch the application of the Kotani principle to the proof of Theorem 3. This says that the decay estimate of Theorem 1 implies the absence of singular continuous spectrum.

**Lemma 5.** *For every $\lambda \neq 1$, the singular continuous part of the measure $\langle \phi, 1_S(H_\lambda)\phi\rangle$ is concentrated on the set of $E$ with $\langle \phi, (H-E)^{-2}\phi\rangle = \infty$.*

Proof: Consider the perturbation identity

$$\langle \phi, H(H_\lambda - E - i\epsilon)^{-1}\phi\rangle = \frac{\lambda^2}{\lambda^2 - 1 - \Gamma}. \tag{28}$$

The singular spectrum is contained in the set of $E$ for which the left hand size becomes infinite as $\epsilon \to 0$. The only way this can happen is for the denominator to vanish. This says that

$$-\frac{1}{\Gamma} = \langle \phi, H(H-E)^{-1}\phi\rangle = -\frac{1}{\lambda^2 - 1}. \tag{29}$$

Now consider an $E$ satisfying (29) and with $\langle \phi, (H-E)^{-2}\phi\rangle < \infty$. This says that $(H-E)^{-1}\phi$ is in the Hilbert space. This implies that such an $E$ must be an eigenvalue of $H_\lambda$. In fact the corresponding eigenvector is

$$u = (\lambda - 1)\langle \phi, H(H-E)^{-1}\phi\rangle\phi + H(H-E)^{-1}\phi. \tag{30}$$

The singular continuous spectrum is concentrated on the subset where the denominator vanishes and where there is no eigenvalue. Thus it is contained in the set of $E$ with $\langle \phi, (H-E)^{-2}\phi\rangle = \infty$. ∎

Note that the eigenvector constructed in the lemma satisfies

$$\langle u, \phi\rangle = \lambda\langle \phi, H(H-E)^{-1}\phi\rangle \tag{31}$$

and

$$\|u\|^2 = E\langle \phi, H(H-E)^{-2}\phi\rangle. \tag{32}$$

This gives a nice formula for the point mass $|\langle u, \phi\rangle|^2/\|u\|^2$.

**Theorem 6.** *If $\langle \phi, (H-E)^{-2}\phi\rangle < \infty$ for almost every $E$ in some interval, then the measure $\langle \phi, 1_S(H_\lambda)\phi\rangle$ has no singular continuous part in this interval, for almost every $\lambda$.*

Proof: Let $S$ be the set where $\langle \phi, (H-E)^{-2}\phi\rangle = \infty$. Apply Theorem 4 to $S$. Since $S$ has Lebesgue measure zero, we conclude that $\langle \phi, 1_S(H_\lambda)\phi\rangle = 0$ for almost every $\lambda$. The conclusion follows from Lemma 5. ∎

Now we may indicate the rest of the proof of Theorem 3. Fix an $\mathbf{x} \in \mathbf{Z}^\nu$. Theorem 1 combined with Theorem 6 shows that $H$ has no singular continuous spectrum in the high frequency interval, for almost every value of $c(\mathbf{x})$. Since $c(\mathbf{x})$ has an absolutely continuous distribution and is independent of the $c(\mathbf{y})$ for $\mathbf{y} \neq \mathbf{x}$, it follows that $H$ has no absolutely continuous spectrum in the interval, with probability one.

## 3. Resonance bounds

In this section we illustrate another application of the Kotani principle. The resonance bounds are the statement that strong resonances are unlikely in moderate sized regions. This is the estimate (iv) in the outline of the proof of the decay estimate in the last section. The resonance bounds follow from a statement about the density of eigenvalues for an operator defined in a bounded region.

As before the operator $H = -c\Delta c$, where $c \geq 0$ is a random real function on $\mathbf{Z}^\nu$. This operator acts in the Hilbert space $\ell^2(\mathbf{Z}^\nu)$. However the strategy is to build this operator out of parts in bounded regions $\Lambda$ of $\mathbf{Z}^\nu$. Thus we define $H^\Lambda$ to be the corresponding operator with Dirichlet boundary conditions acting in $\ell^2(\Lambda)$.

**Theorem 7.** *Let $E_i$ be the random eigenvalues of the random operator $H^\Lambda$. Then for every $E > 0$ and sufficiently small $\kappa > 0$*

$$\mathcal{P}[E - \kappa \leq E_i \leq E + \kappa \text{ for some } i\,] \leq \frac{2b}{E-\kappa}\kappa|\Lambda|, \tag{33}$$

*where $b$ is the measure of the low frequency concentration of the density of the random variables $c(\mathbf{x})^2$ and $|\Lambda|$ is the number of points in $\Lambda$.*

Proof: Fix $\mathbf{x}$ in $\Lambda \subset \mathbf{Z}^\nu$. Let $\delta_\mathbf{x}$ be the Kronecker delta function concentrated at $\mathbf{x}$. Then the conditional expectation of the matrix element $\langle \delta_\mathbf{x}, 1_S(H^\Lambda)\delta_\mathbf{x}\rangle$ given the $c(\mathbf{y})^2$ for $\mathbf{y} \neq \mathbf{x}$ is an integral with respect to the density $\rho(E) \leq b\, dE/E$. It follows from Theorem 4 that

$$\mathcal{E}[\langle \delta_\mathbf{x}, 1_S(H^\Lambda)\delta_\mathbf{x}\rangle \mid c(\mathbf{y})^2, \mathbf{y} \neq \mathbf{x}] \leq b\int_S \frac{dE}{E}. \tag{34}$$

The expectation of the conditional expectation is the expectation, so the expectation satisfies the same bound. If we bound the integral over $S = [E-\kappa, E+\kappa]$, we obtain

$$\mathcal{E}[\langle \delta_\mathbf{x}, 1_{[E-\kappa,E+\kappa]}(H^\Lambda)\delta_\mathbf{x}\rangle] \leq b\frac{2\kappa}{E-\kappa}. \tag{35}$$

Now if we sum over $\mathbf{x} \in \Lambda$ we obtain

$$\mathcal{E}[\operatorname{tr} 1_{[E-\kappa,E+\kappa]}(H^\Lambda)] = \mathcal{E}[\#i, E-\kappa \leq E_i \leq E+\kappa] \leq b\frac{2\kappa}{E-\kappa}|\Lambda|. \tag{36}$$

However the expected number of eigenvalues in the interval is an upper bound for the probability of an eigenvalue in the interval. This proves the theorem. ∎

**Corollary 8.** *For large $\gamma$,*

$$\mathcal{P}[\|(H^\Lambda - E)^{-1}\| \geq \gamma] \leq \frac{2b}{E-\gamma^{-1}}\frac{|\Lambda|}{\gamma}. \tag{37}$$

This corollary is the result that says that large resonances $\gamma$ are improbable in moderate sized regions $\Lambda$. In the inductive construction, if the diameter of $\Lambda$ is bounded by $d_i$, then the probability of a $\gamma_i$ resonance is bounded by $d_i^\nu/\gamma_i$. Thus in the construction is necessary to take $\gamma_i \to \infty$ much faster than $d_i \to \infty$, in such a way that the probability approaches zero rapidly as $i \to \infty$. The work of Fröhlich and Spencer [5] has shown that this is possible.

**References**

1. F. Delyon, Y. Lévy, and B. Souillard, "Anderson localization for multi-dimensional systems at large disorder or large energy," *Comm. Math. Phys.* **100** (1985), 463–470.
2. W. Faris, "Localization estimates for a random discrete wave equation at high frequency," in preparation.
3. W. Faris, "A localization principle for multiplicative perturbations," *J. Funct. Anal.*, to appear.
4. J. Fröhlich, F. Martinelli, E. Scoppola, and T. Spencer, "Constructive proof of localization in the Anderson tight binding model," *Comm. Math. Phys.* **101** (1985), 21–46.
5. J. Fröhlich and T. Spencer, "Absence of diffusion in the Anderson tight binding model for large disorder or low energy," *Comm. Math. Phys.* **88** (1983), 151–184.
6. S. Kotani, "Lyaponov exponents and spectra for one-dimensional random Schrödinger operators," to appear in *Random Matrices and Their Applications*, Proc. 1984 AMS Conference.
7. B. Simon and T. Wolff, "Singular continuous spectrum under rank one perturbations and localization for random Hamiltonians," *Comm. Pure Applied Math.*, to appear.

# SIMULATIONS AND GLOBAL OPTIMIZATION

Basilis Gidas*

Division of Applied Mathematics
Brown University
Providence, RI 02912

## I. Introduction

Let $Z^d$ be the usual d-dimensional cubic lattice. To each site $i \in Z^d$, we associate a random variable ("spin") $x_i$ with values in a spin stable space $X$. We will consider the cases when:

a) $X$ is a finite set

b) $X$ is a compact metric space (e.g. a finite interval in $R$, or a homogeneous space of a compact Lie group with the natural $\sigma$-algebra of Borel subsets).

c) $X$ is a complete separable metric space (e.g. $R^n$, $n > 1$).

The space $C = X^{Z^d}$ is called the configuration space. Under the product topology, if $X$ is a compact metric space then so is $C$, and if $X$ is a complete separable metric space so is $C$. A configuration is a function $x = \{x_i: i \in Z^d\}$ in $C$. For every subset $V \subset Z^d$, we define $\Omega_V = X^V$. This is the set of configurations $x(V) = \{x_i: i \in Z^d\}$ in $V$.

We assume that for every subset $V \subset Z^d$, we have a function $U(x(V)): \Omega_V \to R$. $U$ is called the interaction of the spins in $V$. For a finite subset $\Lambda \subset Z^d$, the energy (or Hamiltonian) in the domain $\Lambda$ is defined by

$$H_\Lambda(x(\Lambda)) = \sum_{V \subset \Lambda} U(x(\Lambda)) \tag{1.1}$$

the interaction is said to be of finite range if there exists an $R_0 > 0$ such that $U(\phi(V)) = 0$ whenever $\operatorname{diam} V > R_0$. For most of our considerations $U$ need not be of finite range.

---

* Partially supported by NSF Grant DMS 85-16230 and U.S. ARO DAAG-29-83-K-0116

Let $d\mu_0$ be a (not necessarily finite) measure on $X$. We assume that the interaction $U$ is such that the partition function

$$Z_\Lambda = Z_\Lambda(T;u) = \int_{\Omega_\Lambda} e^{-\frac{1}{T}H_\Lambda(x(\Lambda))} \prod_{i\in\Lambda} d\mu_0(x_i) \tag{1.2}$$

is finite. Here $T > 0$ is a parameter which plays the role of temperature. Let

$$d\mu_\Lambda(x(\Lambda)) = \frac{e^{-\frac{1}{T}H_\Lambda(x(\Lambda))}}{Z_\Lambda} \prod_{i\in\Lambda} d\mu_0(x_i) \tag{1.3}$$

be a probability measure on $\Omega_\Lambda$. This is the Gibbs distribution (corresponding to the interaction $U$) with free boundary conditions. Our considerations hold also for Gibbs distributions with general boundary conditions [20].

A problem of interest in Statistical Mechanics [18], Gauge Field Theories on a lattice [23], Image Processing [4], Machine Learning [13], Combinatorial Optimization problems [15], etc., is how to simulate from the distribution (1.3). Another problem of interest to the above disciplines is the following optimization problem: Suppose that $H_\Lambda$ has finite global minima on $\Omega_\Lambda$ (of course, this is automatically in the case when $\Omega_\Lambda$ is compact or finite). Then the problem is how to algorithmically construct the global minima of $H_\Lambda$.

The fundamental algorithm for simulating from $d\mu_\Lambda$ when $X$ (and hence $\Omega_\Lambda$) is a finite set, was introduced by Metropolis et al. [17]. The analogue of the Metropolis algorithm in the case when $X$ is a Riemannian manifold (such as $R^m$, or a compact Lie group) is the Langevin equation. We will describe both algorithms below. A modification of these algorithms with a time-dependent temperature $T = T(t)$ which converges to zero at an appropriate rate as $t \to +\infty$, gives rise to global optimization algorithms known as the Annealing Algorithm (AA) in the discrete case, and the Langevin Algorithm (LA) in the manifold case. We will refer to these global optimization algorithms as the Cooling Algorithms.

The main mathematical problems associated with the simulation and global optimization algorithms are:

(i) Convergence

(ii) Speed of Convergence

(iii) The Computational Complexity as $\Lambda \to Z^d$ (in a suitable sense).

The formulation, the convergence, and the speed of convergence of the simulation as well as of the cooling algorithms, do not depend on the fact that $H_\Lambda(x)$ has the form (1.1), i.e. that it is build out of "local" interactions. In fact, the convergence and the speed of convergence for the Metropolis algorithm, are consequences of the theory of stationary Markov chains, while for the Langevin equation are consequences of the theory of stationary Markov processes. Similarly, we will see that the AA can be treated as a special case of non-stationary Markov chains, while the LA can be treated as a special case of non-stationary Markov processes.

The computational complexity of the simulation and cooling algorithms appears to be the most difficult of the above three problems. In contrast to the convergence and speed of convergence, the computational complexity should be tractable only if the Hamiltonian $H_\Lambda$ is of the form (1.1), with some conditions on the interaction $U$. This is related to the fact that the "thermodynamic limit" $\Lambda \to Z^d$ of the Gibbs distribution (1.3) can be controlled only if $H_\Lambda$ has the form (1.1) [20].

Next we define the Metropolis algorithm, the Langevin equation, and the Cooling Algorithm. We fix $\Lambda$ $Z^d$ (finite), and for convenience we denote $\Omega_\Lambda$ by $\Omega$, $H_\Lambda$ by $H$, and $x(\Lambda)$ by $x = \{x_i : i \in \Lambda\}$.

1) <u>Metropolis Algorithms</u>: Here we assume that $\Omega$ (or equivalently $X$) is a finite set. Let $N$ be the cardinality of $\Omega$ (i.e. $N = |X|^{|\Lambda|}$). We denote the points in $\Omega$ by $x^{(i)} \in \Omega$, $i = 1,\dots,N$, and set $H_i = H(x^{(i)})$. Let $R = (R_{ij})$, $i,j = 1,\dots,N$, be a symmetric $(R_{ij} = R_{ji})$ and irreducible transition probability matrix (to be referred as the "proposal matrix") We define the transition rate matrix of a stationary, finite, continuous-time Markov chain $X(t)$ by

$$i \neq j, \quad L_{ij} = L_{ij}(T) = R_{ij}e^{-\frac{1}{T}(H_j - H_i)^+}$$

$$= \begin{cases} R_{ij} & \text{if } H_j < H_i \\ R_{ij}e^{-\frac{1}{T}(H_j - H_i)} & \text{if } H_j > H_i \end{cases} \tag{1.4a}$$

$$\sum_j L_{ij}(T) = 0 \tag{1.4b}$$

The transition matrix $P(t)$ defined by

$$P_{ij}(t) = P\{X(t) = x_j | X(0) = x_i\} \tag{1.5}$$

satisfies

$$\frac{dP(t)}{dt} = -P(t)\, L(T) \tag{1.6}$$

Let

$$\pi_i(T) = \frac{e^{-\frac{1}{T}H_i}}{\sum_j e^{-\frac{1}{T}H_j}}, \quad i = 1,\ldots,N \tag{1.7}$$

Then we have the "detailed balance" relation

$$\pi_i(T)\, L_{ij}(T) = \pi_j(T)\, L_{ji}(T) \tag{1.8}$$

This easily implies that the matrix $L(T)$ is Hermitean on $\ell_2(\Omega)$ with weight $\pi(t)$, i.e. for any two functions $f$ and $q$ on $\Omega$, we have

$$\sum_j \pi_j(T) f_j (Lg)_j = \sum_j \pi_j(T)(Lf)_j g_j \tag{1.9}$$

where $f_i = f(x^{(i)})$, $g_i = g(x^{(i)})$, $i = 1,\ldots,N$. It is a consequence of the theory of stationary Markov chains that (under the above assumptions on $R$)

$$\lim_{t \to +\infty} P_{ij}(t) = \pi_j(T), \quad j = 1,\ldots,N \tag{1.10}$$

Furthermore, the speed of convergence is given by

$$P_{ij}(t) = \pi_j(T) + O(e^{-\lambda_2(T)t}), \quad \text{large } t \tag{1.11}$$

where $\lambda_2(T)$ is the second eigenvalue of $L(T)$ (the first eigenvalue $\lambda_1(T)$ of $L(T)$ is zero, and the corresponding normalized eigenvector is $\phi_1 = (1,1,\ldots,1)$).

The discrete-time Metropolis algorithm is defined in a similar way: one specifies the one-step transition probabilities $P_{ij}(T)$ so that $P_{ij}(T) = L_{ij}(T)$ for $i \neq j$, and $P_{ii}(T) = 1 - \sum P_{ij}(T)$. The t-step transition matrix $P(t)$ is the t-th power of $p = (P_{ij})$. It satisfies (1.10).

The Metropolis algorithm can be generalized by using a larger class of proposal matrices $R$ which are not necessarily symmetric but satisfy certain weaker conditions [10,7,11]. Also, there exist [7,11] several modifications of the Metropolis algorithms obtained by defining the transition rate matrix $L(T)$ (or the one-step transition probabilities $P_{ij}(T)$) suitably. A commonly used such algorithm is the spin-flip algorithm for binary systems.

2) The Langevin Equation: First, we consider the case when $\Omega = R^n$, $n \geq 1$ (which corresponds to the case $X = R$, and $|\Lambda| = n$). Let $w(t)$ be the standard Brownian motion on $R^n$. We define a stationary Markov process $X(t)$ via the Langevin equation

$$dX(t) = \sqrt{2T}\, dw(t) - \nabla H(X(t))dt \tag{1.12}$$

We assume that $H(x)$ is $C^2$ and that

$$Z = Z(T) = \int_{R^n} e^{-\frac{1}{T} H(x)} dx < +\infty$$

Let

$$p(t,x) = P\{X(t) = x | X(0) = x_0\}$$

be the transition functions of the process $X(t)$. It satisfies the "forward" (or Fokker-Planck) equation

$$\frac{\partial p}{\partial t} = T\Delta p + \nabla(\nabla H p) \tag{1.13}$$

$$p(t,x)\big|_{t\downarrow 0} = \delta(x - x_0)$$

Let

$$\pi_T(x) = \frac{e^{-\frac{1}{T}H(x)}}{Z} \tag{1.14}$$

The generator of the diffusion process $X(t)$ is given by

$$L(T) = -T\Delta p + \nabla H \cdot \nabla \tag{1.15}$$

This is a self-adjoint operator on $L_2(\Omega, \pi_T(x)dx)$, and it is unitarily equivalent to the Schrödinger-type operator

$$S(T) = -T\Delta + \frac{T}{(2T)^2}\{|\nabla H|^2 - 2T\Delta H\} \tag{1.16}$$

We have $L(T) = \pi_T^{-1/2} S(T)\, \pi_T^{1/2}$. If we assume that $|\nabla H|^2 - 2T\Delta H \to +\infty$ as $|x| \to +\infty$, then $L(T)$ (and hence $S(T)$) has discrete spectrum [19]. The first eigenvalue $\lambda_1(T)$ is equal to zero, and the corresponding normalized eigenvector of $L(T)$ is equal to $1$ (and of $S(T)$, equal to $\pi_T^{1/2}$). The second eigenvalue $\lambda_2(T)$ is strictly positive.

It is well-known (and easily proven) that

$$|p(t,\cdot) - \pi_T(\cdot)| \underset{t\to+\infty}{\longrightarrow} 0 \tag{1.17}$$

weakly. Furthermore, the rate of convergence is given by $O(e^{-\lambda_2(T)t})$.

The Langevin equation reads the same when $\Omega = I^n$, where $I$ is a bounded interval in $R$, say $I = [-a,a]$ (this corresponds to the case when $X = I$ and $|\Lambda| = n$). However, in order to define precisely the process $X(t)$ one needs to specify boundary conditions on the boundary $\partial\Omega$ of $\Omega$ [2,6]. The result (1.17) holds also in these cases. The Langevin equation is also defined when $\Omega$ is a Riemannian manifold. In this case, equations (1.12) and (1.13), and the operators

$L(T)$ and $S(T)$, read the same provided that one interpretes $\Delta$ and $\nabla$, as the Laplace-Beltrami operator and the covariant gradient, respectively. If $\Omega$ is a compact manifold without boundaries, then (1.17) holds without any further assumptions. But if $\Omega$ is a compact manifold with boundaries, then the process $X(t)$ is specified by imposing appropriate boundary conditions on $\partial\Omega$. And if $\Omega$ is a non-compact, complete (in the sense of having infinitely long geodesics) Riemannian manifold, then one needs to impose certain conditions on $H(x)$ for $x$ near "infinity". In the manifold case, $w(t)$ in (1.12) is the Brownian process on the manifold $\Omega$.

3) Annealing Algorithm (AA): Here we assume that $\Omega$ is a finite set. The AA is obtained from the Metropolis algorithm by considering a time-dependent temperature $T = T(t)$. This gives rise to a non-stationary Markov chain whose transition rate matrix (or one-step transition probabilities, in the discrete-time case) are defined by (1.4) with $T = T(t)$. The transition matrix $P(t)$ satisfies (1.6) with $L(T(t))$. Let $\pi_j(t) = \pi_j(T(t))$, where $\pi_j(T)$ is given by (1.7). In Section II, we will show that if $T(t) \to 0$ as $t \to +\infty$, at an appropriate rate, then

$$\lim_{t \to +\infty} \sum_j |P_{ij}(t) - \pi_j(t)| = 0 \tag{1.18}$$

This also holds for the AA with the more general class of proposal matrices $R$ mentioned before. It also holds [7] for the variant of the AA obtained from the modified Metropolis algorithms (e.g. the spin-flip algorithm).

3) The Langevin Algorithm (LA): Here we assume that $\Omega$ is a Riemannian manifold, as in the case of the Langevin equation. The Langevin equation (1.12) with $T = T(t)$ gives rise to a non-stationary Markov process whose transition function $p(t,x)$ satisfies (1.13) with $T = T(t)$. Let $\pi(t,x) = \pi_{T(t)}(x)$. In Section III, we will show (when $\Omega = R^n$) that if $T(t) \downarrow 0$ as $t \to +\infty$, at an appropriate rate, then in the weak sense

$$|p(t,\cdot) - \pi(t,\cdot)| \to 0 \quad \text{as} \quad t \to +\infty \tag{1.19}$$

Let $\lambda_2(T(t))$ be the second eigenvalue of the transition rate matrix $L(T(t))$ for the AA, or the operator $L(T(t))$ defined by (1.15) for the LA. In Section II and III, we will show that if:

(i) $T(t) \downarrow 0$ monotonically as $t \to +\infty$

(ii) $$\int_0^{+\infty} \lambda_2(T(t))dt = +\infty \tag{1.20}$$

(iii) for the AA: $$-\frac{T'}{T^2}\frac{1}{\lambda_2(T(t))} \to 0 \quad \text{as} \quad t \to +\infty \tag{1.21a}$$

for the LA: $$-\frac{T'}{T}\frac{1}{\lambda_2(T(t))} \to 0 \quad \text{as} \quad t \to +\infty \tag{1.21b}$$

then (1.18) and (1.19) hold. For the AA, conditions (1.20) and (1.21a) hold if

$$T(t) > \frac{\Delta}{\log t}, \quad \text{large } t \tag{1.22}$$

for a specific (optimal) constant $\Delta$. While for the LA conditions (1.20) and (1.21b) hold if

$$T(t) \geq \frac{\Delta}{\log t}, \quad \text{large } t \tag{1.23}$$

again for a specific (optimal) constant $\Delta$. In [8], we have shown that if (i) holds, then (1.20) is a necessary and sufficient condition for the convergence of the AA, and, in certain cases, for the LA. Our techniques are based on methods of Differential equations. In Sections II and III, we present an elementary version of our techniques to establish convergence under the assumptions (1.20) and (1.21).

The first convergence result for the AA was established in [4]. In [7], we treated the annealing algorithms as a special case of non-stationary Markov chains, and we obtained some optimal annealing schedules, an ergodic theorem, and some estimates on the rate of convergence. Optimal annealing schedules for the AA have recently been obtained in [10]. More recently the AA was also treated in [3].

A convergence theorem for the LA in bounded domains of $R^n$, $n \geq 1$, was first obtained in [6]. In [8] (see also [9]) we have used methods from partial differential equations to analyse the Fokker-Planck equation (with $T = T(t)$) in the entire of $R^n$, and have obtained in certain cases optimal temperature schedules. Our methods apply as well to the manifold case. In [16,1], the theory of Large Deviations was employed to obtain convergence of the LA in the entire of $R^n$. Estimates on the rate of convergence for the LA have not yet been worked out.

The computational complexity of the simulation and cooling algorithms is important for applications and an interesting mathematical problem. For the problem of computational complexity we assume that the interaction $U(\phi(V))$ and the single spin state space $X$ are such that we can control the thermodynamic limit $\Lambda \to Z^d$ (in the sense of van Hove [20]) of the Gibbs distribution (1.3). For the simulations algorithms the problem of computational complexity (usually known as the relaxation time problem) is the following: Let $\lambda_2^{(\Lambda)}(T;U)$ be the second eigenvalue of the transition rate matrix for the Metropolis algorithm, or of the diffusion operator for the Langevin equation. The quantity

$$\tau_\Lambda(T;U) = \frac{1}{\lambda_2^{(\Lambda)}(T;U)} \tag{1.24}$$

is called the relaxation time. The problem is to control the behavior of $\tau_\Lambda(T;U)$ as $\Lambda \to Z^d$. For Ising type models, it is empirically known [18] that for $T > T_c$, where $T_c$ is the critical temperature, the relaxation time $\tau_\Lambda(T)$ has a finite limit $\tau(T)$ as $\Lambda \to Z^d$. Furthermore, it is also empirically known that as $T \downarrow T_c$, $\tau(T)$ diverges like $(\xi(T))^z$, where $\xi(T)$ is the correlation length (which diverges like $(T - T_c)^{-\nu}$ as $T \downarrow T_c$), and $z$ is a dynamical critical exponent. For the one dimensional Ising model, Holley [14] has posed the simulation problem directly on $Z^1$ and has shown that there is a finite relaxation length for all $T > 0$. S. Geman has shown [5] that the results of [14] imply the uniform boundedness of $\tau_\Lambda(T)$ for the 1-dimensional Ising model with free boundary conditions. In general, it is expected that in the one phase region of the parameter space $(T,U)$, the relaxation length $\tau_\Lambda(T;U)$ remains finite as $\Lambda \to Z^d$, and diverges like a power of $|\Lambda|$ for $(T,U)$ in the critical region. Recently, we have obtained some preliminary results for the finiteness of $\lim \tau_\Lambda(T;U)$ in the single phase region ("high temperature region") for models on $Z^d$ which are discrete approximations of certain Quantum Field Theories.

For the cooling algorithms, the computational complexity is formulated as follows: Let $D$ be the union of the domains of attraction of the global minima of $H_\Lambda(x(\Lambda))$, i.e. the sets of initial configurations $x^{(0)} \in \Omega_\Lambda$ from which we can reach a global minimum of $H_\Lambda(x)$ by a descent algorithm. Let $\varepsilon$, $0 < \varepsilon < 1$, be an allowable probability of failure. We define

$$\tau_\Lambda(x^{(0)};U;\varepsilon) = \inf\{t: P(X(s) \in D | X(0) = x^{(0)}) > 1 - \varepsilon, \text{ for all } s > t\} \quad (1.25)$$

The computational complexity of the cooling algorithms amounts to controlling $\tau_\Lambda(x^{(0)};U;\varepsilon)$ as $\Lambda \to Z^d$. For spin systems in Statistical Mechanics, lattice Field Theories, and Image processing problems, we make the following

Conjecture: For fixed $\varepsilon$, $0 < \varepsilon < 1$, there exists a temperature schedule $T = T(t)$, a function $\alpha(|\Lambda|) = \alpha(|\Lambda|;U;\varepsilon)$ which grows exponentially with $|\Lambda|$ as $\Lambda \to Z^d$, and constants $c_1, c_2 > 0$, such that

$$c_1 \alpha(|\Lambda|) < \int_{\Omega_\Lambda} \tau_\Lambda(x^{(0)};U;\varepsilon) d\rho_\Lambda(x^{(0)}) < c_2 \alpha(|\Lambda|) \quad (1.26)$$

with probability one. Here $d\rho_\Lambda(x^{(0)})$ is a Gaussian measure on $\Omega_\Lambda$.

This conjecture is consistent with the widely accepted belief that the above systems are NP-complete systems.

## II. Convergence of the Annealing Algorithm

First we consider a non-stationary Markov chain with a finite state space $\Omega = \{x^{(1)},\ldots,x^{(n)}\}$. Let $L(t)$ be the transition matrix of the chain, and $\pi(t) = (\pi_1,\ldots,\pi_n)$ be a probability vector with $\pi_j(t) > 0$, $j = 1,\ldots,n$, satisfying the "detailed balance" relation (1.8) (with $L(T)$ and $\pi(T)$ replaced by $L(t)$ and $\pi(t)$). The transition matrix $P(t)$ of the Markov chain satisfies (1.6) with $L(T)$ replaced by $L(t)$. We assume that the entries of $L(t)$ are bounded, so that there exists a solution $P(t)$ satisfying

$$\sum_j P_{ij}(t) = 1$$

As in the Metropolis algorithm, the matrix $L(t)$ is Hermitean on $\ell_2(\Omega)$ with weight $\pi(t)$, and therefore it has real eigenvalues. The first eigenvalue $\lambda_1(t)$ is zero, and the corresponding normalized eigenvector is $\phi_1 = (1,1,\ldots,1)$. The second eigenvalue $\lambda_2(t)$ is strictly positive. Our basic theorem is

Theorem 2.1

Consider a non-stationary Markov chain as above. Suppose that

$$\left| \frac{d\pi_j(t)}{dt} \right| \leq \gamma(t)\pi_j(t), \quad j = 1,\ldots,n \tag{2.1}$$

where $\gamma(t)$ is independent of $j$. Then, if

$$\int_0^{+\infty} \lambda_2(t)dt = +\infty \tag{2.2}$$

and

$$\lim_{t \to +\infty} \frac{\gamma(t)}{\lambda_2(t)} = 0 \tag{2.3}$$

then

$$\sum_j |P_{ij}(t) - \pi_j(t)| \to 0 \quad \text{as} \quad t \to \infty \tag{2.4}$$

Proof: Set

$$P_{ij}(t) = \pi_j(t)Q_{ij}(t)$$

Then

$$\sum_j |P_{ij}(t) - \pi_j(t)| \leq \left( \sum_j \pi_j(t)(Q_{ij}(t) - 1)^2 \right)^{1/2}$$

We will show that

$$F(t) \equiv \sum_j \pi_j(t)(Q_{ij}(t) - 1)^2 \to 0 \quad \text{as} \quad t \to +\infty \tag{2.5}$$

For fixed $i$, we set $q_j(t) = Q_{ij}(t)$. Then

$$\frac{dq_j}{dt} = -(Lq)_j - \frac{d\pi_j}{dt}\pi_j^{-1}q_j \tag{2.6}$$

This yields

$$\frac{1}{2}\frac{d}{dt}\{\sum_j \pi_j(q_j - 1)^2\} = -\sum_j \pi_j q_j (Lq)_j - \frac{1}{2}\sum_j \frac{d\pi_j}{dt} q_j$$

$$= -\sum_j \pi_j(q_j - 1)(L(q - \phi_1))_j - \frac{1}{2}\sum_j \frac{d\pi_j}{dt} q_j \tag{2.7}$$

Since $\lambda_2(t)$ is the smallest eigenvalue in the orthogonal complement of the subspace spanned by $\phi_1$, we have

$$\sum_j \pi_j(q_j - 1)(L(q - \phi_1))_j \geq \lambda_2(t) \sum_j \pi_j(q_j - 1)^2 \tag{2.8}$$

This together with (2.1) and (2.7) yield

$$\frac{1}{2}\frac{dF(t)}{dt} \leq -\lambda_2(t)F(t) + \frac{1}{2}\gamma(t)F(t) + \frac{1}{2}\gamma(t) \tag{2.9}$$

If (2.2), (2.3) hold, then (2.9) implies (2.5), and completes the proof of the theorem.

Next, we apply Theorem 2.1 to the AA. We assume that $T(t)$ converges to zer monotonically as $t \to +\infty$. From (1.7) with $T = T(t)$, one finds that (2.1) holds with

$$\gamma(t) = -\frac{T'}{T^2}(H_{max} - H_{min})$$

where $H_{max}$, $H_{min}$ are the maximum and minimum values of $H$ on $\Omega$, respectively. Under the conditions on the proposal matrix $R$ stated in the Introduction, there exists a specific constant $\Delta$ such that

$$\lambda_2(T(t)) = \text{const } e^{-\frac{\Delta}{T(t)}} + O(e^{-\frac{2\Delta}{T(t)}}), \quad \text{large } t \tag{2.10}$$

The constant $\Delta$ can be determined by the methods of [22], and is explicitly represented in terms of graphs. If

$$T(t) = \frac{c}{\log t}, \quad \text{large } t \tag{2.11}$$

then (2.2) holds with $c \geq \Delta$, but (2.3) holds only for $c > \Delta$. Thus we see that for temperature schedules of the form (2.11), the elementary proof of Theorem 2.1 gives a constant $c$ near the optimal constant $\Delta$.

## III. Convergence of the Langevin Algorithm

We will treat only the case when $\Omega = \Omega_\Lambda = R^n$, $n \geq 1$. Our techniques apply also to the case when $\Omega$ is a Riemannian manifold.

Under the assumptions of Theorem 3.1 below, the operator $L(T(t))$ defined in (1.15), has discrete spectrum $\lambda_1(T(t)) < \lambda_2(T(t)) \leq \lambda_3(T(t)) \leq \ldots$, with $\lambda_1(T(t)) = 0$. As in the AA, the second eigenvalue $\lambda_2(T(t))$ plays a special role in our analysis.

Theorem 3.1

Suppose that $H(x)$, $x \in R^n$, is a $C^2$ function which grows like const $|x|^\alpha$, $\alpha \geq 2$ at infinity. Suppose that $|\nabla H(x)|$ goes to infinity at infinity, and that $-\frac{\Delta f}{1 + |\nabla f|^2}$ is bounded below. Furthermore, suppose that the critical set of $H(x)$ consists of isolated, non-degenerate, points. Then, if $T(t) \downarrow 0$ monotonically as $t \to +\infty$, and

$$\int_0^{+\infty} \lambda_2(T(t))dt = +\infty \tag{3.1}$$

$$-\frac{T'}{T}\frac{1}{\lambda_2(T(t))} \to 0, \quad \text{as } t \to +\infty \tag{3.2}$$

then (1.19) holds in the weak sense.

Proof: Let

$$p(t,x) = \pi(t,x)q(t,x) \tag{3.3}$$

Then (1.13) with $T = T(t)$, becomes

$$\frac{\partial q}{\partial t} = T(t)\Delta q - \nabla H \cdot \nabla q - \frac{T'}{T^2}(H(x) - \overline{H}(t))q \tag{3.4a}$$

$$= T\pi^{-1}\nabla(\pi\nabla q) - \frac{T'}{T^2}(H(x) - \overline{H}(t))q \tag{3.4b}$$

$$q(t,x)|_{t\downarrow 0} = \frac{\delta(x - x_0)}{\pi(0,x_0)}$$

where $T' = \frac{dT}{dt}$, and

$$\overline{H}(t) = \int \pi(t,x)H(x)dx$$

Let $\eta(x)$ be a smooth function of compact support. Then

$$|\int p(t,x)\eta(x)dx - \int \pi(t,x)\eta(x)dx| \leq (\int \pi\eta^2 dx)^{1/2}(\int \pi(q - 1)^2 dx)^{1/2} \tag{3.5}$$

We will show that

$$F(t) \equiv \int \pi(t,x)(q(t,x) - 1)^2 dx \to 0, \quad \text{as} \quad t \to +\infty \tag{3.6}$$

From (3.4), we obtain

$$\frac{1}{2}\frac{d}{dt}\int \pi(q - 1)^2 dx = -T\int q\nabla(\pi\nabla q)dx - \frac{1}{2}\frac{T'}{T^2}\int \pi(H(x) - \overline{H}(t))q^2 dx \tag{3.7}$$

In [8], we have established the following a priori bound whose proof is indicated below: There exists a $\theta(t)$ tending to zero as $t \to +\infty$, such that

$$q(t,x) \leq ce^{\theta(t)H(x)}, \quad \text{large } t \tag{3.8}$$

for some constant C. This bound together with the estimate $|\nabla q| \in L_2(R^n, \pi dx)$ easily obtained from (3.4), allow us to integrate by parts and obtain

$$-\int q\nabla(\pi\nabla q)dx = \int \pi|\nabla q|^2 dx = \int \pi|q - 1|^2 dx \tag{3.9}$$

the assumptions on $H(x)$ allow us to apply a standard Poincaré inequality for weighted spaces, and obtain

$$T\int \pi|\nabla(q - 1)|^2 dx \geq \lambda_2(T)\int \pi(q - 1)^2 dx \tag{3.10}$$

Here we have used the fact that $\int \pi q\, dx = 1$. Using (3.8), the last term in (3.7) is bounded as follows

$$-\frac{T'}{T^2}\int \pi\,(H(x) - \overline{H}(t))q^2 dx \leq -C^2\frac{T'}{T^2}\int \pi(H - \overline{H}(t))e^{2\theta(t)H(x)}dx \tag{3.11a}$$

$$\leq -\tilde{C}\frac{T'}{T} \tag{3.11b}$$

Here we have used the fact that the integral in the right hand side of (3.11a) goes to zero like $T(t)$. Inserting estimates (3.10), (3.11b) into (3.7), we obtain

$$\frac{1}{2}\frac{dF}{dt} \leq -\lambda_2(T(t))F(t) - \frac{1}{2}\tilde{C}\frac{T'}{T} \tag{3.12}$$

If (3.1) and (3.2) hold, then (3.12) yields (3.6). Now, we indicate the proof of (3.8). Since the coefficient of $q$ in (3.4) is non-negative (at least for large x), we cannot apply the maximum principle directly to the parabolic equation (3.4). However, if we set

$$q(t,x) = \phi(t,x)e^{\theta(t)H(x)} \tag{3.13}$$

Then

$$\frac{\partial\phi}{\partial t} - T(t)\Delta\phi + (1 - \theta T)\nabla H \cdot \nabla\phi + c(t,x)\phi(t,x) = 0 \tag{3.14a}$$

where

$$c(t,x) = \theta(t)|\nabla H|^2 + \theta'(t)H(x) + \frac{T^1}{T^2}(H(x) - \overline{H}(t)) - \theta^2 T|\nabla H|^2 - \theta T\Delta H \tag{3.14b}$$

Using the assumption that the critical points of $H(x)$ are non-degenerate, and that $T(t) \downarrow 0$ monotonically as $t \to +\infty$, one can show that there exists a $\theta(t)$ such that $\theta(t) \to 0$ as $t \to +\infty$, and $c(t,x) > 0$. This allows the applicability of the maximum principle which yeilds (3.8).

Remarks 1) In [8], we have derived sharper upper and lower bounds for $q(t,x)$, which imply in certain cases that (3.1) is necessary and sufficient for (1.19) to holds.

2) The second eigenvalue $\lambda_2(T)$ of the operator (1.15) converges to zero exponentially as $T \downarrow 0$ [12,21]. In [8], we have derived (in certain cases) the asymptotic behavior of $\lambda_2(T)$ as $T \downarrow 0$. Under the assumption on $H(x)$ in Theorem 3.1, the asymptotic behavior of $\lambda_2(T)$ is as follows: Let $a_1,\ldots,a_N$ be the minima of $H(x)$. For simplicity, we assume that $a_N$ is the only global minimum of $H(x)$. We consider paths $\gamma_i(s)$, $s \in [0,1]$, from each $a_i$, $i = 1,\ldots,N-1$, to $a_N$, such that $\gamma_i(0) = a_i$, $\gamma_i(1) = a_N$, and define

$$\Delta_i = \min_{\gamma_i} \max_{s\in[0,1]} f(\gamma_i(s)) - f(a_i) \tag{3.15a}$$

$$\Delta = \max_{i=1,\ldots,N-1} \Delta_i \tag{3.15b}$$

In one dimension ($n = 1$), or in $n \geq 2$ but with $N = 2$, we have

$$\lambda_2(T) = \text{const } e^{-\frac{\Delta}{T}} + O(e^{-\frac{2\Delta}{T}}) \tag{3.16}$$

In these cases, if

$$T(t) \geq \frac{\Delta}{\log t}, \quad \text{large } t \tag{3.17}$$

then (3.1) and (3.2) hold. Thus for temperature schedules of the form (3.17), the above proof of convergence of the LA gives the optimal constant $\Delta$.

## References


1. Chiang, T.S., C.R. Hwang, and S.J. Shen: "Diffusion for Global Optimization in $R^n$", preprint.

2. Dynkin, E.B.: Markov Processes-II, Springer-Verlag (1965).

3. Gelfand, S., and S.K. Mitter: "Analysis of Simulated Annealing for Optimization", preprint, M.I.T.

4. Geman, S., and D. Greman: "Stochastic Relaxation, Gibbs Distribution, and the Bayesian Restoration of Images" IEEE Transactions, PAMI 6 (1984) 721-741.

5. Geman, S.: Private Communication.

6. Geman, S., and C.R. Hwang: "Diffusions for Global Optimization", to appear SIAM Journ. on Control and Optimization (1985).

7. Gidas, B.: "Non-Statinary Markov Chains and Convergence of the Annealing Algorithm", J. Stat. Physics 39 (1985) 73-131.

8. Gidas, B.: "Global Optimization via the Langevin Equation", in preparation.

9. Gidas, B.: "Global Optimization via the Langevin Equation", Conference Proceedings, 24th IEEE Conference on Decision and Control (1985).

10. Hajek, B.: "Cooling Schedules for Optimal Annealing", to appear in Mathematics of Oper. Research.

11. Hastings, W.K.: "Monte Carlo Sampling Methods Using Markov Chains and Their Applications", Biometrika 5 (1970), 97-109.

12. Helffer, B., and I. Sjöstrand: "Multiple Wells in the Semi-Classical Limit I", Comm. Part. Diff. Eqn. 9 (1984), 337-408.

13. Hinton, G., T. Sejnowski, and D. Ackley: "Boltzmann Machine: Constraint Satisfaction Networks that Learn", preprint (1984).

14. Holley, R.: "Rapid Convergence to Equilibrium in One Dimensional Ising Stochstic Models", Ann. of Prop.

15. Kirpatrick, S., C.D. Gelatt, and M. Yecchi: "Optimization by Simulated Annealing", Science 220, 13 May (1983), 621-680.

16. Kushner, H.: "Asymptotic Behavior for Stochastic Approximations and Diffusion with slowly decreasing noice effects: Global Minization via Monte Carlo", preprint (1985).

17. Metropolis, N., A Rosenbluth, M. Rosenbluth, A. Teller, and E. Teller: "Equations of State Calculations by Fast Computing Machines", J. Chem. Phys. 21 (1953) 1087-1091.

18. Monte Carlo Methods in Statistical Physics, Springer-Verlag (1979), ed. K. Binder.

19. Reed, M., and B. Simon: Methods of Modern Mathematical Physics IV: Analysis of Operators, Academic Press (1978).

20. Ruelle, D.: Thermodynamic Formalism, Addison-Wesley (1978

21. Simon, B.: "Semiclassical Analysis of Low Lying Eigenvalues I. Non-Degenerate Minima: Asymptotic Expansions", Ann. Inst. Henri Poincaré 38 (1983), 295-307.

22. Ventel, A.D.: "On the Asymptotics of Eigenvalues of Matrices with Elements of Order $\exp\{-v_{ij}|2\varepsilon^2\}$", Dokl. Akad. Nank SSR 202 (1972) 75-68.

23. Wilson, K.: "Monte Carlo Calculations for Lattice Gauge Theory", in Recent Developments in Gauge Theories, Plenum, N.Y. 1980), eds.: G. 'tHooft, et. al.

# LOW TEMPERATURE BEHAVIOR IN RANDOM ISING MODELS

John Z. Imbrie
Lyman Laboratory of Physics
Harvard University
Cambridge, MA 02138

Abstract

The behavior of Ising models with disorder is considered at low temperature. We discuss the main ideas behind a proof of long-range order in three dimensions for the ground state of the Ising model in a random magnetic field. We comment also on the behavior of interfaces in random field and random bond problems.

## 1. Introduction

My intent in this lecture is to discuss some probabilistic ideas arising in the solution of stochastic minimization problems. These are problems where the goal is to minimize an energy functional in a system with infinitely many degrees of freedom. The stochastic aspect of the problem comes in when the functional itself depends on an infinite number of random variables which are taken as fixed when computing the energy minimum.

Such problems arise naturally in condensed matter physics when a sample is disordered, *i.e.* it has impurities which cause the microscopic structure of the material to vary from point to point. At sufficiently low temperatures, the sample will take on a configuration of minimum energy, so to understand the system we must minimize an energy functional with spacially varying coefficients. For sufficiently large samples, averages over the volume of the sample become identical with averages over the statistical distribution of the impurity parameters.

We will consider specifically the *Ising model in a random magnetic field.* In our Ising model the configuration of the system is obtained by specifying a collection of spins $\{\sigma_i\}$, where $\sigma_i = \pm 1$ for each site $i \in \mathbf{Z}^d$. The integer $d$ is the dimension of the system; it is useful to explore the behavior of the system as $d$ is varied. Usually the energy functional favors alignment of adjacent spins (both

+1 or both -1). In the presence of a random magnetic field, there is in addition a tendency of the spin at a site to be either up (+1) or down (-1), depending on the value of the random magnetic field there. The energy functional is

$$H^{+}(\Lambda) = \sum_{<i,j>} \frac{1}{2}(1-\sigma_i\sigma_j) - \sum_{i\in\Lambda} \frac{1}{2} h_i\sigma_i \quad , \tag{1}$$

where $\Lambda$ is a large rectangle in $\mathbb{Z}^d$ and $\sigma_i$ is fixed at +1 for $i \notin \Lambda$. The sum over i ranges over $\mathbb{Z}^d$, and $<i,j>$ runs over nearest-neighbor pairs in $\mathbb{Z}^d$. The magnetic fields $h_i$ are independent, identically distributed random variables. Each magnetic field is distributed according to a measure $d\mu(h_i)$, which we will take to be a Gaussian with mean zero and variance $\varepsilon^2$. The precise form of $d\mu(h_i)$ is not so important; what is important is that it be a symmetric measure.

We will see that the minimum of $H^{+}(\Lambda)$ behaves nicely as $\Lambda$ increases to all of $\mathbb{Z}^d$, at least for $d \geqslant 3$. The value of the minimum at a particular site converges as $\Lambda \nearrow \mathbb{Z}^d$, and one can almost surely determine the minimum in any finite region S by looking at the magnetic fields in finite regions containing S [1]. This property is likely to fail for other types of randomness--for example when $J_{ij}$, the coefficient of $\sigma_i\sigma_j$, is random (spin glass). The minimization problem is much more difficult in such cases.

We have chosen symmetry-breaking boundary conditions, $\sigma_i = 1$ for $i \notin \Lambda$, because we will be interested in whether or not the minimum energy configuration $\sigma^{min}(\Lambda)$ feels the + boundary conditions even in the limit $\Lambda \nearrow \mathbb{Z}^d$. We ask what is the probability that the spin at the origin agrees with the boundary conditions. If $\mathrm{Prob}(\sigma_0^{min}(\Lambda) = 1) > 1/2$ uniformly as $\Lambda \nearrow \mathbb{Z}^d$, then we say that the system possesses *long-range order*, or that the sample has magnetized.

We will see that there is long-range order for small $\varepsilon$ if $d \geqslant 3$, and furthermore we will find that

$$\mathrm{Prob}(\sigma_0^{min}(\Lambda) = 1) \geqslant 1 - \exp(-O(\varepsilon^{-2})) \quad .$$

(For a complete proof, see [1,2].)

Small values of the disorder parameter $\varepsilon$ mean that large values of $h_i$ are unlikely

When $\varepsilon$ is large, $h_i$ is usually large and then the effect of the boundary condition is lost--there is no long-range order [3].

Lower critical dimension. The question of which values of the dimension $d$ can support long-range order has been controversial during the past several years [4]. Physicists define the lower critical dimension, $d_\ell$, as the borderline dimension above which long-range order can exist. In the ordinary (pure) Ising model, which corresponds to $\varepsilon = 0$, the question is only interesting for the thermal ensemble given by the Gibbs measure $Z^{-1} \exp(-H^+(\Lambda)/T)$ on the set of configurations. Then it is well known that for low enough temperature $T$, long-range order exists in dimensions bigger than 1; hence $d_\ell = 1$. (Long-range order in this case means that $\langle\sigma_0\rangle$, the expectation of the spin at the origin with respect to the Gibbs ensemble, is strictly positive.)

Random Ising models will be ordered in a different set of dimensions as compared with the pure case. Thus the question arises as to the value of the lower critical dimension in this case. The controversy has been between the values $d_\ell = 2$ [5,6] and $d_\ell = 3$ [7]. One can define a lower critical dimension at $T = 0$ or for $T > 0$, but one would expect the value to be the same for both cases, since it is difficult to see how thermal effects could spoil long-range order in the presence of randomness when they do not already do so in the absence of randomness. Physicists are usually willing to work in fractional dimensions, but it happens that the two competing theories for $d_\ell$ predict integral values. Furthermore, dimension $d = 3$ is the deciding case, and our result [1] that there is long-range order in three or more dimensions at $T = 0$ rules out theories that $d_\ell = 3$. (Such theories were based on an idea of dimensional reduction--a relationship between pure systems in $d$ dimensions and random systems in $d + 2$ dimensions [8]. This idea may have some validity in some situations, but evidently not in the random field Ising model in three dimensions.)

Positive temperatures. The behavior of the random field Ising model for small $T > 0$ has not been settled yet. Here one first computes expectations in the Gibbs ensemble at a fixed random field configuration. For example the spin at the origin has expectation

$$\langle\sigma_0\rangle_h = \frac{\sum_{\{\sigma_i\}} \sigma_0 \exp(-H^+(\Lambda,h)/T)}{\sum_{\{\sigma_i\}} \exp(-H^+(\Lambda,h)/T)} . \tag{2}$$

Second, one averages over the field configuration $h$. The result is denoted $\overline{\langle\sigma_0\rangle_h}$, and if it is positive, uniformly in $\Lambda$, then the system is said to have long-range order. One would like to show that there is long-range order for $d \geq 3$, but this has not yet been accomplished. As $T \to 0$, the thermal average (2) approaches $\sigma_0^{min}$, the value of $\sigma_0$ in the ground state, and we are back to the stochastic minimization problem.

The method. We approach the $T = 0$ problem as follows. We attempt to prove central limit theorem type results about the ground state energy $H^+(\sigma^{min})$. That is, we show that $H^+(\sigma^{min})$ is essentially a sum of independent random variables. Defects in the uniformly $\sigma = +1$ configuration can only occur when various sums of these random variables take on large values. Our goal then will be to show that such large deviations have small probability. Then there will be few defects, and most spins will align with the $+1$ boundary conditions.

The dimensionality of the lattice comes into the argument as follows. Let us imagine expressing $H^+(\sigma^{min})$ as a sum of independent random variables,

$$H^+(\sigma^{min}) = -\frac{1}{2}\sum_{i\in V} h_i + \text{(more complicated but local terms)} . \tag{3}$$

We work in a connected region $V \subset \mathbb{Z}^d$ of the lattice. The unspecified terms in the above equation represent the effect of domains of minus spins. Assuming that the field term above is a good guide, the expected deviation from 0 is of the order $[\text{Volume}(V)]^{1/2}$ or $|V|^{1/2}$. If $V$ is a roughly spherical region of linear dimension $L$, the expected deviation is $O(L^{d/2})$. If the deviation is larger than the surface energy involved in making $V$ a minus domain, then it will be favorable to produce a minus domain. The surface energy should be $O(L^{d-1})$. Balancing the two effects gives $d = 2$ as the borderline dimension, with large domains becoming increasingly more unlikely to form for large $L$ if $d > 2$. This is the heart of the Imry-Ma argument for 2 as the lower critical dimension [5].

We would like to use similar arguments to control terms other than the field term in (3). This will be possible if the terms are *local*, in the sense that they depend on the magnetic fields $h_i$ only in a limited region. Only in this case can we hope to have a sum of independent random variables and an $L^{d/2}$ behavior for fluctuations.

An illustrative example: One-dimensional interface with random bonds. Simplified arguments such as this can lead one astray. Consider the problem of finding the path $x(t)$ of least energy when steps of the path have random energies. (This example was also discussed by D. Fisher at this workshop.) More precisely, let $x, t \in \mathbb{Z}$ with $0 \leqslant t \leqslant T$. The path $x(t)$ can be thought of as a one-dimensional interface in the $x-t$ plane. The bonds of the $x-t$ plane have random energies, independently distributed but with positive expected value (see Fig. 1).

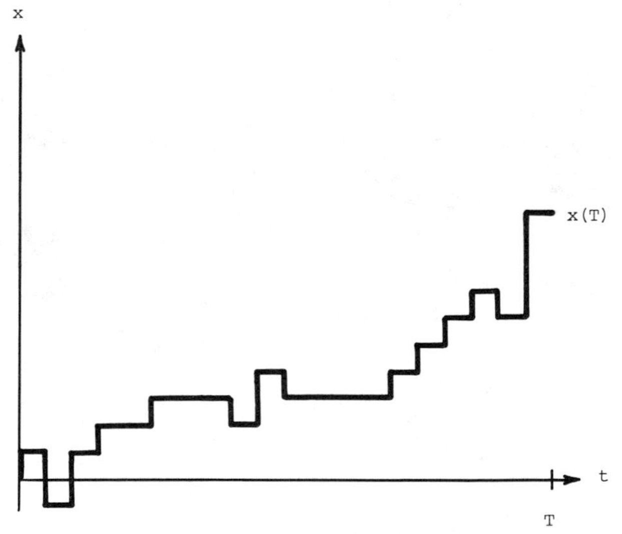


Figure 1. A one-dimensional interface.

Again we have a stochastic minimization problem. We are interested in how the expected value of $|x(T)|$ behaves for large $T$. An argument analogous to the Imry-Ma argument could be made by equating the extra distance (or energy) necessary to reach $x(T)$ with the expected fluctuation in path energy $T^{1/2}$. The extra distance is $\sqrt{T^2 + x(T)^2} - T \approx x(T)^2/T$ for small $x(T)/T$. Equating this with $T^{1/2}$ yields $x(T) \sim T^{3/4}$. This intuitive argument notwithstanding, all numerical and theoretical indications are for $x(T) \sim T^{2/3}$ [9].

This problem is a difficult one to study mathematically because the minimum-energy path can change completely as $T$ increases. In contrast, the ground state of the random field Ising model settles down as the volume $\Lambda$ becomes larger. The most challenging problems in disordered systems theory lack this stability and so behave more like the interface problem above. Spin glasses, for example, have ground states with a high degree of degeneracy or near-degeneracy.

## 2. Long-Range Order in the RFIM

We will explore three basic ideas behind the proof of long-range order in the three-dimensional random field Ising model. The overall strategy, as enunciated above, is to apply central limit theorems to expansions for ground state energies $H^+(V,\sigma^{min})$ in various regions $V$. From now on we will let it be understood that $H^+(V)$ is evaluated at the ground state $\sigma^{min}$ in $V$, and write simply $H^+(V)$.

Telescoping expansions. We obtain an expansion for the ground state energy as a telescopic sum, devised so that random variables will be independent as much as possible. First, we define contours as surfaces in the lattice separating minus domains from plus domains. If $\gamma$ is a (connected) contour, let $|\gamma|$ be its surface area, and let $V(\gamma)$ be the region enclosed by $\gamma$. Let $|V(\gamma)|$ be the volume of $V(\gamma)$. (See Fig. 2.)

If there are no contours, then

$$H^+(V) = -\frac{1}{2}\sum_{i\in V} h_i ,$$

since only the field term contributes. We have an easy central limit theorem for sums of Gaussian random variables $h_i$.

Next, we allow contours, but no contours within contours. The effect of each contour on the ground state energy is then simply described. Each contour adds an

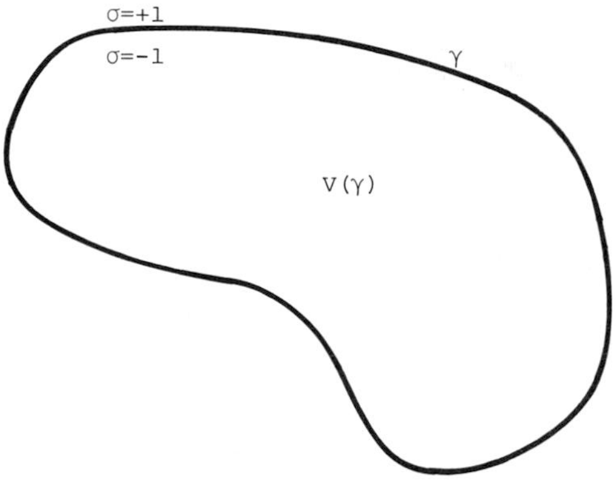


Figure 2. The contour $\gamma$ separates plus spins from minus spins, and encloses the region $V(\gamma)$.

energy $|\gamma|$ associated with the misaligned spins at its surface. Furthermore, the field terms are changed from $-1/2\ h_i$ to $1/2\ h_i$ inside $\gamma$, a change of $\Sigma_{i\in V(\gamma)} h_i$. Thus with one level of contour we have

$$H^+(V) = -\frac{1}{2}\sum_{i\in V} h_i + \sum_{\gamma \text{ in ground state}} \left[|\gamma| + \sum_{i\in V(\gamma)} h_i\right] . \tag{4}$$

In the general case with nested contours, we can only write

$$H^+(V) = -\frac{1}{2}\sum_{i\in V} h_i + \sum_{\gamma \text{ outermost in ground state}} \left[H^+(V(\gamma)) + \sum_{i\in V(\gamma)} \frac{1}{2} h_i\right] . \tag{5}$$

Here the ground state energy replaces $|\gamma| + \Sigma_{i\in V(\gamma)} 1/2\ h_i$ in (4) because we no longer know very much about the structure of the ground state in $V(\gamma)$.

Unfortunately, the event "$\gamma$ is in the ground state" depends on all the $h_i$'s. Thus in (4) and (5) we do not have a sum of independent random variables. To see how to achieve independence, let us go to even simpler cases. We are hoping that large contours are rare, so let us suppose that the only contours are single boxes surrounding one lattice site. Allow only alternate lattice sites. If $h_i < -6$, then

$|\gamma| + h_i < 0$ so it is favorable to insert $\gamma$ into the ground state. (The probability of this happening is $O(e^{-36/\varepsilon^2}) \ll 1$.) So we define random variables

$$r_\gamma(h_i) = \begin{cases} 0 , & \text{if } h_i > -6 \\ |\gamma| + h_i , & \text{if } h_i \leqslant -6 \end{cases} , \tag{6}$$

where $\gamma$ is the contour surrounding the site $i$. Now, as in (4), we have

$$H^+(V) = -\frac{1}{2}\sum_{i\in V} h_i + \sum_\gamma r_\gamma(h_i) . \tag{7}$$

This is a sum of independent random variables, since each $r_\gamma(h_i)$ depends only on one magnetic field.

When we start allowing larger contours (but still no contours within contours), we could define their energy contributions as $\min\{|\gamma| + \Sigma_{i\in V(\gamma)} h_i, 0\}$, analogous to (6). But there is interference between contours; this formula neglects the fact that we have already made some adjustment in the ground state energy in (7) through the $r_\gamma(h_i)$'s with small $\gamma$. The *true effect* of a large contour $\gamma$ is

$$r_\gamma = \min\left\{0, |\gamma| + \sum_{i\in V(\gamma)} h_i - \sum_{\gamma' \text{ inside } \gamma} r_{\gamma'}\right\} , \tag{8}$$

which is our previous expression *minus* the sum of all previous contributions in $V(\gamma)$. Here "$\gamma'$ inside $\gamma$" means $V(\gamma') \subset V(\gamma)$, $V(\gamma') \neq V(\gamma)$. The previous contributions were included in error since we did not realize that a large $\gamma$ would wipe out the smaller contours.

In the real model with nested contours, we generalize (8) to

$$r_\gamma = \min\left\{0, H^+(V(\gamma)) + \sum_{i\in V(\gamma)} \frac{1}{2} h_i - \sum_{\gamma' \text{ inside } \gamma} r_{\gamma'}\right\} , \tag{9}$$

just as in the progression from (4) to (5). This leads to the following "telescopic expansion,"

$$H^+(V) = -\frac{1}{2}\sum_{i\in V} h_i + \sum_{\gamma : V(\gamma)\subset V} r_\gamma(h) . \tag{10}$$

It is telescopic because there are chains of added and subtracted terms implicit in (9), (10). Note that $\gamma$ runs over all possible contours in V, without reference to the ones in any ground state--see (5). This is true also for $\gamma'$ in (9). Note also that $r_\gamma(h)$ depends only on $h_i$ for $i \in V(\gamma)$, so we have achieved independence of random variables, as long as the corresponding contours do not have any common interior. The philosophy is to do the best one can using random variables depending on h's in regions up to some scale; errors can be corrected at a later stage using larger scale variables.

I hope that this description of the telescopic expansion will elucidate the ideas behind the formal definitions in [1]. The definitions here differ in detail from those in [1], but the ideas are the same. Basically, we obtain the ground state in any region as a limit of a sequence of *local ground states* in larger and larger regions.

Symmetrization; large deviation estimates. Formulas (9) and (10) above set the stage for the arguments discussed earlier. In particular, our task now is to show that large deviations in sums of random variables $r_\gamma$ are unlikely. This will lead to the diluteness of large contours and the existence of long-range order. These estimates are accomplished in an inductive fashion. We assume probability estimates on $r_\gamma$'s for smaller $\gamma$'s, and derive estimates on sums of such $r_\gamma$'s. These estimates then imply statements about the distribution of $r_\gamma$'s for larger $\gamma$'s.

It is important to see how a *symmetrized version*

$$\delta r_\gamma(h) = r_\gamma(h) - r_\gamma(-h) \tag{11}$$

of $r_\gamma$ appears in the formulas. Only for this random variable, which has zero mean and a distribution symmetric under $h \to -h$, can we expect to prove that large deviations from 0 are unlikely.

In (9), let us suppose that $\gamma$ is in the ground state in $V(\gamma)$, so that the minimum is attained at the second term. In this case, it is clear that all the spins in $V(\gamma)$ adjacent to $\gamma$ must be -1 (this defines $\gamma$). Let us delete these sites from $V(\gamma)$ and call the result $\tilde{V}(\gamma)$. (See Fig. 3.) The ground state energy

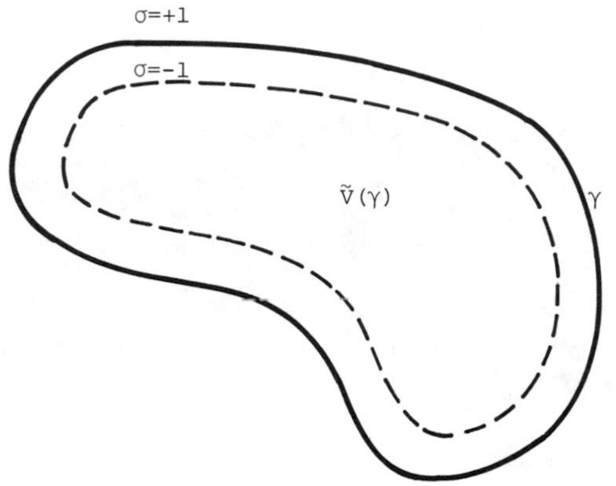


Figure 3. Spins just insice the contour $\gamma$ are equal to -1. The region $\tilde{V}(\gamma)$ is obtaining by deleting this collar from the interior of $\gamma$.

$H^+(V(\gamma))$ can be divided into a contribution from $\gamma$, a contribution from the spins in $V(\gamma)\backslash\tilde{V}(\gamma)$, and a contribution $H^-(\tilde{V}(\gamma))$ from $\tilde{V}(\gamma)$. Here we take note of the fact that $\tilde{V}(\gamma)$ has minus boundary conditions. Altogether, we have

$$H^+(V(\gamma)) = |\gamma| + \sum_{i\in V(\gamma)\backslash\tilde{V}(\gamma)} \frac{1}{2} h_i + H^-(\tilde{V}(\gamma)) \quad . \tag{12}$$

Now $H^-(\tilde{V}(\gamma))$ possesses an expansion analogous to (10), but with $h \to -h$ to reflect the flipped boundary conditions. Thus

$$H^-(\tilde{V}(\gamma)) = \sum_{i\in\tilde{V}(\gamma)} \frac{1}{2} h_i + \sum_{\gamma':V(\gamma')\subset\tilde{V}(\gamma)} r_\gamma(-h) \quad . \tag{13}$$

Inserting (12) and (13) into (9), we obtain

$$r_\gamma = |\gamma| + \sum_{i\in V(\gamma)} h_i + \sum_{\gamma':V(\gamma')\subset\tilde{V}(\gamma)} r_\gamma(-h) - \sum_{\gamma' \text{ inside } \gamma} r_{\gamma'}(h). \tag{14}$$

Now (14) can serve as the basis for an induction since only $\gamma'$ inside $\gamma$ appear on the right-hand side. Furthermore, we have sums of symmetrized variables, either $h_i$ or $\delta r_{\gamma'}(h) = r_{\gamma'}(h) - r_{\gamma'}(-h)$ [except for a few terms where $\gamma'$ is inside $\gamma$ but $V(\gamma') \not\subset \tilde{V}(\gamma)$]. These sums must exceed $|\gamma|$ in magnitude before the right-hand side of (14) can be negative. This is important, because if $\gamma$ is in the ground state, it must reduce the energy at least a little, so that $r_\gamma < 0$. If we accept that the sums are unlikely to exceed $|\gamma|$, then we see immediately that $\gamma$ is unlikely to occur.

It turns out that the probability distribution of $r_\gamma$ can essentially be estimated as

$$\mathrm{Prob}(r_\gamma < -R) \leqslant \exp\left( - \frac{(|\gamma|+R)^2}{\varepsilon^2 |V(\gamma)|} \right) \tag{15}$$

for $R \geqslant 0$. The dimensional form of the estimate ($\mathrm{Area}^2/\mathrm{Volume}$) is the same as the naive one obtained from the field terms alone. As before, we conclude that for $d > 2$, $|\gamma|^2$ dominates $|V(\gamma)|$ for large $\gamma$, and thus large contours are unlikely to occur in the ground state. (The true estimate gives away a factor logarithmic in the size of $\gamma$ in the exponent in (15).)

<u>Coarse-grained contours.</u> The third idea that is important in this problem has to do with the "entropy" or counting problem for contours. The problem is that there are too many contours to be compatible with the individual estimates (15) on the likelihood of their appearing in the ground state. Quantitatively, in three dimensions, there are $\exp(O(L^2))$ contours of area $L^2$ surrounding the origin, while each has its probability estimated by $\exp(-(L^2)^2/\varepsilon^2 L^3) = \exp(-L/\varepsilon^2)$. Clearly the probability is not small enough to compensate for the "entropy" of contours.

This crude estimate ignores the fact that the events "$r_\gamma < 0$" are highly dependent when two contours enclose approximately the same volume. Fisher, Fröhlich, and Spencer [10] devised a scheme for exploiting the dependence, and used it to prove long-range order in the approximation of no contours within contours. (See also [11].) The idea is to approximate a contour $\gamma$ on different length scales using contours $\gamma^{(j)}$ on a lattice of spacing $2^j$. Associated to these "coarse-grained contours" is a decomposition of $V(\gamma)$ into annular regions between $\gamma^{(j)}$ and $\gamma^{(j+1)}$. The "entropy-energy" or "counting-small probability" balance is considered independently for each annular region. The first region counts as $\gamma$ does, giving $\exp(O(L^2))$ choices. However, the probability of a large deviation is smaller because the volume of the annulus is only of order $L^2$. Thus the probability is only $\exp(-(L^2)^2/L^2\varepsilon^2) = \exp(-L^2/\varepsilon^2)$, which controls the entropy for small $\varepsilon$. The balance is maintained for the j-th annulus, with entropy reduced from $L^2$ to $(L/2^j)^2$ and energy reduced only to $L^2/2^j\varepsilon^2$.

As a result of this type of analysis, we obtain statements that it is unlikely for *any* contour surrounding the origin to occur in the ground state (as opposed to the weaker statement about a particular contour). Thus the origin agrees with the boundary conditions (i.e. $\sigma_0^{min} = +1$) with probability $1 - \exp(-O(\varepsilon^{-2}))$.

## 3. Related Problems: Interfaces and Positive Temperatures

Interfaces. We can view the problem of a d-dimensional interface in $d+1$ dimensions as similar to a d-dimensional bulk problem. In the Ising model, the interface is a surface separating plus spins above from minus spins below, the overall geometry being imposed by boundary conditions. The question of whether the interface is rigid or not is analogous to the bulk problem of long-range order. A very rigid interface remains at its height at $\infty$ over most of its surface, while with long-range order the spin agrees with its value at $\infty$ over most of the volume. When the interface steps away from its values at $\infty$, the places where it jumps can be thought of as a contour (see Fig. 4). The energy increases by the (d-1)-dimensional area of the contour.

Using this analogy with bulk problems, we should expect that the methods used in the $d \geq 3$ random field Ising model would be applicable to proving rigidity for the interface in $d+1 \geq 4$ dimensions. (The two-dimensional bulk problem and the three-dimensional interface problem are borderline cases, and should be quite difficult to handle rigorously. It is expected that long-range order/rigidity does not occur even at $T = 0$.) In fact we can consider either the random field problem or the random bond problem for the interface. Provided the average bond energy

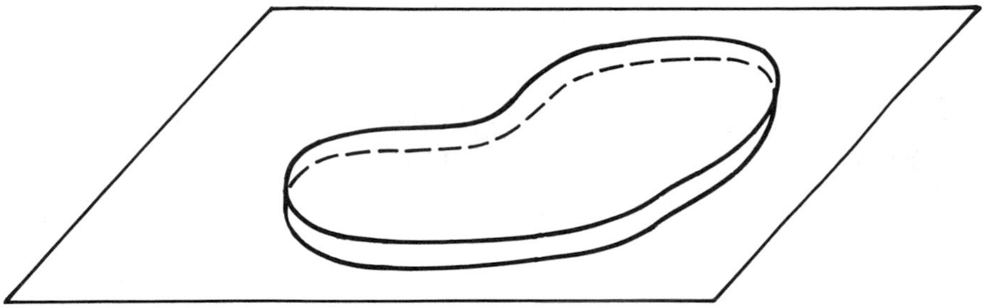

Figure 4. A step in an interface which is mainly rigid. Its excess energy is the area of the vertical surface.

is positive, both problems behave energetically like the bulk random field problem. For the random field interface, the energy change resulting from a step/contour $\gamma$ is $|\gamma| \pm \Sigma_{i\in V(\gamma)} h_i$, depending on whether the step is up or down. For the random bond interface, the energy change is $|\gamma| + \Sigma_{i\in V(\gamma)} (J_{i,i+\hat{e}_z} - J_{i,i-\hat{e}_z})$, where $\hat{e}_z$ is the unit vector perpendicular to the interface. Since we have the difference between the old bond energies and the new bond energies, we are again in a situation with symmetric random variables.

Rough interfaces in either the random bond or the random field problem present very difficult problems. As in the one-dimensional interface problem discussed above, we expect that the height grows as some power of the width of the system [6]. Roughness can also set in at intermediate temperatures or amounts of disorder in dimensions where there is presumably a rigid interface at low temperature and weak disorder [12].

Positive temperatures. To approach the random field Ising model at positive temperatures, it is necessary to work with low temperature expansions. Each energy in the $T = 0$ problem must be replaced with a free energy (or logarithm of a partition function) at positive temperatures. Free energies can be effectively computed as a perturbation series in powers of $e^{-1/T}$ (low temperature expansion).

In disordered systems, perturbation expansions usually fail to converge due to Griffiths singularities, even in circumstances when one would otherwise expect convergence. This is because domains in the lattice can be badly behaved, from the point of view of the expansion--a high temperature region could occur when one is doing a low temperature expansion, for example.

A method for devising convergent expansions despite the existence of such domains was developed in [3]. Essentially, the expansion must be performed away from the "singular" domains. It is also necessary to expand the interactions between the domains on a sequence of increasing length scales, before one can take the logarithm of the partition function.

In the random field Ising model, the following construction should give a good definition of a singular region. We say $V(\gamma)$ is singular if

$$\mathrm{Prob}_{\text{Gibbs in } V(\gamma)}(\gamma) \geqslant \exp\left(-\frac{1}{2T}|\gamma|\right) . \tag{16}$$

That is, we use the Gibbs measure on configurations in $V(\gamma)$ with plus boundary conditions, and if the probability of $\gamma$ occurring is too large, then $V(\gamma)$ is singular. The total singular region is the union of all $V(\gamma)$ for $\gamma$ satisfying (16). This is a useful definition, because if $V(\gamma)$ is not singular, the opposite inequality in (16) is sufficient to guarantee convergence of a low temperature expansion. On the other hand, as in (12) we have that

$$\mathrm{Prob}_{\text{Gibbs in } V(\gamma)}(\gamma) = \frac{\exp\left[-\frac{1}{T}\left(|\gamma| + \sum_{i\in V(\gamma)} h_i\right)\right] Z^-(\tilde{V}(\gamma))}{Z^+(V(\gamma))} ,$$

where we have normalized the partition functions with plus or minus boundary conditions as follows:

$$Z^{\pm}(V) = \sum_{\{\sigma\restriction V\},\sigma=\pm 1 \text{ in } V^c} \exp\left[-\frac{1}{T}\left(\sum_{<i,j>} \frac{1}{2}(1-\sigma_i\sigma_j) - \sum_{i\in V} \frac{1}{2} h_i(\sigma_i \mp 1)\right)\right] .$$

Thus (16) implies that there is a large deviation, either in $\Sigma_{i\in V(\gamma)} h_i$ or in $\log Z^-(\tilde{V}(\gamma)) - \log Z^+(V(\gamma))$, since at least half of the $|\gamma|/T$ in the exponent has to be cancelled. If the difference of logs can be written as an expansion in local random variables analogous to $\delta r_\gamma(h)$, then the large deviation means small probability in $d\mu(h)$. The structure of the argument here is quite analogous to our obtaining small probability from the fact that $r_\gamma < 0$, using (14).

Thus we see that singular regions are rare. Using these ideas, which combine the expansion method of [3] with the telescoping expansion and probabilistic estimates described for the $T = 0$ problem, I hope that long-range order can be obtained at positive temperatures as well.

## Acknowledgements

My understanding of these problems owes a great deal to conversations with T. Spencer and J. Fröhlich. This work was supported in part by the National Science Foundation under grants PHY-82-03669 and PHY-84-13285.

## References

1. Imbrie, J.: The ground state of the three-dimensional random-field Ising model. *Commun. Math. Phys.* 98, 145-176 (1985).

2. Imbrie, J.: Lower critical dimension of the random-field Ising model. *Phys. Rev. Lett.* 53, 1747-1750 (1984).

3. Fröhlich, J., Imbrie, J.: Improved perturbation expansion for disordered systems: Beating Griffiths singularities. *Commun. Math. Phys.* 96, 145-180 (1984).

4. Von Foerster, T.: Random magnetic fields reduce critical dimensionality. *Physics Today* 36, No. 7, 17-19 (July 1983); Imry, Y.: Random external fields. *J. Stat. Phys.* 34, 849-862 (1984); Grinstein, G.: On the lower critical dimension of the random field Ising model. *J. Appl. Phys.* 55, 2371-2376 (1984).

5. Imry, Y., Ma. S.: Random-field instability of the ordered state of continuous symmetry. *Phys. Rev. Lett.* 35, 1399-1401 (1975).

6. Grinstein, G., Ma, S.: Roughening and lower critical dimension in the random-field Ising model. *Phys. Rev. Lett.* 49, 685-688 (1982); Grinstein, G., Ma, S.: Surface tension, roughening, and lower critical dimension in the random-field Ising model. *Phys. Rev.* B28, 2588-2601 (1983); Villain, J.: Commensurate-incommensurate transition with frozen impurities. *J. Physique* 43, L551-L558 (1982).

7. Pytte, E., Imry, Y., Mukamel, D.: Lower critical dimension and the roughening transition of the random-field Ising model. *Phys. Rev. Lett.* 46, 1173-1177 (1981); Mukamel, D., Pytte, E.: Interface fluctuations and the Ising model in a random field. *Phys. Rev.* B25, 4779-4786 (1982).

8. Parisi, G., Sourlas, N.: Random magnetic fields, supersymmetry, and negative dimensions. *Phys. Rev. Lett.* 43, 744-745 (1979). Supersymmetric field theories and stochastic differential equations. *Nucl. Phys.* B206, 321-332 (1982).

9. Huse, D., Henley, C.: Pinning and roughening of domain walls in Ising systems due to random impurities. *Phys. Rev. Lett.* 54, 2708-2711 (1985); Kardar, M., Nelson, D.: Commensurate-incommensurate transitions with quenched random impurities. *Phys. Rev. Lett.* 55, 1157-1160 (1985); Fisher, D.: Exact exponents for an interface in a two-dimensional random exchange Ising model. *Phys. Rev. Lett.*, to appear; Burgers, J., *The Nonlinear Diffusion Equations*. Boston: Reidel, 1974.

10. Fisher, D., Fröhlich, J., Spencer, T.: The Ising model in a random magnetic field. *J. Stat. Phys.* 34, 863-870 (1984).

11. Chalker, J.: On the lower critical dimensionality of the Ising model in a random field. *J. Phys.* C16, 6615-6622 (1983).

12. Natterman, T.: On the interface stiffness of the random-field Ising model. *Z. Phys.* B54, 247-252 (1984).

# SOME RECENT RESULTS ON WAVE EQUATIONS, PATH INTEGRALS, AND SEMICLASSICAL APPROXIMATIONS

by

John R. Klauder
AT&T Bell Laboratories
Murray Hill, New Jersey 07974

**Abstract**

Following a review of standard semiclassical approximations to wave equations in the configuration-space formulation, an approach based on a coherent-state representation is fully developed. Unlike direct configuration-space approaches, it is shown that the coherent-state formulation leads to a global, uniform, semiclassical configuration-space approximation, which has distinct advantages when one is faced with numerous caustics.

## 1. Introduction, and Review of Configuration-Space Approach

The ubiquitous relevance of wave equations makes their continued study of fundamental interest. Many wave equations of interest can be written in the general form

$$i\hbar \frac{\partial \phi(x,t)}{\partial t} = \mathscr{H}(-i\hbar \frac{\partial}{\partial x}, x)\, \phi(x,t) , \tag{1.1}$$

where $\mathscr{H}$ is a self-adjoint operator on $L^2(\mathbf{R})$. For illustrative purposes we have chosen the one-dimensional ($x \in \mathbf{R}$) Schrödinger equation of quantum mechanics where $\mathscr{H}$ denotes the Hamiltonian operator. Our remarks are readily generalized to higher dimensions and to operators $\mathscr{H}$ that additionally depend on the independent variable t. Equally well, our remarks apply to other wave equations of a similar form such as the constant frequency paraxial approximation to the hyperbolic wave equation in a medium of varying index of refraction. In such a (one dimensional) case the relevant equation is of the form

$$i\lambda\!\!\!^{-} \frac{\partial \chi(x,z)}{\partial z} = \mathscr{H}(-i\lambda\!\!\!^{-} \frac{\partial}{\partial x}, x, z)\, \chi(x,z) , \tag{1.2}$$

where $\lambda\!\!\!^{-} = \lambda/2\pi$ denotes the reduced wave length, and z denotes the direction of wave propagation (no backscattering). In a number of cases, such as wave propagation in a random medium (e.g., fluctuations in sound speed in ocean acoustics, or electrons in a disordered crystal), it is appropriate to regard part of $\mathscr{H}$ as a stochastic variable. Interest then commonly centers on various correlation functions of the wave-equation solution or other aspects (such as the spectrum) that can be derived therefrom. At the end of our main discussion we shall make some brief remarks in Sec. 3 on propagation in a random medium.

To set our discussion in proper context it is useful to begin with a review of one standard approach to treat (1.1) based on a path integral representation of the solution and its semiclassical approximation [1]. Equation (1.1) admits a solution expressed in the form

$$\phi(x'',T) = \int J(x'',T;x',0)\,\phi(x',0)\,dx' , \tag{1.3}$$

where $\phi(x',0)$ is the field amplitude at time $t = 0$ and J denotes the propagator. Consistency of this expression requires, of course, that

$$\lim_{T\to 0} J(x'',T;x',0) = \delta(x''-x') . \tag{1.4}$$

The propagator J is conveniently represented as a phase-space path integral. In a heuristic and formal fashion one writes

$$J(x'',T;x',0) = \mathscr{M}\int e^{(i/\hbar)\int[p\dot{x}-H(p,x)]dt}\,\Pi\,dp(t)\,dx(t) , \tag{1.5}$$

which purports to be an integral over all paths $x(t)$ pinned at $t = 0$ and $t = T$ so that $x(0) = x'$ and $x(T) = x''$, and over all paths $p(t)$. The factor $\mathscr{M}$ represents a formal normalization factor. The integrand is $\exp(iI/\hbar)$, where

$$I = \int [p(t)\,\dot{x}(t) - H(p(t), x(t))]dt \tag{1.6}$$

is the classical action for the paths x and p. The classical Hamiltonian $H(p,x)$ is related to the Hamiltonian operator $\mathscr{H}$ by Weyl quantization, namely

$$\mathscr{H}(-i\hbar\,\partial/\partial x, x) = \mathscr{W}H(p,x)\Big|_{p\,=\,-i\hbar\partial\partial x} , \tag{1.7}$$

where $\mathscr{W}$ is an ordering operation which may be deduced from

$$\mathscr{W}\exp[i(ap+bx)]\Big|_{p\,=\,-i\hbar\partial/\partial x} = \exp\{i[a(-i\hbar\frac{\partial}{\partial x}) + bx]\} . \tag{1.8}$$

Equation (1.5) can be better defined by means of the introduction of a lattice-space regularization of the time followed by a limit as the lattice spacing goes to zero. However, for purposes of an approximate evaluation by means of a stationary-phase approximation it is sufficient to proceed formally. Such an approximation leads to a stationary point whenever

$$\dot{x}(t) = \partial H/\partial p(t) , \qquad \dot{p}(t) = -\partial H/\partial x(t) \tag{1.9}$$

for $0 \leqslant t \leqslant T$ subject to the boundary conditions $x(0) = x'$ and $x(T) = x''$. We denote a solution to these equations by $\hat{p}(t)$ and $\hat{x}(t)$, and we set

$$S(x'',x') = \int_0^T [\hat{p}\dot{\hat{x}} - H(\hat{p},\hat{x})]\, dt \tag{1.10}$$

evaluated for an extremal path. The contribution in the neighborhood of such an extremal path (assumed to be well separated from other such extremals) is given, in the first approximation, by

$$e^{iS/\hbar}\mathcal{M}\int e^{(i/\hbar)\int[u\dot{v} - \frac{1}{2}a u^2 - buv - \frac{1}{2}cv^2]dt}\Pi\, du(t)dv(t)$$

$$\equiv E\, e^{iS/\hbar} . \tag{1.11}$$

Here we have set $p(t) = \hat{p}(t) + u(t)$, $x(t) = \hat{x}(t) + v(t)$, and

$$a(t) = \partial_p^2 H(p,x) , \quad b(t) = \partial_p \partial_x H(p,x) , \quad c(t) = \partial_x^2 H(p,x) \tag{1.12}$$

all evaluated for the extremal path $\hat{p}(t)$ and $\hat{x}(t)$. The integral over u and v is such that $v(0) = v(T) = 0$ while u is unrestricted. It follows, in Dirac notation, that the amplitude factor is alternatively given by

$$E = \hbar^{-\frac{1}{2}} \langle \underline{0}|\, \mathbf{T} \exp\{-i\int[\tfrac{1}{2}aP^2 + bP\circ Q + \tfrac{1}{2}cQ^2]dt\}\, |\underline{0}\rangle \tag{1.13}$$

where **T** denotes time ordering, P and Q are ($\hbar-$less) abstract Heisenberg operators, $[Q,P] = i$, $P\circ Q = (PQ + QP)/2$, and $|\underline{0}\rangle$ is a formal eigenvector of Q, $Q|\underline{0}\rangle = 0$. Observe that E is unbounded since $\langle\underline{0}|\underline{0}\rangle = \infty$.

There are several standard ways to calculate E, and we now outline one of them. The three (symmetric) operators $\frac{1}{2}P^2$, $\frac{1}{2}Q^2$, and $P\circ Q$ form a closed Lie algebra for which

$$[\tfrac{1}{2}Q^2, \tfrac{1}{2}P^2] = i\, P \circ Q , \tag{1.14}$$

$$[\tfrac{1}{2}Q^2, P \circ Q] = i\, Q^2 ,$$

$$[\tfrac{1}{2}P^2, P \circ Q] = -i P^2 .$$

A nonsymmetric but nevertheless faithful representation is given by the $2 \times 2$ matrices

$$\tfrac{1}{2}P^2 = \begin{pmatrix} 0 & 1 \\ 0 & 0 \end{pmatrix} , \quad \tfrac{1}{2}Q^2 = \begin{pmatrix} 0 & 0 \\ 1 & 0 \end{pmatrix} , \quad P \circ Q = \begin{pmatrix} i & \\ & -i \end{pmatrix} . \tag{1.15}$$

Thus the operator in question is represented by

$$T \exp[\int \begin{pmatrix} b & -ia \\ -ic & -b \end{pmatrix} dt] \equiv \begin{pmatrix} A & B \\ C & D \end{pmatrix} . \tag{1.16}$$

We seek an alternative form given by

$$e^{i\lambda Q^2/2}\, e^{i\mu P^2/2}\, e^{-i\nu P \cdot Q}$$

$$= \begin{pmatrix} 1 & 0 \\ i\lambda & 1 \end{pmatrix} \begin{pmatrix} 1 & i\mu \\ 0 & 1 \end{pmatrix} \begin{pmatrix} e^{\nu} & \\ & e^{-\nu} \end{pmatrix} = \begin{pmatrix} e^{\nu} & i\mu e^{-\nu} \\ i\lambda e^{\nu} & (1-\mu\lambda)e^{-\nu} \end{pmatrix} . \tag{1.17}$$

The quantity of interest, namely (1.13), is now given by

$$E = \hbar^{-1/2} \langle \underline{0} |\, e^{i\lambda Q^2/2}\, e^{i\mu P^2/2}\, e^{-i\nu P \cdot Q} | \underline{0} \rangle$$

$$= \frac{e^{\nu/2}}{2\pi\sqrt{\hbar}} \int dk\, e^{i\mu k^2/2} = \frac{e^{\nu/2}}{\sqrt{-2\pi\hbar i\mu}} = \frac{1}{\sqrt{-2\pi\hbar B}} . \tag{1.18}$$

We note also that the auxiliary, classical, linearized Hamiltonian

$$H_2(\tilde{p}, \tilde{x}, t) = \tfrac{1}{2} a\tilde{p}^2 + b\tilde{p}\tilde{x} + \tfrac{1}{2} c\tilde{x}^2 , \tag{1.19}$$

leads to the equations of motion

$$\dot{\tilde{x}}(t) = b(t)\tilde{x}(t) + a(t)\tilde{p}(t) , \tag{1.20}$$

$$\dot{\tilde{p}}(t) = -c(t)\tilde{x}(t) - b(t)\tilde{p}(t) ,$$

or if $\tilde{r}(t) \equiv -i\tilde{x}(t)$, then

$$\dot{\tilde{r}}(t) = b(t)\,\tilde{r}(t) - i\,a(t)\,\tilde{p}(t)\ , \tag{1.21}$$

$$\dot{\tilde{p}}(t) = -i\,c(t)\,\tilde{r}(t) - b(t)\,\tilde{p}(t)\ .$$

Thus we recognize that

$$\begin{pmatrix} \tilde{r}\ (T) \\ \tilde{p}\ (T) \end{pmatrix} = \begin{pmatrix} A & B \\ C & D \end{pmatrix} \begin{pmatrix} \tilde{r}\ (0) \\ \tilde{p}\ (0) \end{pmatrix} . \tag{1.22}$$

Hence if $\tilde{p}(0) = 1$ and $\tilde{x}(0) = 0$, then $\tilde{r}(T) = B$, or $\tilde{x}(T) = iB$. Consequently, we determine that the semiclassical approximation is given by

$$J_{sc} = \frac{1}{\sqrt{i\,2\pi\hbar\tilde{x}(T)}}\ e^{iS(x'',x')/\hbar}\ , \tag{1.23}$$

where $\tilde{x}(T)$ is a solution of (1.20) subject to the initial conditions $\tilde{x}(0) = 0$, $\tilde{p}(0) = 1$. If there are several such extremal paths, which is generally the case, then

$$J_{sc} = \Sigma\ \frac{1}{\sqrt{i\,2\pi\hbar\tilde{x}(T)}}\ e^{iS(x'',x')/\hbar} \tag{1.24}$$

with the sum extending over all such paths. Clearly, whenever $\tilde{x}(T) = 0$, $T > 0$, for one or more paths this level of approximation may fail, as is the case at various caustics. The standard remedy of including higher-order deviations from the extremal path is well documented in standard texts [1] and need not be repeated here.

Instead we outline in the same spirit the proposal of Maslov [2] based on the observation that

$$J\,(x'',T;x',0) = \frac{1}{\sqrt{2\pi\hbar}} \int e^{ix''p''/\hbar} dp''\, M\,(p'',T;x',0)\ , \tag{1.25}$$

where M admits a formal path integral expression given by

$$M(p'',T;x',0) = \mathscr{M} \int e^{(i/\hbar)\{-p''x(T) + \int [p\dot{x} - H(p,x)]dt\}}\, \Pi\, dp(t)\,dx(t) \tag{1.26}$$

integrated over all variables subject to $x(0) = x'$ and $p(T) = p''$. A stationary-phase approximation derived in complete analogy to that given previously leads to

$$M_{sc}(p'',T;x',0) = \Sigma \frac{1}{\sqrt{2\pi\hbar\tilde{x}(T)}} e^{iT(p'',x')/\hbar} \tag{1.27}$$

where, for each term in the sum,

$$T(p'',x') = -p''\hat{x}(T) + \int_0^T [\hat{p}\dot{\hat{x}} - H(\hat{p},\hat{x})]dt , \tag{1.28}$$

$$\dot{\hat{x}} = \partial H/\partial \hat{p}(t) , \quad \dot{\hat{p}} = -\partial H/\partial \hat{x}(t) ,$$

subject to $\hat{x}(0) = x''$ and $\hat{p}(T) = p''$. In this case $\tilde{x}(T)$ is a solution of (1.20) — with a, b, and c based on the new extremal solution, of course — subject in the present case to the initial conditions $\tilde{x}(0) = 1$ and $\tilde{p}(0) = 0$. Now the proposal is to introduce

$$\hat{J}_{sc}(x'',T;x',0) \equiv \frac{1}{\sqrt{2\pi\hbar}} \int e^{ix''p''/\hbar} dp'' M_{sc}(p'',T;x',0) \tag{1.29}$$

as a replacement for the ill-defined expression (1.24) at the caustic. As a new solution, $\tilde{x}(T) \neq 0$ at the caustic. Normally, in a Taylor series expansion, $T(p'',x')$ contains terms linear and quadratic in $p''$, but at the caustic the quadratic term is absent; the term cubic in $p''$ then becomes important leading to the expected Airy function. The general result (1.29) yields a uniform expression, but one which is only locally valid. That is, as $x''$ or $x'$ is repositioned there will generally arise caustic points where $\tilde{x}(T) = 0$, which thus invalidates (1.27). Some improvements on a Maslov-type of construction have been proposed by Littlejohn [3].

## 2. Coherent-State Formulation

We now turn our attention to a rather different procedure of construction, which, as we shall see, provides a global, uniform semiclassical approximation to wave-equation solutions. This procedure involves a unitary transformation of the field amplitude $\phi(x,t)$ known in the physics literature as the coherent-state representation [4]. Specifically, for all $\phi \in L^2(\mathbf{R})$ we introduce

$$\psi(p,q,t) = (\pi\hbar)^{-1/4} \int e^{-x^2/2\hbar - ipx/\hbar} \phi(x+q,t)\, dx . \tag{2.1}$$

Clearly $\psi$ is bounded and continuous for all $\phi$, and it follows directly that

$$\int |\phi(x,t)|^2 dx = \int |\psi(p,q,t)|^2 \, (dpdq/2\pi\hbar) . \tag{2.2}$$

The set of image functions $\{\psi\}$ cannot comprise all of $L^2(\mathbf{R}^2)$, but instead represents a proper subspace. It is most easy to characterize that subspace as a reproducing kernel Hilbert space for which the reproducing kernel is given by

$$\mathscr{K}(p'',q'';p',q') \equiv \exp\{\frac{i}{2\hbar}(p''+p')(q''-q') - \frac{1}{4\hbar}[(p''-p')^2 + (q''-q')^2]\} . \quad (2.3)$$

Every element $\psi$ of the relevant subspace satisfies the equation

$$\psi(p'',q'',t) = \int \mathscr{K}(p'',q'';p',q')\, \psi(p',q',t)\, (dp'dq'/2\pi\hbar) , \quad (2.4)$$

and it is straightforward to show that (2.4) is satisfied if and only if $\psi$ is given by (2.1).

The inverse transformation to that of (2.1) is, in fact, not unique. However, the expression we shall adopt is the simplest of these expressions, and is given by

$$\phi(q,t) = \frac{(\pi\hbar)^{1/4}}{2\pi\hbar} \int \psi(p,q,t)\, dp . \quad (2.5)$$

It is perhaps useful to observe that $\psi$ may also be represented in terms of the Fourier amplitude

$$\chi(k,t) \equiv \frac{1}{\sqrt{2\pi\hbar}} \int e^{-ikx/\hbar}\, \phi(x,t)\, dx . \quad (2.6)$$

In particular, one finds that

$$\psi(p,q,t) = (\pi\hbar)^{-1/4}\, e^{ipq/\hbar} \int e^{-k^2/2\hbar + iqk/\hbar}\, \chi(k+p,t)\, dk , \quad (2.7)$$

for which, of course,

$$\int |\chi(k,t)|^2 dk = \int |\psi(p,q,t)|^2\, (dpdq/2\pi\hbar) . \quad (2.8)$$

The transformation inverse to (2.7) is again not unique, but it is most simply given by

$$\chi(p,t) = \frac{(\pi\hbar)^{1/4}}{2\pi\hbar} \int e^{-ipq/\hbar} \psi(p,q,t)\, dq . \quad (2.9)$$

Equations (2.5) and (2.9) tell us that both $\phi$ and its Fourier transform $\chi$ can be easily projected out of the coherent-state amplitude $\psi$. The phase factor in the integrands of (2.5) and (2.9) can be redistributed, but it cannot be eliminated from both expressions simultaneously. For convenience, we have adopted a phase convention that makes (2.5) appear in its simplest guise. For other applications alternative phase conventions may be appropriate.

The wave equation (1.1) is readily transformed to

$$i\hbar \frac{\partial\psi(p,q,t)}{\partial t} = \mathscr{H}(-i\hbar \frac{\partial}{\partial q}, q + i\hbar \frac{\partial}{\partial p}) \psi(p,q,t) \tag{2.10}$$

suitable for the amplitude $\psi$. Indeed, if, for example, $\mathscr{H}$ is a polynomial then (1.1) is easily recovered from (2.10) simply by integrating over p and integrating by parts where necessary. Premultiplication of (2.10) by $e^{-ipq/\hbar}$ followed by integration over q, again with appropriate integration by parts, leads, with (2.9), to the wave equation (1.1) expressed in terms of the Fourier amplitude $\chi$.

The solution of (2.10) admits the representation

$$\psi(p'',q'',T) = \int K(p'',q'',T;p',q',0)\, \psi(p',q',0)\, (dp'dq'/2\pi\hbar) \tag{2.11}$$

in terms of the initial value amplitude and the propagator K. The expression for the propagator K is not unique but can be made so by choosing its form at T = 0. For example, it is possible to choose

$$\lim_{T\to 0} K(p'',q'',T;p',q',0) = \mathscr{K}(p'',q'';p',q') , \tag{2.12}$$

the reproducing kernel, or, analogous to (1.4), as

$$(2\pi\hbar)\, \delta(p''-p')\, \delta(q''-q') . \tag{2.13}$$

We shall adopt (2.12) from which it follows that K is continuous and uniformly bounded,

$$|K(p'',q'',T;p',q',0)| \leqslant 1 . \tag{2.14}$$

Given the propagator K it is possible to recover the configuration-space propagator J [cf. (1.3)] from the expression

$$J(q'',T;q',0) = \frac{(\pi\hbar)^{1/2}}{(2\pi\hbar)^2} \int K(p'',q'',T;p',q',0)\, dp''dp' . \tag{2.15}$$

This relation follows, in turn, from (2.5) and thus

$$\phi(q'',T) = \frac{(\pi\hbar)^{1/4}}{2\pi\hbar} \int dp''\ K(p'',q'',T;p',q',0)\ \psi(p',q',0)\ (dp'dq'/2\pi\hbar)$$

$$\equiv \frac{(\pi\hbar)^{1/4}}{2\pi\hbar} \int dp''\ L(p'',q'',T;q',0)\ \phi(q',0)\, dq' , \tag{2.16}$$

where the last line follows from the isometry afforded by (2.2). Here L denotes the image of K in $L^2(\mathbf{R})$, namely

$$L(p'',q'',T;q',0) = \frac{(\pi\hbar)^{1/4}}{2\pi\hbar} \int K(p'',q'',T;p',q',0)\, dp' . \tag{2.17}$$

When combined with (2.16) and the uniqueness of J, then Eq. (2.15) follows.

We now take up a path-integral representation of the propagator K. Recently, and for a wide class of Hamiltonian operators $\mathscr{H}$, it has been proved [5] that

$$K(p'',q'',T;p',q',0)$$

$$= \lim_{\nu\to\infty} 2\pi\, e^{\hbar\nu T/2} \int e^{(i/\hbar)\int[pdq - h(p,q)dt]} d\mu_w^\nu(p)\, d\mu_w^\nu(q) , \tag{2.18}$$

where $d\mu_w^\nu(p)$ [or, with $p \to q$, $d\mu_w^\nu(q)$] denotes a pinned Wiener measure, such that $p(0) = p'$, $p(T) = p''$, with total weight

$$\int d\mu_w^\nu(p) = (2\pi\hbar\nu T)^{-1/2} \exp[-(p''-p')^2/2\hbar\nu T] . \tag{2.19}$$

If $\mathbf{E}[\cdot]$ denotes expectation in the normalized measure $d\mu_w^\nu/\int d\mu_w^\nu$, then, for $t_1 \leqslant t_2$,

$$\mathbf{E}[p(t_1)p(t_2)] - \mathbf{E}[p(t_1)]\,\mathbf{E}[p(t_2)] = \hbar\nu t_1 (1-t_2/T) , \tag{2.20}$$

a relation which clearly shows that $\hbar\nu$ is the diffusion constant. The expression $\int pdq$ is a well-defined stochastic integral, while the function $h(p,q)$ in (2.18) is related to the function $H(p,q)$ through the operation

$$h(p,q) = e^{-(\hbar/4)(\partial_p^2+\partial_q^2)}\, H(p,q) ; \tag{2.21}$$

note that this is the opposite of a smoothing operation. If H is a polynomial in p and q so too is h, and that may be a sufficient class for many purposes. However, regarding the approximation to come, we may consider very general Hamiltonians since we will only need to deal with

$$h_1(p,q) \equiv H(p,q) - \tfrac{1}{4}\,\hbar\,(\partial_p^2+\partial_q^2)\,H(p,q)\,, \tag{2.22}$$

which just represents the lowest-order $\hbar$ correction to H.

The technical conditions for (2.18) to hold are summarized as:

(i) $\int |h(p,q)|^2\, e^{-\alpha(p^2+q^2)} dp\,dq < \infty$ , for all $\alpha > 0$;

(ii) $\int |h(p,q)|^4\, e^{-\beta(p^2+q^2)} dp\,dq < \infty$, for some $\beta$, $0 < \beta < \tfrac{1}{2}$;

(iii) $\mathscr{H}(-i\hbar \frac{\partial}{\partial x}, x)$ is essentially self adjoint on the finite linear span of vectors of the form $\exp[ipx/\hbar - (x-q)^2/2\hbar]$, $(p,q) \in \mathbf{R}^2$. For the proof of (2.18) we refer the reader to Ref. [5].

We next intend to evaluate (2.18) approximately by stationary-phase methods to the same level of accuracy as treated previously. For this purpose we follow our earlier example and introduce a formal, heuristic path-integral expression for (2.18), somewhat in the spirit of (1.5), namely

$$\begin{aligned} &K(p'',q'',T;p',q',0) \\ &\approx \lim_{\nu\to\infty} \mathscr{M} \int \exp\{(i/\hbar)\int [p\dot{q} - H(p,q) + \tfrac{1}{4}\,\hbar\,(\partial_p^2+\partial_q^2)\,H(p,q)]\,dt\} \\ &\qquad \times \exp\{-(1/2\hbar\nu)\int [\dot{p}^2+\dot{q}^2]\,dt\}\ \Pi\ dp(t)dq(t)\ . \end{aligned} \tag{2.23}$$

Note that since we intend to employ a stationary-phase approximation we have already approximated $h(p,q)$ by $h_1(p,q)$. The extremal equations that follow from (2.23) are given by

$$\dot{q}(t) - \partial H/\partial p(t) = i\ddot{p}(t)/\nu\,, \tag{2.24}$$

$$\dot{p}(t) + \partial H/\partial q(t) = -\,i\ddot{q}(t)/\nu\,,$$

subject to the boundary conditions $q(0) = q'$, $p(0) = p'$, $q(T) = q''$, and $p(T) = p''$. (We observe that the order $\hbar$ correction to the Hamiltonian does not figure in the extremal equations, but instead it appears first as a modification of the next-order term, the amplitude coefficient.) Our interest centers on a solution of (2.24) for large $\nu$ which is continuous for all $T \geqslant 0$. The structure of that solution readily follows by analogy of (2.24) with the Navier-Stokes equations [6] for an incompressible fluid with a small viscosity (large Reynolds number). The solution of interest has two very thin boundary layers, one near $t = 0$ and the other near $t = T$, with a laminar region in between; in particular, the solution of (2.24), for $\nu T >> 1$ and $\nu >> |\partial^2 H/\partial y_1 \partial y_2|$, $y_j = p$ or $q, j = 1,2$, is given by [7]

$$\hat{q}(t) = (q' - \overline{q}')\, e^{-\nu t} + \overline{q}(t) + (q'' - \overline{q}'')\, e^{-\nu(T-t)} , \tag{2.25}$$

$$\hat{p}(t) = (p' - \overline{p}')\, e^{-\nu t} + \overline{p}(t) + (p'' - \overline{p}'')\, e^{-\nu(T-t)} .$$

Here $\overline{q}(t)$, $\overline{p}(t)$ denotes a generally complex solution of the classical equations

$$\dot{\overline{q}}(t) = \partial H/\partial \overline{p}(t) , \quad \dot{\overline{p}}(t) = -\, \partial H/\partial \overline{q}(t) , \tag{2.26}$$

which hold in the middle (laminar) region where it is acceptable to set $\nu = \infty$ (zero viscosity). In addition, $\overline{q}' = \overline{q}(0)$, $\overline{p}' = \overline{p}(0)$, $\overline{q}'' = \overline{q}(T)$, and $\overline{p}'' = \overline{p}(T)$. To satisfy the coupled equations of motion in the boundary layers, where the dynamics makes a negligible contribution, it is necessary that

$$q' + i p' = \overline{q}' + i \overline{p}' , \tag{2.27}$$

$$q'' - i p'' = \overline{q}'' - i \overline{p}'' ,$$

which are simple consequences of the equation $q(t) = i\dot{p}(t)/\nu + \text{const.}$ that holds in each boundary layer.

In seeking the extremal solution it is convenient to first set

$$\overline{q}' = q' + w , \quad \overline{p}' = p' + i w , \tag{2.28}$$

where w is a free complex parameter, and then to choose w so that the time-evolved solution $\overline{q}''$ and $\overline{p}''$, according to (2.26), satisfies the one final complex boundary condition.

When the path integral is approximately evaluated by stationary phase, and the limit $\nu \to \infty$ is taken, the result is given by

$$K_{sc}(p'', q'', T; p', q', 0) = E\, e^{i F(p'', q''; p', q')/\hbar} , \tag{2.29}$$

where

$$F = \tfrac{1}{2}\,(p''q''-p'q') - \tfrac{1}{2}\,(p''\bar{q}''-q''\bar{p}'' + \bar{p}'q'-\bar{q}'p')$$

$$+\int_0^T [\tfrac{1}{2}\,(\bar{p}\dot{\bar{q}}-\bar{q}\dot{\bar{p}}) - H(\bar{p},\bar{q})]\,dt\,. \tag{2.30}$$

The factor E incorporates the $\hbar$ correction to the Hamiltonian plus, in the first approximation, the Gaussian integral contribution of the quadratic deviations from the extremal trajectory. With $q(t) = \hat{q}(t) + v(t)$, $p(t) = \hat{p}(t) + u(t)$, it follows that E is formally given by

$$E = \lim_{\nu\to\infty} \mathcal{M}\int \exp\{(i/\hbar)\int [u\dot{v}-\tfrac{1}{2}\,\alpha u^2-\beta uv-\tfrac{1}{2}\gamma v^2 + \tfrac{1}{4}\,\hbar(\alpha+\gamma)]\,dt\}$$

$$\times\, \exp\{-(1/2\hbar\nu)\int [\dot{u}^2+\dot{v}^2]\,dt\}\,\Pi\, du(t)\,dv(t)\,, \tag{2.31}$$

subject to the boundary conditions $u(0) = v(0) = u(T) = v(T) = 0$. We have also introduced

$$\alpha(t) = \partial_{\bar{p}}^2 H(\bar{p},\bar{q})\,, \qquad \beta(t) = \partial_{\bar{p}}\partial_{\bar{q}} H(\bar{p},\bar{q})\,, \qquad \gamma(t) = \partial_{\bar{q}}^2 H(\bar{p},\bar{q})\,, \tag{2.32}$$

evaluated, as indicated, for the extremal path. Since the extremal solution $\bar{p}$, $\bar{q}$ is generally complex it follows that $\alpha$, $\beta$, and $\gamma$ are generally complex functions. For the time being we assume that E, as given by (2.31), is finite. Exceptions to this assumption are discussed subsequently.

To evaluate E it is convenient to reexpress it in the abstract Dirac form used previously, in which case

$$E = \langle 0|\, \mathbf{T}\exp[-i\int (\tfrac{1}{2}\alpha P^2 + \beta P\circ Q + \tfrac{1}{2}\gamma Q^2)\,dt]\,|0\rangle\,, \tag{2.33}$$

where most of the notation is similar to that in (1.13). The principal difference in the two expressions lies in the fact that here $|0\rangle$ denotes a unit vector, $\langle 0|0\rangle = 1$, specifically the harmonic oscillator ground state for which $(P^2+Q^2)\,|0\rangle = |0\rangle$.

An explicit value for E may be obtained with the aid of the nonsymmetric operator representation (1.15). It follows that the operator of present interest is represented by

$$\mathbf{T}\exp[\int \begin{pmatrix} \beta & -i\alpha \\ -i\gamma & -\beta \end{pmatrix} dt] \equiv \begin{pmatrix} A & B \\ C & D \end{pmatrix}. \tag{2.34}$$

In this case we seek an alternative expression in the form

$$e^{i\xi ½ (P^2+Q^2)}\; e^{i\eta P\cdot Q}\; e^{i\zeta ½ (P^2+Q^2)}$$

$$= \begin{pmatrix} \cos\xi & i\sin\xi \\ i\sin\xi & \cos\xi \end{pmatrix} \begin{pmatrix} e^{-\eta} & \\ & e^{\eta} \end{pmatrix} \begin{pmatrix} \cos\zeta & i\sin\zeta \\ i\sin\zeta & \cos\zeta \end{pmatrix} . \tag{2.35}$$

In this form the quantity of interest is given by

$$<0|\; e^{i\xi ½ (P^2+Q^2)}\; e^{i\eta P\cdot Q}\; e^{i\zeta ½ (P^2+Q^2)}\; |0> \;=\; e^{i ½ (\xi+\zeta)}\; <0|\; e^{i\eta P\cdot Q}\; |0> . \tag{2.36}$$

Now

$$<0|\; e^{i\eta P\cdot Q}\; |0> = \pi^{-½}\, e^{\eta/2} \int e^{-(1+e^{2\eta})y^2/2}\, dy$$

$$= 1/\sqrt{\cosh\eta}\;, \tag{2.37}$$

and thus

$$E = [\cosh\eta\; e^{-i(\xi+\zeta)}]^{-½} . \tag{2.38}$$

From the two 2 × 2 matrices involved it follows that

$$½\;(A + D) = \cosh\eta\;\; \cos(\xi+\zeta)\;, \tag{2.39}$$

$$½\;(B + C) \;=\; i\cosh\eta\;\; \sin(\xi+\zeta)\;,$$

and therefore

$$E = 1/\sqrt{(A + D - B - C)/2}\,. \tag{2.40}$$

We may reinterpret this result in terms of solutions to the linearized classical Hamiltonian equations of motion

$$\dot{\tilde{q}}(t) = \beta(t)\; \tilde{q}(t) + \alpha(t)\; \tilde{p}(t)\;, \tag{2.41}$$

$$\dot{\tilde{p}}(t) = -\gamma(t)\; \tilde{q}(t) - \beta(t)\; \tilde{p}(t)\;.$$

If we set $\tilde{r}(t) = -i\tilde{q}(t)$, then it follows that

$$\dot{\tilde{r}}(t) = \beta(t)\,\tilde{r}(t) - i\alpha(t)\,\tilde{p}(t)\,, \tag{2.42}$$

$$\dot{\tilde{p}}(t) = -i\gamma(t)\,\tilde{r}(t) - \beta(t)\,\tilde{p}(t)\,.$$

Hence we recognize

$$\begin{pmatrix}\tilde{r}\,(T)\\ \tilde{p}\,(T)\end{pmatrix} = \begin{pmatrix}A & B\\ C & D\end{pmatrix}\begin{pmatrix}\tilde{r}\,(0)\\ \tilde{p}\,(0)\end{pmatrix}. \tag{2.43}$$

With $\tilde{r}(0) = -1/2$, $\tilde{p}(0) = 1/2$ interest centers on $\tilde{p}(T) - \tilde{r}(T) = (A+D-B-C)/2$. In other words,

$$E = \frac{1}{\sqrt{\tilde{p}(T) + i\tilde{q}(T)}} \tag{2.44}$$

based on a solution of (2.41) subject to the initial conditions $\tilde{p}(0) = 1/2$, $\tilde{q}(0) = -i/2$.

The similarity of this amplitude to that in (1.23) is striking, yet it differs importantly because of the complex argument. In (1.23), $\tilde{x}(T)$, being real, cannot pass from positive to negative values without crossing zero; that undesirable feature is avoided in the complex parameterization.

Putting our results together we have determined a semiclassical approximation for K given by

$$K_{sc}(p'',q'',T;p',q',0) = \frac{1}{\sqrt{\tilde{p}(T) + i\tilde{q}(T)}}\, e^{iF(p'',q'';p'q')/\hbar}\,. \tag{2.45}$$

We emphasize that there is *no* sum over extremal paths in this formula; there is just one relevant extremal path connecting any pair of phase-space points $(p',q')$ and $(p'',q'')$, and in general the required path is complex as noted above.

It is instructive to examine this expression first for the simple case of a harmonic oscillator with $H(p,q) = \frac{1}{2}\,(p^2+q^2)$. In that case $\tilde{p}(T) + i\tilde{q}(T) = e^{iT}$, and

$$K_{sc}(p'',q'',T;p',q',0) \tag{2.46}$$

$$= e^{-iT/2}\exp\{\frac{i}{2\hbar}(p''q''-p'q') + \frac{i}{2\hbar}(p'_Tq''-q'_Tp'') - \frac{1}{4\hbar}[(p''-p'_T)^2 + (q''-q'_T)^2]\}\,,$$

where

$$q'_T = \cos(T)\, q' + \sin(T)\, p' \,, \tag{2.47}$$

$$p'_T = -\sin(T)\, q' + \cos(T)\, p' \,.$$

For this example the approximation is exact, i.e., $K_{sc} = K$, which is not really too surprising since only Gaussian integrals are involved. More important, almost, is what this result teaches us about the solution in more general cases. Here the "amplitude" factor E only contributes a phase, which is really nothing but the zero-point energy [8]. The true amplitude information is contained in F leading to an exponential reduction whenever $p'' \neq p'_T$ and $q'' \neq q'_T$. Observe that $(p'_T, q'_T)$ is just the image of the phase-space point $(p',q')$ in time T under the action of $H = ½\,(p^2+q^2)$. Therefore $|K_{sc}| = 1$ if and only if $p'' = p'_T$ and $q'' = q'_T$, otherwise $|K_{sc}| < 1$ with an exponential decrease. Since $K_{sc}$ is exact in this case it follows that $J_{sc}$ given by

$$J_{sc}(q'',T;q',0) = \frac{(\pi\hbar)^{½}}{(2\pi\hbar)^2} \int K_{sc}(p'',q'',T;p',q',0)\, dp''dp' \tag{2.48}$$

will be the exact expression in configuration space. In fact $K_{sc}$ and consequently $J_{sc}$ will be exact for an arbitrary quadratic Hamiltonian, namely, if

$$H(p,q) = ½\, X(t)\, p^2 + Y(t)\, pq + ½\, Z(t)\, q^2 \tag{2.49}$$

for arbitrary, piece-wise smooth X, Y and Z.

For nonquadratic Hamiltonians $K_{sc}$ is not exact, but its general structure is similar to that discussed above. In the general case, the amplitude factor E contributes to the phase and, as we shall see, introduces relatively modest modulations of the amplitude of $K_{sc}$. The major contributions to $K_{sc}$, therefore, come from

$$K^{o}_{sc}(p'',q'',T;p',q',0) \equiv e^{iF(p'',q'';p',q')/\hbar} \,. \tag{2.50}$$

For "special" pairs of phase-space points, $(p'',q'')$ and $(p',q')$, it occurs that $p'' = p'_T$ and $q'' = q'_T$, where $(p'_T, q'_T)$ denotes the image in time T under evolution by the general classical Hamiltonian $H(p,q)$. In that case the extremal solution discussed in (2.28) has $w = 0$ and is entirely real; moreover there are no rapidly changing boundary layers. As a consequence, at special points,

$$F(p'',q'';p',q') = \int [\bar{p}\dot{\bar{q}} - H(\bar{p},\bar{q})]dt \tag{2.51}$$

which is also real, and in fact $F(p'',q'';p',q') = S(q'',q')$ for that particular classical path with the indicated initial and final momenta. The upshot is that for special points, which are connected in time T by the classical dynamics,

$$|K^{o}_{sc}(p'',q'',T;p',q',0)| = 1 \ . \tag{2.52}$$

Moreover, for special points the functions $\alpha, \beta$, and $\gamma$ in (2.32) and (2.33) are *real*, so for special points it always follows that

$$|E| \leqslant 1 \ . \tag{2.53}$$

This bound is also implied by (2.14).

For "general" pairs of phase-space points, $(p'',q'')$ and $(p',q')$, either $p'' \neq p'_T$ or $q'' \neq q'_T$, or more generally both. In that case $w \neq 0$, the solution is complex, and there are boundary layers near both $t = 0$ and $t = T$. It is true that the combination $z_+(t) \equiv q(t) + ip(t)$ has no boundary layer at $t = 0$, thanks to the boundary condition there, but it will have a boundary layer at $t = T$ for general pairs of phase-space points. Likewise $z_-(t) \equiv q(t) - ip(t)$ has no boundary layer at $t = T$, but generally has one at $t = 0$. There is no combination of paths that is devoid of all boundary layers (except in the case of special points where there are no boundary layers at all). Therefore, at general points the solution $\bar{p},\bar{q}$ is complex, and as a consequence F is complex. It follows, in the case of general points, that

$$|K^{o}_{sc}(p'',q'',T;p',q',0)| < 1 \ , \tag{2.54}$$

namely, F contributes an *exponential damping factor*, $\mathrm{Im}\, F > 0$. It cannot be otherwise for the simple reason that $|K| \leqslant 1$ holds as a general rule [cf. (2.14)] and $K_{sc}$ given by (2.45) becomes all the more accurate as $\hbar$ decreases. The amplitude factor E cannot affect this bound as it does not depend on $\hbar$, unless, of course, it happens that $E = 0$. However, it follows that $|E| > 0$ holds as a general rule since as a solution of a linear system of equations (2.41) there is no way in which $|\tilde{p}(T) + i\tilde{q}(T)|$ could diverge in a finite time, $T < \infty$.

For general points, where $\alpha, \beta$, and $\gamma$ are rather general complex functions, the amplitude factor given by (2.33) is no longer constrained by (2.53). However, the following general remarks hold. For general points in the neighborhood of a special point, where, in some sense w remains small and the solution remains nearly real, then it follows that $|E| \lesssim 1$. This relation holds just by continuity. When w is not small and $\alpha, \beta$, and $\gamma$ are not nearly real, then $|E|$ may become large. But in such cases, $|E\, K^{o}_{sc}| << 1$ thanks to a dominating exponential damping factor in $K^{o}_{sc}$. If, for certain $\alpha, \beta$, and $\gamma$, it should happen that E diverges there is no cause for alarm. A divergent E simply means a breakdown in the quadratic approximations to deviations from the extremal path in (2.31); inclusion of high-order terms will render E finite. But the precise finite value of E is actually unimportant since invariably $|E\, K^{o}_{sc}| << 1$, and so the true magnitude of E is irrelevant. There is no real loss of accuracy in this scheme if E is replaced, for example, by $E' = E \min(1, 100/|E|)$, thus rendering E uniformly bounded.

Thus we arrive at a qualitative picture of the semiclassical approximation $K_{sc}$ for a general Hamiltonian. Put simply, the function $K_{sc}(p'',q'',T;p',q',0)$ is nonzero, but exponentially small for all pairs of phase-space points, $(p'',q'')$ and $(p',q')$, save at (and near) those special points where the classical evolution of $(p',q')$ in time T coincides with $(p'',q'')$. This picture helps us understand how an appropriate configuration-space approximation, generally different from $J_{sc}$, emerges from the definition

$$\tilde{J}_{sc}(q'',T;q',0) \equiv \frac{(\pi\hbar)^{1/2}}{(2\pi\hbar)^2} \int K_{sc}(p'',q'',T;p',q',0)\, dp''dp' \,. \tag{2.55}$$

In this integral the integrand $K_{sc}$ will generally be negligible save in the vicinity of special points connected by a genuine, real classical trajectory. Each such trajectory leads to a contribution the strength of which is determined by just how fast $K_{sc}$ decreases in amplitude and/or changes in phase in the neighborhood of that trajectory. The result will be a sum of such weighted contributions, one for each of the appropriate classical trajectories. This is the qualitative picture appropriate to a situation having well-separated classical trajectories and gives rise to an expression substantially similar to the usual result given in (1.24). On the other hand, (2.55) has a far wider range of applicability. For example, (2.55) is directly applicable in cases where paths are not well separated, as arises in the case of general caustics, or even in the case of a "thicket" of classical paths by which we mean that for a finite time interval two or more paths remain so close that a sensible exponential separation between them does not exist. Such a situation could occur for rays making very small angles with the axis of the sound channel in long-distance propagation ocean acoustics problems.

Equation (2.55) offers a global, uniform, semiclassical approximation to the configuration-space wave equation, Eq. (1.1). As such it constitutes the principal result of this paper [9]. However, it is also useful to proceed one step further and discuss, at least in outline form, an approximate, stationary-phase evaluation of (2.55) itself. Assume we deal with a single, well-separated classical trajectory connecting $q'$ and $q''$ in time T. The particular path may be distinguished by the value of its initial and final momenta, $p' = p'_c$ and $p'' = p''_c$. We fix the amplitude factor E at its value for the special point, and set

$$F = F(p''_c + (p''-p''_c), q''; p'_c + (p'-p'_c), q') . \qquad (2.56)$$

If this expression is expanded out to terms linear and quadratic in $(p''-p''_c)$ and $(p'-p'_c)$, then with this approximation for F it becomes possible to evaluate (2.55) analytically [8]. This procedure is suitable provided that the matrix of quadratic deviations is nonsingular. When this is not the case at least one term cubic in $(p''-p''_c)$ or $(p'-p'_c)$ is needed giving rise to the expected Airy function. In that sense (2.55) offers a solution not unlike the Maslov solution, but unlike the Maslov solution the approximation offered by (2.55), or by its analytic stationary-phase approximation is globally applicable. This would be a nontrivial improvement for a problem in which so many rays contribute that two or more of them would almost always form a caustic in such a way as to invalidate the Maslov-type approximation. In such cases the advantages of the global, uniform, semiclassical approximation offered here become manifest.

## 3. Remarks on Propagation in a Random Medium

Techniques for inclusion of a random perturbation in the configuration-space approach of Sec. 1 are well known [10, 11]. In view of the general representation of the semiclassical amplitude as a sum over contributions for each classical trajectory [cf. (1.24)] it is especially convenient to perform ensemble averages over the random medium before making a semiclassical approximation. Such an average is commonly carried out on a path-integral representation of the amplitude, or on a path-integral representation of products of several amplitude factors. Invariably, the result of such an average leads to a multi-time path integral that is typically difficult to deal with analytically unless one makes the so-called Markov approximation which, in essence, reduces the problem to a single-time path integral.

In the coherent-state representation the handling of a random medium can proceed in either of two ways. In the first way, analogous to the discussion just given, it is possible to average before making a semiclassical approximation. This average can be accomplished with ease in a path-integral representation, but further analytic progress would seem to depend on invoking the Markov approximation, just as in the configuration case. However, there is also a second way, which consists of averaging after making a semiclassical approximation. This possibility is open to us now since there is only a single term in the semiclassical approximation rather than a large number of terms. Moreover, in this approach there is no longer any need to assume a Markov approximation for the correlation function. The associated configuration-space moment approximation is obtained [cf. (2.55)] by integrating out all the initial and final momentum values. We do not carry out the indicated analysis here, but instead reserve that to a subsequent article.

**Acknowledgements**

It is a pleasure to thank M. Berry for a discussion regarding the Maslov approach. The research reported here has been largely motivated by conversations with A. H. Carter, R. Holford, and R. S. Patton.

## REFERENCES

1. See, e.g., L. Schulman, *Techniques and Applications of Path Integration* (Wiley & Sons, New York, 1981).
2. V. P. Maslov, *Thèorie des Perturbations et Méthodes Asymptotiques* (Dunod, Paris, 1972).
3. R. G. Littlejohn, Phys. Rev. Letters *54*, 1742 (1985).
4. See, e.g., J. R. Klauder and B.-S. Skagerstam, *Coherent States* (World Scientific, Singapore, 1985).
5. J. R. Klauder and I. Daubechies, Phys. Rev. Letters *52*, 1161 (1984); I. Daubechies and J. R. Klauder, J. Math. Phys. *26*, 2239 (1985).
6. See, e.g., R. S. Brodkey, *The Phenomena of Fluid Motions* (Addison-Wesley, Reading, Mass., 1967), Chap. 9.
7. J. R. Klauder, in *Path Integrals and their Applications in Quantum, Statistical and Solid-State Physics,* eds. G. J. Papadopoulis and J. T. Devreese (Plenum, New York, 1978), p. 5; Phys. Rev. D*19*, 2349 (1978); in *Quantum Fields-Algebras, Processes,* ed. L. Streit (Springer, Vienna, 1980), p. 65.
8. J. R. Klauder, "Path Integrals and Semiclassical Approximations to Wave Equations," (AT&T Bell Laboratories preprint).
9. An announcement of this result appears in J. R. Klauder, "Global, Uniform, Semiclassical Approximation to Wave Equations," (AT&T Bell Laboratories preprint).
10. P. L. Chow, J. Math. Phys. *13*, 1224 (1972); in *Multiple Scattering and Waves in Random Media,* eds. P. L. Chow, W. E. Kohler, and G. C. Papanicolaou (North Holland, Amsterdam, 1981), p. 89.

11. R. Dashen, J. Math. Phys. *20*, 892 (1979); R. S. Patton, "Second Moments of the Pressure Field Near a Smooth Caustic," in *Adaptive Methods in Underwater Acoustics*, ed. H. G. Urban (D. Reidel, Dordrecht, Holland, 1985).

# EFFECTIVE CONDUCTIVITIES OF RECIPROCAL MEDIA*

Joseph B. Keller
Departments of Mathematics
and Mechanical Engineering
Stanford University
Stanford, California 94305

## 1. The Concept of Effective Conductivity

The conductivity $\sigma$ of a medium is the coefficient of proportionality between a force $E$ and a flux $J$ in the linear relation

$$J = \sigma E. \tag{1.1}$$

When $E$ is the electric field intensity and $J$ is the electric current density, $\sigma$ is called the electrical conductivity. When $E$ is minus the temperature gradient and $J$ is the heat flux, $\sigma$ is called the thermal conductivity, etc. In an isotropic medium $\sigma$ is a scalar while in an anisotopic medium it is a tensor.

A heterogeneous medium is one with a conductivity $\sigma(x)$ which varies with position $x$, and this variation makes it difficult to analyze fields and currents in such media. Therefore it is desirable, to replace the function $\sigma(x)$ by a constant tensor or scalar $\Sigma$ to simplify such analyses. The constant tensor or scalar $\Sigma[\sigma(\cdot)]$, which must depend upon the function $\sigma(\cdot)$, is called the effective conductivity of the medium. Its introduction is useful only if it leads to the same overall or large scale fields and currents as would the actual conductivity.

There is such a $\Sigma$ when $\sigma(x)$ is a periodic function and when $\sigma(x)$ is a stationary random function. In these two cases $\Sigma$ is defined by the relation

$$\langle J \rangle = \Sigma \langle E \rangle. \tag{1.2}$$

The angular brackets denote the average value of the enclosed quantity. For a periodic medium the average is the spatial average over a period cell, while for a stationary random medium it is the stochastic average. In both cases a precise problem must be specified to characterize the functions $E$ and $J$ which enter into (2).

* Research supported by the Office of Naval Research, the Air Force Office of Scientific Research, the Army Research Office, and the National Science Foundation.

For time-independent fields, $E = \nabla\varphi$ where $\varphi$ is a scalar potential function, and in a source-free region $\nabla \cdot J = 0$. Upon combining these two relations with (1.1) we get

$$\nabla \cdot (\sigma \nabla \varphi) = 0. \tag{1.3}$$

This is the equation governing $\varphi$. It must be solved in all of space subject to one of the conditions

$$\varphi(x + a) - \varphi(x) = E_0 \cdot a, \tag{1.4a}$$

$$\langle \nabla \varphi(x) \rangle = E_0. \tag{1.4b}$$

In (1.4a), which holds for periodic media, $a$ is any period vector and $E_0$ is a prescribed constant vector. In (1.4b), which holds for stationary random media, the average of $\nabla\varphi$ must be equal to the constant $E_0$ at all points. In terms of $\varphi$ (1.2) becomes

$$\langle \sigma(x) \nabla \varphi(x) \rangle = \Sigma[\sigma(\cdot)] \langle \nabla \varphi(x) \rangle. \tag{1.5}$$

## 2. An Example

The preceding considerations lead to a well-defined effective conductivity tensor $\Sigma[\sigma(\cdot)]$, but the problem of calculating it is a formidable one. One of the first such calculations was done by Maxwell for a periodic medium consisting of perfectly conducting spheres of diameter $b$ arranged in a simple cubic lattice of spacing $a$. The medium surrounding the spheres has the constant scalar conductivity $\sigma_1$. When $b/a$ is small, the effective conductivity $\Sigma(\sigma_1, \infty)$ has the form

$$\Sigma(\sigma_1, \infty) = \sigma_1[1 + c_1(\tfrac{b}{a}) + c_2(\tfrac{b}{a})^2 + \cdots] \tag{2.1}$$

Maxwell calculated $c_1$, Rayleigh calculated $c_2$, Runge calculated $c_3$, and Meredith and Tobias [1] calculated $c_4$.

It is clear on physical grounds that $\Sigma(\sigma_1, \infty) = \infty$ at $b = a$ because then the perfectly conducting spheres touch one another. However this behavior is not revealed by any finite number of terms in the expansion (2.1). To reveal this behavior and to determine the nature of the singularity I considered the case in which $a - b$ is small, and calculated the potential in the narrow gap between two adjacent spheres. From it I found [2] the following asymptotic result, which is valid for $a - b \ll a$:

$$\Sigma(\sigma_1, \infty) \sim -\frac{\sigma_1 \pi}{2} \log(\tfrac{b}{a} - 1) + \cdots \qquad \text{(spheres)} \tag{2.2}$$

I repeated the calculation for a square array of perfectly conducting cylinders and obtained for the conductivity in the plane normal to the cylinders

$$\Sigma(\sigma_1, \infty) \sim \frac{\sigma_1 \pi}{2^{\frac{1}{2}}} (\tfrac{a}{b} - 1)^{-\frac{1}{2}} + \cdots \qquad \text{(cylinders).} \tag{2.3}$$

For a square array of nonconducting cylinders I obtained instead

$$\Sigma(\sigma_1, 0) \sim \frac{\sigma_1 2^{\frac{1}{2}}}{\pi}\left(\tfrac{a}{b} - 1\right)^{\frac{1}{2}} + \cdots \qquad \text{(cylinders).} \tag{2.4}$$

Comparison of (2.3) and (2.4) shows that $\Sigma(\sigma_1, \infty)/\sigma_1$ becomes infinite at $a/b = 1$ while $\Sigma(\sigma_1, 0)/\sigma_1$ becomes zero there, and they behave exactly as reciprocals of one another.

## 3. Reciprocal Theorems for Two Component Media

The preceding asymptotic result suggested that the reciprocal relation might hold for all values of $a/b$. I was able to prove this by using conjugate harmonic functions.

**Theorem 1 (Keller 1963)** Let $\Sigma(\sigma_1, \infty)$ and $\Sigma(\sigma_1, 0)$ be the effective conductivities of two composite media obtained by imbedding in a medium of scalar conductivity $\sigma_1$ a square lattice of circular cylinders of conductivities infinity and zero respectively. Then

$$\frac{\Sigma(\sigma_1, \infty)}{\sigma_1} = \frac{\sigma_1}{\Sigma(\sigma_1, 0)}. \tag{3.1}$$

The conductivities apply to the plane normal to the cylinders.

Generalizing this theorem, still using conjugate harmonic functions, led to the following [3]:

**Theorem 2 (Keller 1964)** Let $\Sigma_x(\sigma_1, \sigma_2)$ be the effective conductivity in the $x$ direction of a rectangular lattice of identical parallel cylinders of scalar conductivity $\sigma_2$ in a medium of scalar conductivity $\sigma_1$, with the cylinders parallel to the $z$-axis. Let $\Sigma_y(\sigma_2, \sigma_1)$ be the effective conductivity in the $y$ direction for the same lattice with $\sigma_1$ and $\sigma_2$ interchanged. If each cylinder is symmetric in the $x$ and $y$ axes, which are the lattice axes, then

$$\frac{\Sigma_x(\sigma_1, \sigma_2)}{\sigma_1} = \frac{\sigma_2}{\Sigma_y(\sigma_2, \sigma_1)}. \tag{3.2}$$

An immediate corollary of Theorem 2 is

**Corollary 1** When the lattice is square and each cylinder is symmetric in the line $x = y$

$$\frac{\Sigma(\sigma_1, \sigma_2)}{\sigma_1} = \frac{\sigma_2}{\Sigma(\sigma_2, \sigma_1)}. \tag{3.3}$$

The subscripts $x$ and $y$ have been omitted because in this case the effective medium is isotropic in the $x, y$ plane, i.e., $\Sigma_x(\sigma_1, \sigma_2) = \Sigma_y(\sigma_1, \sigma_2)$.

This corollary applies in particular to a checkerboard pattern of $\sigma_1$ and $\sigma_2$. Furthermore the pattern is unchanged when $\sigma_1$ and $\sigma_2$ are interchanged, so $\Sigma(\sigma_1, \sigma_2) = \Sigma(\sigma_2, \sigma_1)$. Thus (3.3) yields for the checkerboard

$$\Sigma(\sigma_1, \sigma_2) = (\sigma_1 \sigma_2)^{\frac{1}{2}}. \tag{3.4}$$

In the same way (3.4) follows for various other patterns.

Theorem 2 was extended by Mendelson [4]. He proved

**Theorem 3 (Mendelson 1975)** Let $\Sigma_x(\sigma_1, \sigma_2)$ be the effective conductivity in the $x$ direction of a two dimensional medium composed of constituents with scalar conductivities $\sigma_1$ and $\sigma_2$. Let $\Sigma_y(\sigma_2, \sigma_1)$ be the effective conductivity in the $y$ direction for the corresponding medium with $\sigma_1$ and $\sigma_2$ interchanged. If the $x$ and $y$ axes are principal axes of the two media then (3.2) holds.

Since Theorem 3 applies to any geometrical arrangement of the constituents, it applies in particular to statistically stationary random media. When the media are also statistically isotropic, then (3.3) holds. If in addition the interchange of $\sigma_1$ and $\sigma_2$ leaves the medium statistically unchanged, then (3.4) holds.

Schulgasser [5] considered three dimensional media and proved

**Theorem 4 (Schulgasser 1976)** Let $\Sigma_x(\sigma_1, \sigma_2)$ be the effective conductivity in the $x$ direction of a statistically homogeneous three dimensional medium composed of two constituents with scalar conductivities $\sigma_1$ and $\sigma_2$. Let $\Sigma_y(\sigma_2, \sigma_1)$ be the effective conductivity in the $y$ direction with $\sigma_1$ and $\sigma_2$ interchanged. Then

$$\Sigma_x(\sigma_1, \sigma_2)\Sigma_y(\sigma_2, \sigma_1) \geq \sigma_1\sigma_2. \tag{3.5}$$

His proof utilized the variational formulation of the conductivity problem and the representation of $\Sigma_x$ as the minimum of a certain quadratic functional. He also showed that equality in (3.5) does not hold in general by considering explicit cases, and also by a general argument for symmetric configurations.

## 4. Generalizations

The preceding results for periodic media were generalized by Hansen [6] in two ways. First of all he considered the full conductivity tensor $\Sigma$, and he also considered any periodic conductivity $\sigma(x)$.

**Theorem 5 (Hansen 1978)** Let $\Sigma[\sigma(\cdot)]$ be the effective conductivity tensor of a two dimensional periodic medium with scalar conductivity $\sigma(\cdot)$. Let $\Sigma[\frac{\sigma_0^2}{\sigma(\cdot)}]$ pertain to the medium with conductivity $\sigma_0^2/\sigma(\cdot)$ where $\sigma_0^2$ is a positive constant. Then

$$\Sigma\left[\frac{\sigma_0^2}{\sigma(\cdot)}\right] = \frac{\sigma_0^2}{\det\Sigma[\sigma(\cdot)]}\Sigma[\sigma(\cdot)]. \tag{4.1}$$

From this he obtained the

**Corollary** If $\Sigma[\sigma(\cdot)]$ is a scalar and if $\sigma_0^2/\sigma(\cdot)$ is obtainable from $\sigma(\cdot)$ by a rigid body motion, then

$$\Sigma[\sigma(\cdot)] = \sigma_0 I. \tag{4.2}$$

Here $I$ is the identity.

By using this corollary he calculated the effective conductivity of various periodic media.

The results for stationary random media were generalized by Kohler and Papanicolau [7], who proved the following two theorems.

**Theorem 6 (Kohler and Papanicolau 1982)** Let $\Sigma_x[\sigma(\cdot)]$ be the effective conductivity in the $x$ direction of a two dimensional statistically stationary random medium. Then

$$\Sigma_x[\sigma(\cdot)] = \frac{1}{\Sigma_y[\frac{1}{\sigma(\cdot)}]}. \tag{4.3}$$

**Theorem 7 (Kohler and Papanicolau 1982)** For a three dimensional statistically stationary random medium

$$\Sigma_x[\sigma(\cdot)] \geq \frac{1}{\Sigma_y[1/\sigma(\cdot)]}. \tag{4.4}$$

These theorems were proved by using the variational method. The analysis was simplified by a direct formulation of the problem for the entire space $R^3$, rather than for a bounded region followed by passage to the limit of an infinite region.

## 5. Anisotropic Media

The most recent extension, by Nevard and Keller [8], concerns the case in which $\sigma(x)$ is a symmetric positive definite tensor. They proved the following two theorems:

**Theorem 8 (Nevard and Keller 1985)** Let $\sigma(x)$ be the piecewise smooth conductivity tensor of a two dimensional medium which is either periodic or stationary random. Then for any positive constant scalar $k$

$$\Sigma\left[\frac{k\sigma(\cdot)}{\det\sigma(\cdot)}\right] = \frac{k\Sigma[\sigma(\cdot)]}{\det\Sigma[\sigma(\cdot)]}. \tag{5.1}$$

**Theorem 9 (Nevard and Keller 1985)** Let $\sigma(x)$ be a piecewise smooth positive definite symmetric tensor and let $\rho$ be a constant tensor. Assume that for any coordinate directions in the $x, y$ plane they are given by

$$\sigma(x) = \begin{pmatrix} \sigma_{xx} & \sigma_{xy} & 0 \\ \sigma_{xy} & \sigma_{yy} & 0 \\ 0 & 0 & \sigma_{zz} \end{pmatrix}, \qquad \rho = \begin{pmatrix} 0 & -1 & 0 \\ 1 & 0 & 0 \\ 0 & 0 & 1 \end{pmatrix}. \tag{5.2}$$

Suppose that $\sigma$ is statistically stationary, or periodic in the $x, y$ plane and periodic in $z$. Then for any constant scalar $k > 0$,

$$\Sigma_{xx}[\sigma(\cdot)]\Sigma_{yy}[k\rho\sigma^{-1}(\cdot)\rho^{-1}] \geq k. \tag{5.3}$$

Various consequences of these theorems are described in [8]. An analogue of Theorem 2 was proved for the effective shear modulus of an elastic composite by Flaherty and Keller [9].

## References


[1] R.E. Meredith and C.W. Tobias, *J. Appl. Phys.* **31**, 1270 (1960).

[2] J.B. Keller, "Conductivity of a Medium Containing a Dense Array of Perfectly Conducting Spheres or Cylinders, or Nonconducting Cylinders," *J. Appl. Phys.* **34**, 991 (1963).

[3] J.B. Keller, "A Theorem on the Conductivity of a Composite Medium," *J. Math Phys.* **5**, 548 (1964).

[4] K. Mendelsohn, *J. Appl. Phys.* **46**, 917 (1975).

[5] K. Schulgasser, "On a phase interchange relationship for composite materials, *J. Math Phys.* **17**, 378 (1976).

[6] E.B. Hansen, "Generalization of a Theorem on the Conductivity of a Composite Medium," unpublished.

[7] W. Kohler and G.C. Papanicolau, "Macroscopic Properties of Disordered Media," edited by R. Burridge, S. Childress and G. Papanicolau, *Lect. Notes Phys.* **154**,Springer (1981), p. 111.

[8] J. Nevard and J.B. Keller, "Reciprocal relations for effective conductivities of anisotropic media," *J. Math. Phys.* **26**, 2761 (1985).

[9] J.B. Keller and J.E. Flaherty, "Elastic behavior of composite media," *Comm. Pure Appl. Math.* **26**, 565 (1973).

# REFLECTION FROM A LOSSY, ONE-DIMENSIONAL OCEAN SEDIMENT MODEL

Werner E. Kohler

Department of Mathematics
Virginia Tech
Blacksburg, Virginia

ABSTRACT: The interplay of random layering, small attenuation and refraction is studied in the context of an idealized one-dimensional model of ocean sediments. Appropriate scalings are defined and the Fokker-Planck equation for the reflection coefficient is asymptotically analyzed.

## 1. Introduction

Abyssal plains constitute a significant portion of the world's ocean bottom. These plains are typically regions of thick sediment layering (nominally hundreds of meters thick) covering a rock basement. As the sediments have formed over the ages, the various physical processes contributing to their formation have imparted a very fine layering to the overall sediment structure as a function of depth. This fine-scale layering causes the acoustic properties of the sediment layer (e.g. sound speed and density) to fluctuate rapidly with depth; these rapid variations, in turn, are superposed upon mean or average values of these acoustic parameters that gradually increase with depth due to compactification. This slow-scale (nominally linear) increase with depth of mean sound speed and density tends to refract penetrating acoustic energy back towards the surface (i.e. the acoustic rays are bent back upward towards the water-sediment surface). As the acoustic energy is refracted through the sediment material, it interacts with the fine-scale layering via multiple scattering. Our goal is to understand the interplay of refraction and multiple scattering within this highly-laminated material.

We shall adopt an idealized one-dimensional model in which a thick sediment layer lies beneath an isovelocity water column and rests on top of a homogeneous semi-infinite rock basement. Within the sediment layer itself, slow-scale transverse variations will be neglected and acoustic properties will be assumed to vary only as a function of depth. The fine-scale fluctuations in these properties will be modeled in terms of a single (deposition) stochastic process; these random fluctuations will be added to slowly-increasing mean values. We shall assume that the sediment layer is illuminated from the water column by a monochromatic plane acoustic wave.

Parameter values characteristic of the abyssal plains will be adopted. A value of acoustic frequency that results in both significant sediment penetration and multiple scattering will be chosen. We shall assume that acoustic energy impinges upon the sediment layer in such a way that complete internal reflection is achieved. An understanding of the interplay of random scattering and refraction in the presence of a turning point is our goal.

This problem was discussed in [1] for the case of a lossless sediment and basement. The analysis in that case focussed upon the asymptotics of a suitably-scaled Fokker-Planck equation for the angle process of the (unimodular) reflection coefficient. In this work we extend this study to include small, deterministic dissipation characteristic of actual ocean sediments; this is accomplished through the artifice of making the mean sound speed complex. In Section II we formulate the stochastic two-point boundary value problem being considered, specify typical parameter values and introduce the appropriate scaling. Section III discusses the centering trajectory associated with the averaged problem (the geoacoustic model); Section IV derives the Riccati equation for the (stochastic) reflection coefficient and introduces a white noise approximation. The main result of this work, the formal asymptotic analysis of the ensuing Fokker-Planck equation for the polar coordinates of the reflection coefficient, is presented in Section V. Section VI presents the results of some computer simulations, evaluates the asymptotic analysis in the light of these simulations and presents conclusions.

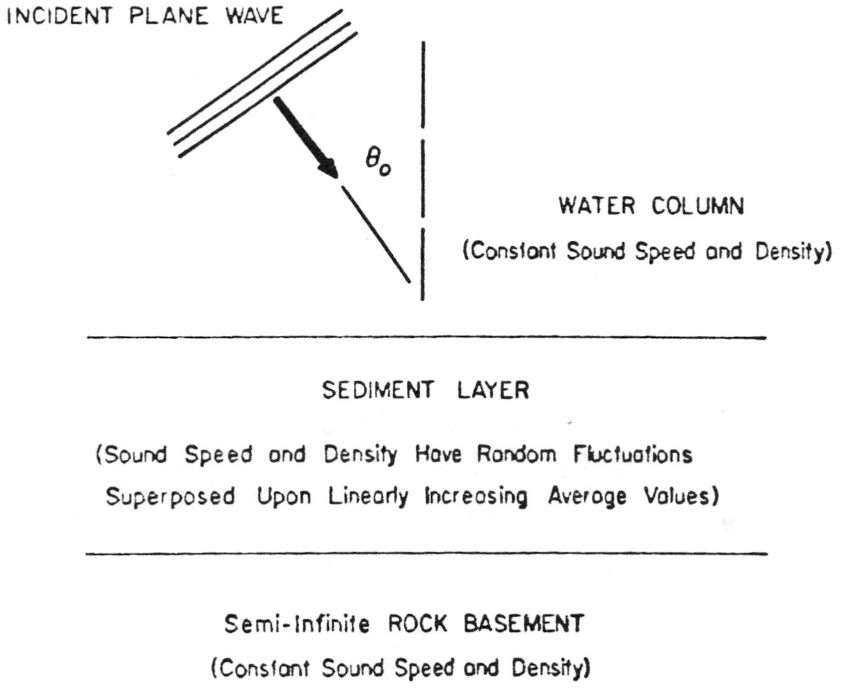


Figure 1 : PROBLEM CONFIGURATION

## II. Problem Formulation:

Our aim is to adopt as simple a model as possible to study the interplay of refraction and random scattering. Since measured acoustic response of thick ocean sediments do not exhibit shear wave effects [2], we shall adopt the following simplified description:

$$\begin{aligned} \partial_t p + \rho c^2 \nabla \cdot \underline{v} &= 0 \\ \partial_t \underline{v} + \rho^{-1} \nabla p &= \underline{0}. \end{aligned} \tag{1}$$

In (1), p, $\underline{v}$, $\rho$ and c denote pressure, particle velocity, density and sound speed, respectively. In the water column $(z < 0)$, density and sound speed will have constant values $\rho_0$ and $c_0$. The rock basement $(z > L)$ will be characterized either by constant constitutive parameters, i.e. density $\rho_r$ and sound speed $c_r$, or by an interface reflection coefficient. Within the sediment layer itself $(0 \leqslant z \leqslant L)$, density and sound speed will vary randomly with depth.

We shall assume that a monochromatic plane acoustic wave impinges on the sediment layer at incidence angle $\theta_0$ (measured from the normal) as shown in Figure 1. Then, pressure and particle velocity can be written as:

$$p(\underline{r},t) = e^{i(kz-\omega t)}\hat{p}(z) \tag{2}$$

$$\underline{v}(\underline{r},t) = e^{i(kz-\omega t)}[\hat{v}_x(z)\underline{x}_0 + \hat{v}_z(z)\underline{z}_0]$$

where $k \equiv \omega/c_0$ and $\underline{x}_0$, $\underline{z}_0$ denote unit vectors. When representations (2) are substituted into (1), a first order system of ordinary differential equations for $\hat{p}$ and $\hat{v}_z$ is obtained. It proves convenient to introduce new dependent (scattering) variables a, b defined as:

$$a \equiv \frac{1}{2}[\hat{p} + (\rho_0 c_0 \sec\theta_0)\hat{v}_z], \quad b \equiv \frac{1}{2}[\hat{p} - (\rho_0 c_0 \sec\theta_0)\hat{v}_z]. \tag{3}$$

In terms of these scattering variables, the problem of interest reduces to the following two-point boundary value problem:

$$\frac{d}{dz}\begin{bmatrix} a \\ b \end{bmatrix} = \frac{ik}{2}\begin{bmatrix} A_{11} & A_{12} \\ -A_{12} & -A_{11} \end{bmatrix}\begin{bmatrix} a \\ b \end{bmatrix}, \quad 0 < z < L \tag{4a}$$

$$a(0) = 1; \quad b(L) = \Gamma a(L) \tag{4b}$$

where

$$A_{\substack{11\\12}} \equiv \pm \tilde{\rho} \cos\theta_0 + \tilde{\rho}^{-1} \sec\theta_0(\tilde{c}^{-2} - \sin^2\theta_0) \tag{4c}$$

$$\tilde{\rho} \equiv \rho/\rho_0 \quad , \quad \tilde{c} \equiv c/c_0$$

The dependent variable $a$ represents an incident or downward-propagating wave while $b$ represents a reflected or upward-propagating wave. Note that a unit strength wave is incident from the water column; $b(0)$ represents the (unknown) water-sediment interface reflection coefficient. $\Gamma$ is the (assumed known) sediment-rock interface reflection coefficient.

Within the sediment layer, the density and sound speed in the lossless case considered in [1] were assumed to vary as:

$$\begin{aligned} \rho(z) &= \rho_1(1 + \eta z) + \delta\rho(z/\ell). \\ c(z) &= c_1(1 + \eta z) + \delta c(z/\ell) \end{aligned} \tag{5}$$

where $\rho_1$, $c_1$ are constants, $\eta$ is a common (constant) gradient parameter and $\delta\rho(\cdot)$, $\delta c(\cdot)$ are zero mean, unit correlation length random fluctuations. The constant $\ell$, then, represents the actual correlation length. Parameter values typical of the Hatteras abyssal plain were noted in [1] to be:

$$\begin{aligned} &\rho_0 = 1 \text{ gm/cm}^3, \quad \rho_1 = 1.5 \text{ gm/cm}^3 \\ &c_0 = c_1 = 1500 \text{ m/sec}, \quad \eta^{-1} \simeq 1500 \text{ m} \approx L, \quad \ell \cong 0.15 \text{ ,} \end{aligned} \tag{6}$$

A realistic sediment model, however, should incorporate small levels of dissipation. Gilbert [3] reports attenuation levels of approximately 0.1 dB/m/kHz. We shall introduce loss into the model by adding a small imaginary component to the mean sound speed [4]; the second of equations (5) will be changed to:

$$c(z) = c_1[(1 + \eta z) - i\nu] + \delta c(z/\ell), \quad \nu \cong 0.00275 \tag{7}$$

As discussed in [1], the frequency regime where the richest interplay between refraction and multiple scattering occurs is on the order of hundreds of Hertz. At significantly lower frequencies scattering is not very pronounced. At significantly higher frequencies, dissipation and intense multiple scattering both act to localize the impinging acoustic energy to a region near the water-sediment interface; appreciable penetration and refraction does not occur.

We shall assume a nominal frequency $f \sim 10^2$Hz; therefore acoustic wavelength $\lambda \equiv c_0/f \sim 15$m. We shall introduce a small dimensionless parameter $\varepsilon$ by defining:

$$\varepsilon^2 \equiv \eta\ell. \tag{8}$$

Note that $\varepsilon = 10^{-2}$ for the parameters specified. Let $\xi \equiv z/\lambda$. Since $\eta\lambda = \ell/\lambda = 10^{-2}$ and $\nu \sim 0(10^{-2})$, the following scaled version of equation (40) will be considered:

$$\frac{d}{d\xi}\begin{bmatrix} a \\ b \end{bmatrix} = i\kappa\, A(\varepsilon\xi, \xi/\varepsilon)\begin{bmatrix} a \\ b \end{bmatrix} - \kappa\varepsilon\bar{a}\, J \begin{matrix} a \\ b \end{matrix}, \quad 0 < \xi < \tau/\varepsilon \tag{9a}$$

where $\kappa$, $\tau \equiv \eta L$ are $0(1)$ constants and:

$$\bar{a} \equiv 2\tilde{\nu}\tilde{\rho}_1^{-1}\tilde{c}_1^{-2}\sec\theta_0(1+\varepsilon\xi)^{-4}$$

$$\tilde{\nu} \equiv \nu/\varepsilon, \quad \tilde{\rho}_1 \equiv \rho_1/\rho_0, \quad \tilde{c}_1 \equiv c_1/c_0 \tag{9b}$$

$$J \equiv \begin{bmatrix} 1 & 1 \\ -1 & -1 \end{bmatrix}$$

Equation (9a) represents a linearized approximation of the original equation (4), with $c(z)$ defined by (7). Coefficient matrix $A$ in (9a) is thus identical to the lossless case (i.e. (4) with $\rho$, $c$ defined by (5)); the dependence of this matrix upon the slow-scale mean refraction variation $(\eta z \to \varepsilon\xi)$ and the rapid-scale random fluctuations $(z/\ell \to \xi/\varepsilon)$ is explicitly noted. Small deterministic dissipation enters the model through the last term (the $\varepsilon\bar{a}$ term) in (9a).

We shall rescale the independent variable in (9), letting $\varepsilon\xi \to \xi$. We shall also assume that coefficient matrix $A$ in (9a) can be written as the sum of a slowly-varying mean $\bar{A}$ and a zero mean random fluctuation $a$. Thus, the scaled version of (4) that we shall ultimately consider is:

$$\frac{d}{d\xi}\begin{bmatrix} a \\ b \end{bmatrix} = \frac{i\kappa}{\varepsilon}[\bar{A}(\xi) + \mathcal{a}(\xi,\xi/\varepsilon^2)]\begin{bmatrix} a \\ b \end{bmatrix} - \kappa\bar{\mathcal{a}}\, J\begin{bmatrix} a \\ b \end{bmatrix}, \quad 0 < \xi < \tau$$

$$a(0) = 1, \quad b(\tau) = \Gamma a(\tau) \tag{10}$$

Let $<\cdot>$ denote expected value. We shall assume (as in [1]) that $<\tilde{\rho}><\tilde{\rho}^{-1}(\tilde{c}^{-2} - \sin^2\theta_0)>$ is a monotonically decreasing function of $\xi$, positive on $[0,\xi_0)$, negative on $(\xi_0,\tau]$ with nonvanishing (negative) derivative at $\xi_0$. Since the eigenvalues of the averaged matrix $\bar{A}$ are $\pm 2[<\tilde{\rho}><\tilde{\rho}^{-1}(\tilde{c}^{-2} - \sin^2\theta_0)>]^{1/2}$, $\xi_0$ represents a turning point. The averaged problem thus exhibits the same general qualitative features as the geoacoustic model in the absence of random fluctuations.

## III. The Averaged Problem:

The averaged problem (or geoacoustic model) will consist of the two point boundary value problem:

$$\frac{d}{d\xi}\begin{bmatrix} \bar{a} \\ \bar{b} \end{bmatrix} = \frac{i\kappa}{\varepsilon}\bar{A}(\xi)\begin{bmatrix} \bar{a} \\ \bar{b} \end{bmatrix} - \kappa\bar{\mathcal{a}}\, J\begin{bmatrix} \bar{a} \\ \bar{b} \end{bmatrix}, \quad 0 < \xi < \tau \tag{11a}$$

$$\bar{a}(0) = 1, \quad \bar{b}(\tau) = \Gamma\bar{a}(\tau) \tag{11b}$$

$$\bar{A}_{\substack{11\\12}} = \pm<\tilde{\rho}>\cos\theta_0 + \sec\theta_0<\tilde{\rho}^{-1}(\tilde{c}^{-2} - \sin^2\theta_0)> \tag{11c}$$

Let $\bar{\Phi}$ denote a unimodular fundamental matrix for (11a); note, however, that $\bar{\Phi} \notin SU(1,1)$ if $\nu \neq 0$. Using the ansatz of Ludwig [5], we can again construct a WKB approximant of $\bar{\Phi}$. For brevity, define:

$$\begin{aligned}
\alpha^2 &\equiv \kappa<\tilde{\rho}\cos\theta_0> = \kappa\tilde{\rho}_1(1+\xi)\cos\theta_0 \\
\beta^2 &\equiv \kappa<\tilde{\rho}^{-1}\sec\theta_0(\tilde{c}^{-2} - \sin^2\theta_0)> \\
\Lambda(\xi) &\equiv 2\int_0^\xi \alpha\beta\, d\sigma, \quad 0 < \xi < \xi_0 \\
\gamma(\xi) &\equiv \kappa\int_0^\xi \bar{\mathcal{a}}\,\frac{\alpha}{\beta}\, d\sigma, \quad 0 < \xi < \xi_0
\end{aligned} \tag{12}$$

Thus $\beta^2$ is positive above the turning point $(0 < \xi < \xi_0)$ and negative below it $(\xi_0 < \xi < \tau)$. In the region above the turning point, the ansatz of Ludwig leads naturally to the WKB approximant:

$$\Phi \sim \begin{bmatrix} \frac{1}{2}[(\frac{\alpha}{\beta})^{1/2}+(\frac{\alpha}{\beta})^{-1/2}]e^{[i\varepsilon^{-1}\Lambda(\xi)-\gamma(\xi)]} & \frac{1}{2}[(\frac{\alpha}{\beta})^{1/2}-(\frac{\alpha}{\beta})^{1/2}]e^{[-i\varepsilon^{-1}\Lambda(\xi)+\gamma(\xi)]} \\ \frac{1}{2}[(\frac{\alpha}{\beta})^{1/2}-(\frac{\alpha}{\beta})^{-1/2}]e^{[i\varepsilon^{-1}\Lambda(\xi)-\gamma(\xi)]} & \frac{1}{2}[(\frac{\alpha}{\beta})^{1/2}+(\frac{\alpha}{\beta})^{-1/2}]e^{[-i\varepsilon^{-1}\Lambda(\xi)+\gamma(\xi)]} \end{bmatrix}$$

$$0 < \xi < \xi_0 \qquad (13)$$

In the transition layer at the turning point $\xi_0$:

$$\alpha^2(\xi) \cong \alpha^2(\xi_0) \equiv \alpha_0^2, \quad \beta^2(\xi) \cong -\mu^2(\xi - \xi_0) \qquad (14a)$$

Define:

$$\begin{aligned} a_0 &\equiv (2\alpha_0\mu/\varepsilon)^{2/3}, \quad b_0 \equiv -i\varepsilon^{1/3}\kappa\bar{\alpha}(\xi_0)(2\alpha_0/\mu^2)^{2/3} \\ \psi_1^{(\pm)}(\cdot) &\equiv \frac{1}{2}[Ai(\cdot) \pm \frac{i\varepsilon a_0}{2\alpha_0^2} Ai'(\cdot)] \\ \psi_2^{(\pm)}(\cdot) &\equiv \frac{1}{2}[Bi(\cdot) \pm \frac{i\varepsilon a_0}{2\alpha_0^2} Bi'(\cdot)] \\ C_1 &\equiv \pi^{1/2}2^{1/6}\alpha_0^{2/3}\mu^{-1/3}\varepsilon^{-1/6}e^{[i\varepsilon^{-1}\Lambda(\xi_0) + i\pi/4 - \gamma(\xi_0)]} \\ C_2 &\equiv C_1^* e^{2\gamma(\xi_0)} \end{aligned} \qquad (14b)$$

The customary linearization of the coefficient matrix $\overline{A}$ and the matching of the solution with (13), leads to the transition layer approximant [6]:

$$\Phi \sim \begin{bmatrix} \psi_1^{(-)}(a_0(\xi-\xi_0) + b_0) \\ \psi_1^{(+)}(a_0(\xi-\xi_0) + b_0) \end{bmatrix} \begin{bmatrix} \psi_2^{(-)}(a_0(\xi-\xi_0) + b_0) \\ \psi_2^{(+)}(a_0(\xi-\xi_0) + b_0) \end{bmatrix} \begin{bmatrix} -iC_1 & iC_2 \\ C_1 & C_2 \end{bmatrix} \qquad (15)$$

In the region below the turning pont $(\xi_0 < \xi < \tau)$, define:

$$\tilde{\Lambda}(\xi) \equiv 2 \int_{\xi_0}^{\xi} \alpha|\beta| d\sigma$$

$$\tilde{\gamma}(\xi) \equiv \kappa \int_{\xi_0}^{\xi} \bar{a} \frac{\alpha}{|\beta|} d\sigma$$

$$k(\xi) = \frac{1}{2} [ ( \frac{\alpha}{|\beta|} )^{1/2} + i ( \frac{\alpha}{|\beta|} )^{-1/2} ] \tag{16}$$

$$K_1 \equiv e^{[i\varepsilon^{-1}\Lambda(\xi_0) + i\pi/4 - \gamma(\xi_0)]}$$

$$K_2 \equiv K_1^* e^{2\gamma(\xi_0)}$$

A WKB approximant for $\Phi$ in this region below the turning point is then:

$$\Phi \sim \begin{bmatrix} k(\xi)e^{[-\varepsilon^{-1}\tilde{\Lambda}(\xi)+i\tilde{\gamma}(\xi)]} & k^*(\xi)e^{[\varepsilon^{-1}\tilde{\Lambda}(\xi)-i\tilde{\gamma}(\xi)]} \\ j^*(\xi)e^{[-\varepsilon^{-1}\tilde{\Lambda}(\xi)+i\tilde{\gamma}(\xi)]} & k(\xi)e^{[\varepsilon^{-1}\tilde{\Lambda}(\xi)-i\tilde{\gamma}(\xi)]} \end{bmatrix} \begin{bmatrix} -\frac{i}{2}K_1 & \frac{i}{2}K_2 \\ K_1 & K_2 \end{bmatrix} \tag{17}$$

## IV. Riccati Equation for the Reflection Coefficient:

We shall formulate the two-point boundary value problem (4) as an initial value problem for the reflection coefficient. Let $r \equiv b/a$; also, let $\sigma \equiv \tau - \xi$ so that scaled distance is measured upward from the sediment-basement interface. Viewed as a function of $\sigma$, $r$ satisfies the following nonlinear stochastic initial value problem:

$$\frac{dr}{d\sigma} = \frac{i\kappa}{\varepsilon}[2A_{11}r + A_{12}(1 + r^2)] - \kappa\bar{a}(1 + r)^2, \quad 0 < \sigma < \tau$$

$$r(0) = \Gamma. \tag{18a}$$

The introduction of polar coordinates, i.e. $r \equiv Re^{i\theta}$ and $\Gamma \equiv R_0e^{i\theta_0}$, leads to the system:

$$\frac{d}{d\sigma} R = \frac{\kappa}{\varepsilon} A_{12}\sin\theta(1 - R^2) - \kappa\bar{a}(2R + \cos\theta(1 + R^2))$$

$$\frac{d}{d\sigma}\theta = 2\frac{\kappa}{\varepsilon}A_{11} + \frac{\kappa}{\varepsilon}A_{12}\cos\theta(R^{-1} + R) + \kappa\bar{a}\sin\theta(R^{-1} - R) \tag{18b}$$

$$R(0) = R_0, \ \theta(0) = \theta_0.$$

We shall also consider the reflection coefficient for the averaged problem (11). Let $\overline{r} \equiv \overline{R}e^{i\overline{\theta}} \equiv \overline{b}/\overline{a}$, where $\overline{a}$ and $\overline{b}$ form the solution of (11). Then, $\overline{R}$ and $\overline{\theta}$ are solutions of the problem:

$$\frac{d}{d\sigma}\overline{R} = \frac{\kappa}{\varepsilon}\overline{A}_{12}\sin\theta(1 - \overline{R}^2) - \kappa\overline{a}(2\overline{R} + \cos\overline{\theta}(1 + \overline{R}^2))$$

$$\frac{d}{d\sigma}\overline{\theta} = 2\frac{\kappa}{\varepsilon}\overline{A}_{11} + \frac{\kappa}{\varepsilon}\overline{A}_{12}\cos\theta\,(\overline{R}^{-1} + \overline{R}) + \kappa\overline{a}\sin\overline{\theta}(\overline{R}^{-1} - \overline{R}) \tag{19a}$$

$$\overline{R}(0) = R_0, \quad \overline{\theta}(0) = \theta_0.$$

The reflection coefficient $\overline{r}$ can also be determined from the fundamental matrix $\overline{\Phi}(\xi) = (\overline{\phi_{ij}}(\xi))$, whose WKB approximation was discussed in Section III. One can show that:

$$\overline{r} = \frac{\overline{\phi}_{21}(\xi)[\overline{\phi}_{22}(\tau) - r\overline{\phi}_{12}(\tau)] - \overline{\phi}_{22}(\xi)[\overline{\phi}_{21}(\tau) - r\overline{\phi}_{11}(\tau)]}{\overline{\phi}_{11}(\xi)[\overline{\phi}_{22}(\tau) - r\overline{\phi}_{12}(\tau)] - \overline{\phi}_{12}(\xi)[\overline{\phi}_{21}(\tau) - r\overline{\phi}_{11}(\tau)]} \tag{19b}$$

Both points of view, i.e. (19a) and (19b), will prove convenient.

As noted in [1], the frequency regime is such that the correlation length of the random fluctuations is much less than acoustic wavelength. We thus introduce the "white noise" approximation:

$$\varepsilon^{-1}\, a_{1j}d\sigma \to \mathcal{B}_{1j}(\sigma)dw, \; j = 1,2 \tag{20}$$

where $\mathcal{B}_{1j}$ is an $O(1)$ deterministic function of $\sigma$ and $w$ is Brownian motion. The equations for the reflection coefficient polar coordinates then become:

$$dR = [\frac{\kappa}{\varepsilon}\overline{A}_{12}\sin\theta(1 - R^2) - \kappa\overline{a}(2R + \cos\theta(1 + R^2))]d\sigma +$$
$$\kappa\mathcal{B}_{12}\sin\theta(1 - R^2)dw \equiv F_1 d\sigma + G_1 dw$$
$$d\theta = [2\frac{\kappa}{\varepsilon}\overline{A}_{11} + \frac{\kappa}{\varepsilon}\overline{A}_{12}\cos\theta(R^{-1} + R) + \kappa\overline{a}\sin\theta(R^{-1} - R)]d\sigma +$$
$$[2\kappa\mathcal{B}_{11} + \kappa\mathcal{B}_{12}\cos\theta(R^{-1} + R)]dw \equiv F_2 d\sigma + G_2 dw \tag{21}$$
$$R(0) = R_0, \quad \theta(0) = \theta_0$$

Equations (21) will be interpretted in the sense of Stratonovich [7]. We thus obtain the following Fokker-Planck equation for the density function $P(\sigma,R,\theta)$ of the $(R,\theta)$ process:

$$\partial_\sigma P = -\partial_R[(F_1 + H_1)P] - \partial_\theta[(F_2 + H_2)P] + \frac{1}{2}\partial^2_{RR}[G_1^2 P]$$
$$+ \partial^2_{R\theta}[G_1 G_2 P] + \frac{1}{2}\partial^2_{\theta\theta}[G_2^2 P]$$

$$P(0,R,\theta) = \delta(R - R_0)\delta(\theta - \theta_0) \tag{22}$$

$$H_j \equiv \frac{1}{2}[G_1 \partial_R G_j + G_2 \partial_\theta G_j], \quad j = 1,2.$$

Our ultimate goal is to solve this initial value problem over the interval $0 < \sigma \leq \tau$. Once $P(\tau,R,\theta)$ is known, the response of the sediment layer (as measured in the water column) can be statistically characterized.

## V. Asymptotic Analysis of the Fokker-Planck Equation:

Before analyzing (22) asymptotically, we shall review the underlying physics of the problem to determine the type of behavior that should be expected. The small level of dissipation incorporated into this model should not qualitatively alter the discussion of [1]. Recall that a turning point is assumed to exist at $\sigma_0 \equiv \tau - \xi_0$. Note also that as the solution $P(\sigma,R,\theta)$ of (22) is evolved upward from the basement toward the water-sediment interface, it can be physically interpretted at each value of $\sigma$ as statistically characterizing reflection from the sediment-rock half-space lying beneath that particular level $\sigma$.

At the sediment-basement interface $(\sigma = 0)$, the probability mass is concentrated at the polar coordinates of the basement reflection coefficient. As $\sigma$ increases, an $O(\varepsilon)$ boundary layer (i.e. an initial layer) is encountered. The reflection coefficient "loses communication" with the basement and the probability mass should rapidly coalesce (with little diffusive spreading) near the boundary of the unit disk at a point depending only upon the local sediment parameters and not upon the basement.

Remark:

For simplicity, a deterministic rock basement model has been adopted. One could easily incorporate a random basement (statistically independent of the sediment) into the discussion. In that case, the initial condition in (22) would be spread out over the disk. In this case as well, one should expect the probability mass to coalesce in the same way as described above as $\sigma$ undergoes an $O(\varepsilon)$ increase from zero.

In the outer region, above the initial layer but below the turning point, the sediment layer is in a cutoff state. Wave propagation, and the associated randomization due to multiple scattering, can not occur. We expect the probability mass to evolve, with little diffusion, as $\sigma$ increases to the turning point value of $\sigma_0$. Once the transition through the turning point is made, the region of multiple scattering is reached. In the outer layer above the turning point ($\sigma_0 < \sigma < \tau$), multiple scattering (tempered by the dissipation) should cause significant diffusive spreading of the probability mass.

In order to quantitatively describe this behavior, consider first the reflection factor $\overline{r} = \overline{R}e^{i\overline{\theta}}$ for the average problem. The two scales of variation of $\overline{r}$ become apparent if one uses (17) in (19b); within an $O(\varepsilon)$ initial layer, $\overline{r}$ adjusts from its initial value of $\Gamma$ to a value - call it $\overline{r}_\infty$ - that depends only upon the local average sediment properties and thus varies on the slower $\sigma$ length scale. To leading order, $\overline{r}_\infty = k^*/k$, where $k$ is defined by (16). This limiting behavior was found in [1] for the lossless model. The effects of dissipation enter through $O(\varepsilon)$ corrections. Let $\overline{r}_\infty \equiv \overline{R}_\infty e^{i\overline{\theta}_\infty}$; then a multiscale analysis of (19a) leads to:

$$\overline{R}_\infty = 1 - \varepsilon\overline{a}|\overline{A}_{12} \sin \overline{\theta}_\infty^{(0)}|^{-1}(1 + \overline{A}_{11}/|\overline{A}_{12}|) + \cdots$$

$$\overline{\theta}_\infty = \overline{\theta}_\infty^{(0)} - \varepsilon(2\kappa|\overline{A}_{12} \sin \overline{\theta}_\infty^{(0)}|)^{-1} \frac{d}{d\sigma} (\overline{A}_{11}/\overline{A}_{12}) + \cdots \qquad (23)$$

$$0 < \sigma < \sigma_0$$

$$\overline{\theta}_\infty^{(0)} \equiv -2 \,\mathrm{Tan}^{-1}(|\beta|/\alpha)$$

In the region below the turning point, $\overline{A}_{12} < 0$ and $|\overline{A}_{12}| > |\overline{A}_{11}|$. Thus, $\overline{R}_\infty < 1$; note also that $\overline{\theta}_\infty^{(0)} = \arg(k^*/k)$. Representation (23) remains valid until the turning point is approached. As $\sigma \uparrow \sigma_0$, $\overline{\theta}_\infty^{(0)} \to 0$ since $|\beta| \to 0$ and approximation (23) breaks down.

Consider now initial value problem (22) in the region below the turning point. In the lossless case discussed in [1], the angle process $\theta$ remained centered about trajectory $\overline{\theta}$ with little diffusive spreading not only through the initial layer but also through the outer layer below the turning point. In the dissipative case, the joint $(R,\theta)$ process behaves in a qualitatively similar way. In particular, consider equation (22) in the region where (23) holds. We introduce fast and slow length scales, $\sigma$ and $\zeta \equiv \sigma/\varepsilon$, as well as scaled variables $\rho$ and $\psi$ defined as follows:

$$\rho \equiv (R - \overline{R}_\infty)/\varepsilon^\mu, \quad \psi \equiv (\theta - \overline{\theta}_\infty)/\varepsilon^\nu \tag{24}$$

where in this context $\mu$ and $\nu$ are exponents that must be determined. Dominant balancing considerations lead to the prescription $\mu \equiv 3/2$ and $\nu \equiv 1/2$. The leading order equation obtained from (22) in this scaling then takes on the form:

$$\begin{aligned}\partial_\zeta P = {} & 2\partial_\rho[(\tilde{a}\rho + [\tilde{b} + \tilde{c}]\psi P] + 2\partial_\psi[(\tilde{a}\psi)P] + 2\partial^2_{\rho\rho}[\tilde{d}^2 P] \\ & - 4\partial^2_{\rho\psi}[(\tilde{d}\tilde{e})P] + 2\partial^2_{\psi\psi}[\tilde{e}^2 P]\end{aligned} \tag{25a}$$

where:

$$\begin{aligned} &\tilde{a} \equiv \kappa\overline{A}_{12} \sin \overline{\theta}_\infty^{(0)} && \tilde{d} \equiv -\kappa\, \mathcal{B}_{12}\overline{R}_\infty^{(1)} \sin \overline{\theta}_\infty^{(0)} \\ &\tilde{b} \equiv -\kappa\overline{A}_{12}\overline{R}_\infty^{(1)} \cos \overline{\theta}_\infty^{(0)} && \tilde{e} \equiv \kappa[\mathcal{B}_{11} + \mathcal{B}_{12} \cos \overline{\theta}_\infty^{(0)}] \\ &\tilde{c} \equiv -\kappa\mathcal{a} \sin \overline{\theta}_\infty^{(0)} && \overline{R}_\infty^{(1)} \equiv \mathcal{a}|\overline{A}_{12}\sin \overline{\theta}_\infty^{(0)}|^{-1}(1 + \overline{A}_{11}/|\overline{A}_{12}|)\end{aligned} \tag{25b}$$

Note that all of the quantities defined in (25b) are positive and depend only upon the slow variable $\sigma$. Equation (25a) can be shown to have the $\zeta$-invariant jointly normal solution:

$$P \sim \exp\{-(\tilde{A}x^2 + \tilde{B}xy + \tilde{C}y^2)\}$$

$$x \equiv (\tilde{d} + \tilde{e})^{-1}(\rho - \psi), \quad \tilde{y} \equiv (\tilde{d} + \tilde{e})^{-1}(\tilde{e}\rho + \tilde{d}\psi)$$

$$\tilde{A} \equiv \tilde{a} \tag{26}$$

$$\tilde{B} \equiv -2\tilde{a}(\tilde{a} + \tilde{e}\tilde{f})/\tilde{e}^2\tilde{f}, \quad \tilde{f} \equiv (\tilde{b} + \tilde{c})/(\tilde{d} + \tilde{e})$$

$$\tilde{C} \equiv \tilde{a}(2\tilde{a}(\tilde{a} + \tilde{e}\tilde{f}) + (\widetilde{ef})^2)/\tilde{e}^2\tilde{f}$$

To leading order, therefore, the probability mass remains localized about the centering trajectory in the region below the turning point. The centering trajectory, in turn, varies as a function of $\sigma$ according to the dictates of the mean local sediment properties (c.f. (23)). Again, however, approximation (26) breaks down near the turning point; $\tilde{A}$, $\tilde{B}$ and $\tilde{C}$ all tend to zero as $\sigma \uparrow \sigma_0$.

Consider now the transition region about the turning point. In this region the reflection coefficient for the averaged problem can be determined by using (15) and (19b); the result is:

$$r \sim \frac{\psi_1^{(+)}}{\psi_1^{(-)}} = \frac{Ai + i\varepsilon(a_0/2\alpha_0^2)Ai'}{Ai - i\varepsilon(a_0/2\alpha_0^2)Ai'} \tag{27a}$$

where the argument in (27) is:

$$\alpha_0(\sigma_0 - \sigma) + b_0 = (2\alpha_0\mu/\varepsilon)^{2/3}(\sigma_0 - \sigma) - i\varepsilon^{1/3}\overline{\kappa a}(\xi_0)(2\alpha_0/\mu^2)^{2/3} \tag{27b}$$

In the transition layer beneath the turning point, Ai and Ai' are basically nonoscillatory. In this region, the polar coordinates of $\bar{r}$ can be found from (27) to behave as:

$$\bar{R} \sim 1 - \varepsilon^{2/3}\bar{\rho}, \quad \bar{\theta} \sim -\varepsilon^{1/3}\bar{\psi} \tag{28}$$

We also introduce the transition region scaled length variable $\eta \equiv (\sigma - \sigma_0)/\varepsilon^{2/3}$. Note that (23) "matches up" with (28) in this scaling since $|\beta|$ becomes $O(\varepsilon^{1/3})$. We again introduce scaled variables and invoke dominant balance considerations to determine the appropriate scales; the scaled variables appropriate to this lower half of the transition region (and again called $\rho$ and $\psi$) are

$$\rho \equiv (\overline{R} - R)\varepsilon^{2/3}, \quad \psi = (\theta - \overline{\theta})/\varepsilon^{1/3}. \tag{29}$$

Using (28) and (29), the leading order behavior of (22) in this region becomes:

$$\partial_\eta P = -\partial_\rho[\kappa\overline{A}_{12}(-2\overline{\psi}\rho + 2\overline{\rho}\psi)P] - \partial_\psi[\kappa\overline{A}_{12}(-\rho + 2\overline{\psi}\psi)P]$$
$$+ \partial^2_{\psi\psi}[2\kappa^2(\mathcal{B}_{11} + \mathcal{B}_{12})^2 P] \tag{30}$$

where $\overline{A}_{12}$, $\mathcal{B}_{11}$ and $\mathcal{B}_{12}$ are evaluated at the turning point. Note that (30) is degenerate, involving diffusion only in angle.

Once the turning point is reached, actual wave propagation and multiple scattering can begin to occur. This multiple scattering, caused by the random layering, tends to diffuse the probability mass; dissipation, to the extent that it tends to repress or modify this activity, also plays a more important role above the turning point. Above the turning point, it becomes preferable to remove the rapid oscillations by a change of dependent variable. Let $\overline{\Phi}_0$ denote the fundamental matrix obtained from $\overline{\Phi}$ by letting the dissipation $\nu$ vanish; the WKB approximant for $\overline{\Phi}_0$ can be obtained from (12)-(17) by setting $\overline{\mathcal{A}} \equiv 0$. We introduce the new dependent variables $\hat{a}$, $\hat{b}$ by:

$$\begin{bmatrix} a \\ b \end{bmatrix} \equiv \overline{\phi}_0 \begin{bmatrix} \hat{a} \\ \hat{b} \end{bmatrix} . \tag{31a}$$

Note that $\phi_0 \in SU(1,1)$. Along with these "slowly-varying" scattering variables $\hat{a}$, $\hat{b}$, we introduce a new reflection coefficient:

$$\hat{r} \equiv e^{-i2\varepsilon^{-1}\Lambda(\xi_0)}(\hat{b}/\hat{a}) \equiv \hat{R}e^{i\hat{\theta}}. \tag{31b}$$

Reflection coefficient $\hat{r}$ satisfies a Riccati equation similar to (18). After again introducing white noise approximation (20), the polar coordinates of $\hat{r}$ are found to satisfy the equations:

$$dR = [-2\chi_{11}\hat{R} + (\psi_{12}\sin\hat{\theta} - \chi_{12}\cos\hat{\theta})(1+\hat{R}^2)]d\sigma +$$

$$[(\mu_{12}\sin\hat{\theta} + \nu_{12}\cos\hat{\theta})(1-\hat{R}^2)]dw \equiv F_1 d\sigma + G_1 dw \tag{32a}$$

$$d\hat{\theta} = [(\chi_{12}\sin\hat{\theta} + \psi_{12}\cos\hat{\theta})(\hat{R}^{-1} - \hat{R})]d\sigma +$$

$$[2\mu_{11} + (\mu_{12}\cos\hat{\theta} - \nu_{12}\sin\hat{\theta})(\hat{R}^{-1} + \hat{R})]dw \equiv F_2 d\sigma + G_2 dw$$

where:

$$\mu_{11} \equiv \kappa[\mathcal{B}_{11}(|\bar{\phi}_{11_0}|^2 + |\bar{\phi}_{12_0}|^2) + 2\mathcal{B}_{12}\mathrm{Re}\{\bar{\phi}_{11_0}\bar{\phi}_{12_0}\}]$$

$$\mu_{12} + i\nu_{12} \equiv \kappa[\mathcal{B}_{11}(2\bar{\phi}^*_{11_0}\bar{\phi}_{12_0}) + \mathcal{B}_{12}((\bar{\phi}^*_{11_0})^2 + \bar{\phi}^2_{12_0})]$$

$$\chi_{11} \equiv \kappa\bar{a}[|\bar{\phi}_{11_0}|^2 + |\bar{\phi}_{12_0}|^2 + 2\mathrm{Re}\{\bar{\phi}_{11_0}\bar{\phi}_{12_0}\}] \tag{32b}$$

$$\chi_{12} + i\psi_{12} \equiv \kappa\bar{a}[(\bar{\phi}^*_{11_0})^2 + \bar{\phi}^2_{12_0} + 2\bar{\phi}^*_{11_0}\bar{\phi}_{12_0}].$$

Interpretting these equations in the sense of Stratonovich leads again to a Fokker-Planck equation having the structure of (22). Consider first the outer region above the turning point, $\sigma_0 < \sigma < \tau$. Using approximant (13) with $\gamma \equiv 0$ for $\bar{\Phi}_0$ leads to a Fokker-Planck equation with rapidly-oscillating coefficients; the functions $F_j$, $G_j$ and $H_j$, $j = 1,2$, involve trigonometric dependence upon the argument $\hat{\theta} + 2\varepsilon^{-1}(\Lambda(\xi_0) - \Lambda(\xi))$. Let $\hat{P}(\sigma,\hat{R},\hat{\theta})$ denote the density function for the $\hat{r}$ process. Use of the Khasminskii averaging approximation [8] leads to the following simplified description:

$$\partial_\sigma\hat{P} = \partial_{\hat{R}}[[2\kappa\bar{a}\, e^{X}\hat{R} - \frac{\kappa^2}{4}\eta_1^2\hat{R}^{-1}(1-\hat{R}^2)^2]\hat{P}] + \partial^2_{\hat{R}\hat{R}}[\frac{\kappa^2}{4}\eta_1^2(1-\hat{R}^2)^2\hat{P}] +$$

$$+ \partial^2_{\hat{\theta}\hat{\theta}}[[2\kappa^2\eta_2^2 + \frac{\kappa^2}{4}\eta_1^2(\hat{R}^{-1} + \hat{R})^2]\hat{P}]$$

$$e^X \equiv \alpha/\beta \tag{33}$$

$$\eta_1 \equiv \mathcal{B}_{11}\sinh X + \mathcal{B}_{12}\cosh X$$

$$\eta_2 \equiv \mathcal{B}_{11}\cosh X + \mathcal{B}_{12}\sinh X$$

By using a Fourier series decomposition, $\hat{P} \equiv \sum_{m=-\infty}^{\infty} P_m(\hat{\sigma},\hat{R})e^{im\hat{\theta}}$, one can further reduce (33) to a decoupled system of equations for each of the Fourier components $\hat{P}_m$. Note that dissipative and random scattering effects both influence the drift term in (33). Dissipation tends to drive the probability mass toward the center of the disk. Random scattering counteracts this tendency within the drift term; it also acts to diffuse the mass through the disk.

Within the transition layer above the turning point, one can use (14), (15), with $\bar{a}$ and $\gamma$ set equal to zero, for $\bar{\Phi}_0$ in (32). The resulting Fokker-Planck equation, however, is unwieldy; no significant simplification can be achieved.

The asymptotic analysis presented in this section is not self-contained in the sense of providing a complete, simplified approximate description. It does, however, delineate the scaling regimes of the problem and provide the basis for combined asymptotic-numerical approach. One can use (26) as an initial condition for a numerical solution through the transition region about the turning point. The decoupled system of averaged equations for the Fourier components $\{\hat{P}_m(\sigma,\hat{R})\}$ can then be solved numerically to evolve the probability density through the outer layer above the turning point to the sediment-water interface at $\sigma = \tau$.

## VI. Conclusions:

Numerical implementation of the program outlined in the prior section will permit one to statistically characterize the water-sediment interface reflection coefficient. Thus, one will be able to characterize the reflected signal in the water column. These computations will not provide insight, however, on the statistical nature of the fields themselves within the sediment layer.

Figures 2-4 present the results of numerical simulations performed to study the behavior of both reflection coefficient and internal fields (c.f. [1] for details of the simulations themselves.) These figures show the response of a 500 meter randomly-layered sediment to a 150 Hz. acoustic plane wave incident at 60° from the normal. The turning point occurs at a nominal depth of 232 meters from the water-sediment interface. Results are given for both the lossless and dissipative cases to illustrate the effect of small, realistic amounts of attenuation.

In general, the dissipation tends to substantially suppress the fluctuations that manifest themselves in the lossless case. Note in Figure 2 that the reflection factor fluctuations $\langle|r - \langle r\rangle|^2\rangle$ increase somewhat abruptly above the turning point; this is true in both lossless and dissipative cases and is consistent with our analysis. Figures 3 and 4, on the other hand, point out that this knowledge of $r = b/a$ does not provide any real insight into the behavior of $a$ and $b$ individually. In each random realization, field intensity increases as the acoustic rays approach the turning point. Because of the random layering, the actual turning point depth will vary slightly from realization to realization. It is obvious from the figures that dissipation acts to markedly suppress this peaking of the field strength.

It would be very desirable to have a stochastic theory that could quantitatively characterize the fields themselves within the sediment layer (in both the lossless and slightly dissipative cases). At present, such a theory does not exist.

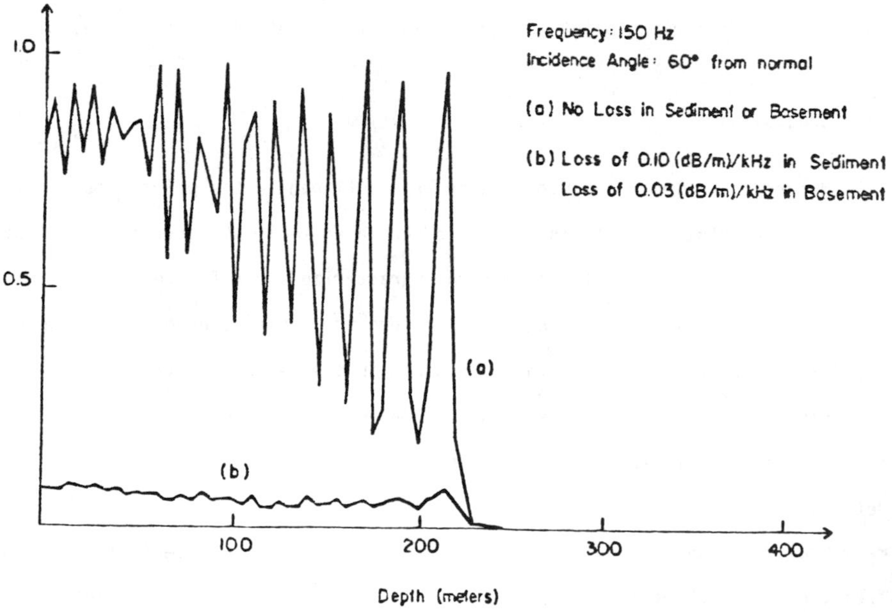


Figure 2 : CENTERED SECOND MOMENT OF THE REFLECTION COEFFICIENT v.s. DEPTH

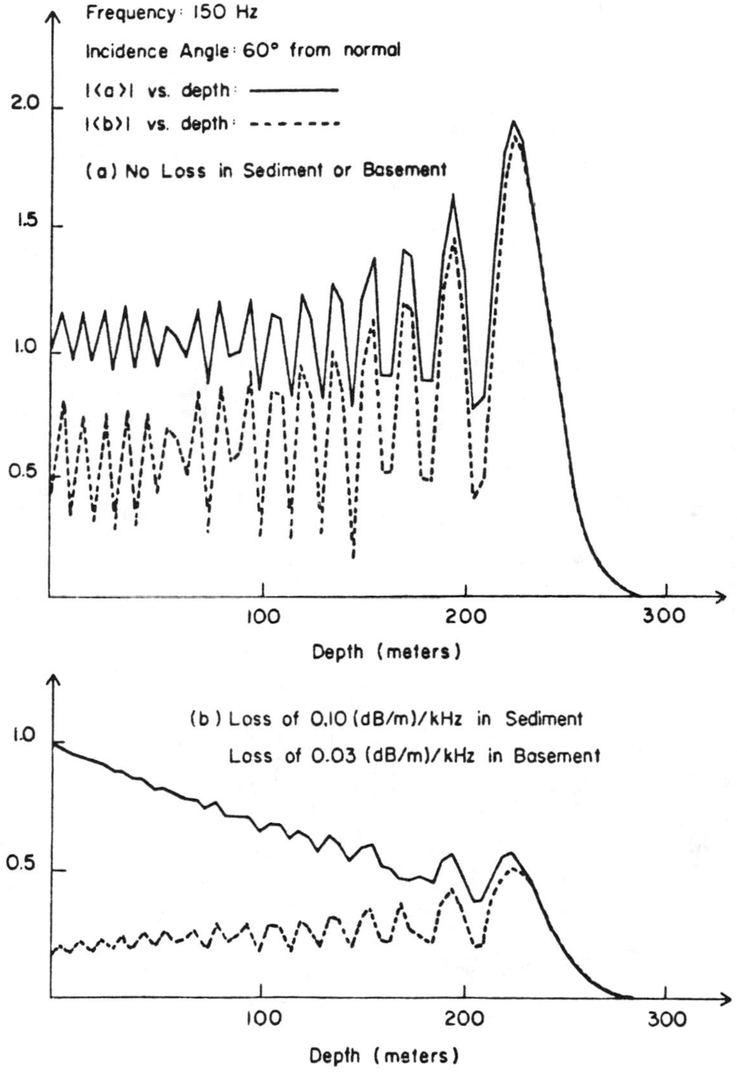


Figure 3 : COHERENT FIELD MODULUS vs DEPTH

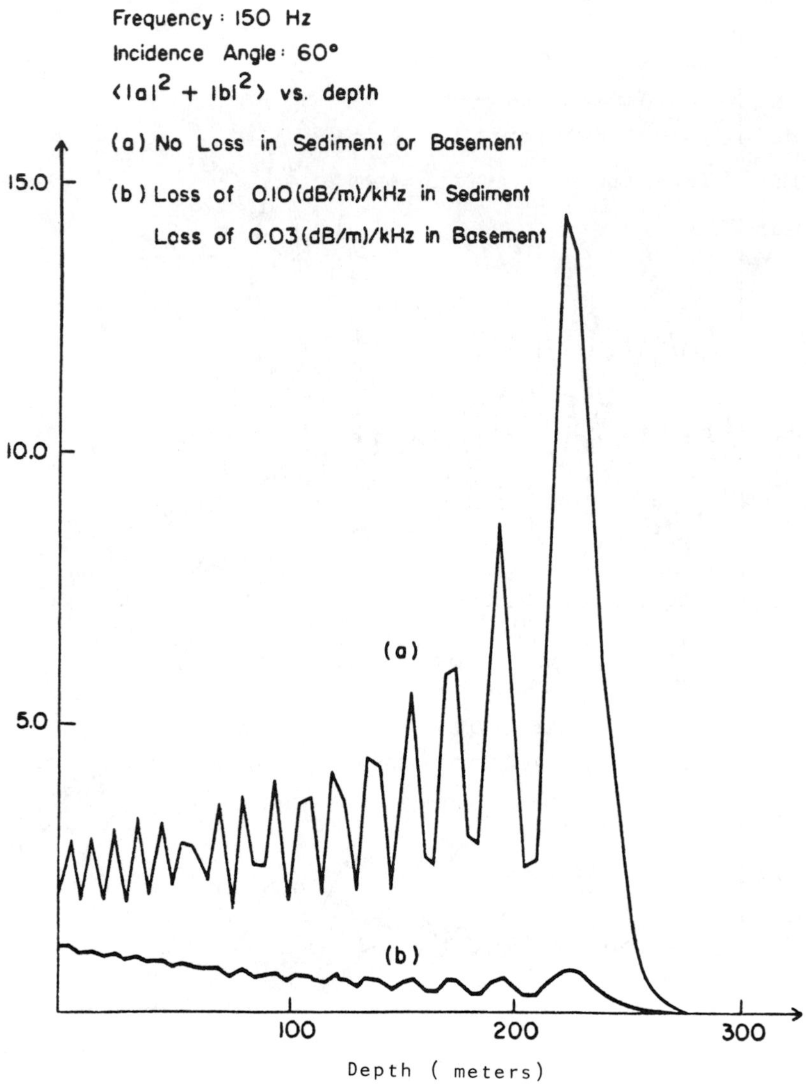


Figure 4 : TOTAL FIELD INTENSITY vs DEPTH

## Acknowledgement

This research was supported by the Office of Naval Research under Contract Nos. N00014-76-C-0056P00004 and N 00014-85-K-0629. The author gratefully acknowledges the computing assistance of Sam Robinson.


## References


1. W.E. Kohler, Reflection from a One-Dimensional, Totally Refracting Random Multilayer, SIAM J. Appl. Math., 46 (1986), pp. 464-482.

2. H. Freese, private communication.

3. K.E. Gilbert, Reflection of Sound from a Randomly Layered Ocean Bottom, J. Acoust. Soc. Amer., 68 (1980), pp. 1454-1458.

4. H.E. Morris, Bottom-Reflection-Loss Model with a Velocity Gradient, J. Acoust. Soc. Amer., 48 (1970), pp. 1198-1202.

5. D. Ludwig, Modified W.K.B. Method for Coupled Ionospheric Equations, J. Atmos. Terr. Phys., 32 (1970), pp. 991-998.

6. C.M. Bender and S.A. Orzag, Advanced Mathematical Methods for Scientists and Engineers, McGraw-Hill, New York (1978).

7. L. Arnold, Stochastic Differential Equations: Theory and Applications, J. Wiley and Sons, New York (1974).

8. R.Z. Khasminskii, Principle of Averaging for Parabolic and Elliptic Differential Equations and for Markov Processes with Small Diffusion, Theory of Prob. and Applic., 8 (1963), pp. 1-21.

# PROBLEMS OF ONE-DIMENSIONAL RANDOM SCHRÖDINGER OPERATORS

Shinichi Kotani

Kyoto University

In this note the author collects the known facts and presents several open problems in the study of one-dimensional random Schrödinger operators. Let $\Omega = L^2_{real}(R, \frac{dx}{1+|x|^3})$. Then for $q \in \Omega$ we can define a Schrödinger operator:

$$L(q) = -\frac{d^2}{dx^2} + q$$

in $L^2(R, dx)$ as self-adjoint operator. For a probability measure $P$ on $\Omega$ we consider the following three conditions:

(1) shift invariance: $P(T_x A) = P(A)$ for any $x \in R$,

(2) integrability: $\int_\Omega \{ \int_0^1 q(x)^2 dx\} P(dq) < +\infty$,

(3) ergodicity: $P(T_x A \ominus A) = 0$ For any $x \in R \Rightarrow P(A) = 0$ or $1$, where $T_x : \Omega \to \Omega$ is defined by $T_x q(\cdot) = q(\cdot + x)$. Let

$$\mathcal{P} = \{P;\ \text{probability measure on } \Omega \text{ satisfying (1) and (2)}\},$$

$$\mathcal{P}_e = \{P \in \mathcal{P};\ (3)\}.$$

Problems we have in mind are various spectral properties of $L(q)$ which are valid with probability one under various probability measures from $\mathcal{P}_e$. If we introduce the unitary operator $U_x$ in $L^2(R, dx)$ by $U_x f(\cdot) = f(\cdot + x)$, then we see

$$L(T_x q) = U_x L(q) U_x^{-1} .$$

Therefore we easily obtain

<u>Theorem 1</u> (L.A. Pastur [4]) For $P \in \mathcal{P}_e$, there exist closed sets $\Sigma$, $\Sigma_{a.c}$, $\Sigma_{s.c}$, $\Sigma_p$ of $R$ such that for a.e. $q \in \Omega$ with respect to $P$, $\Sigma(q) = \Sigma$, $\Sigma_j(q) = \Sigma_j$ for $j = a.c, s.c, p$. $\Sigma_j(q)$ denote the spectrum, the absolutely continuous spectrum, the singular continuous spectrum and the point spectrum for $L(q)$ respectively.

To describe the absolutely continuous spectrum we introduce "Floquet exponent" $w(\lambda)$. Let $g_\lambda(x,y,q)$ be the Green kernel for $L(q) - \lambda$ with $\lambda \in \mathcal{C}_+$. For $P \in \mathcal{P}_e$, it is known (see [3]) that for $\lambda \in \mathcal{C}_+$ fixed

$$g_\lambda(x,y,q) \sim e^{|x-y|w(\lambda) + o(|x-y|)}$$

is valid as $|x-y| \to \infty$ for a.e. $q \in \Omega$. R. Johnson - J. Moser [5] showed that

$$w(\lambda) = \int_\Omega \{-2g_\lambda(0,0,q)\}^{-1} P(dq),$$

$$w'(\lambda) = \int_\Omega g_\lambda (0,0,q)P(dq) \quad .$$

Therefore if we introduce a class $W$ of holomorphic functions on $\mathcal{C}_+$ by

$$W = \{w;\ \mathrm{Im}w(\lambda) > 0,\ \mathrm{Re}w(\lambda) < 0,\ I_m w'(\lambda) > 0 \text{ on } \mathcal{C}_+ \}.$$

Then we easily see from the above relations that the $w(\lambda)$ coming from $(L(q),P)$ satisfies $w \subset W$. Then we can show the following existence of the limits:

$$\gamma(\xi) = -\mathrm{Re}\, w(\xi+i0), \quad \pi n(\xi) = \mathrm{Im}\, w(\xi+i0).$$

for $\xi \in \mathbf{R}$, and they are called "Lyapunov exponent" and "Integrated density of states" for the random system $(L(q), P)$. We have

Theorem 2. (L.A. Pastur [4]) For $P \in \mathcal{P}_e$

$$\Sigma = \mathrm{Supp}\, dn.$$

Theorem 3. (K. Ishii [1], L.A. Pastur [4], S. Kotani [2]).

For $P \in \mathcal{P}_e$, $\Sigma_{a,c} = \overline{\{\xi \in \mathbf{R};\ \gamma(\xi) = 0\}}^\mu$,

where $\overline{A}^\mu = \{x \in \mathbf{R};\ \mu(U \cap A) > 0$ for any neighbourhood $U$ of $x\}$ and $\mu$ is Lebesgue measure.

Now even for $P \in \mathcal{P}$, we define

$$w(\lambda) = w_P(\lambda) = \int_\Omega \{-2g_\lambda(0,0,q)\}^{-1} P(dq).$$

Then we have.

Theorem 4. (S. Kotani [3]) For a $\in$ w W to be $w = w_P$ with some $P \in \mathcal{P}$, it is necessary and sufficient that w satisfies

$$-i\sqrt{\lambda}w(\lambda) = \lambda + m + O\left(\frac{1}{|\lambda|}\right) \quad \text{as} \quad \lambda \to \infty$$

with $m \in \mathbf{R}$.

Theorem 5 (S. Kotani [3]) If $w \in W$ satisfies $\gamma(\xi) = 0$ a.e., with respect to $d\eta$ besides the condition mentioned in Theorem 4, then we can choose $P \in \mathcal{P}_e$ such that $w = w_p$.

Now we briefly sketch future problems in the above context. Since it may be hopeless to get information about $\Sigma_{s.c}$ or $\Sigma_p$ only from w, taking Th.2 and Th.3 into account, we had better consider the following subclass of W:

$$W_0 = \{w \in W;\ \gamma(\xi) = 0 \text{ a.e. } d\eta,\ - i\sqrt{\lambda}w(\lambda) = \lambda + m + O\left(\frac{1}{|\lambda|}\right)\}.$$

First we remark that $\Sigma$ = supp $d\eta$ uniquely determines $w \in W_0$. This is because; if we have two $w_1, w_2 \in W_0$ with $\Sigma = \text{supp } d\eta_1 = \text{supp } d\eta_2$, then $\gamma_1(\xi) = \gamma_2(\xi) = 0$ on $\Sigma$, modulo logarithmic capacity zero. (This comes from the subharmonicity of $\gamma_1, \gamma_2$ and the smooth property of $d\eta_1, d\eta_2$ in the sense of potential theory.) Hence $\gamma_1(\xi) = 0$ a.e., $d\eta_2$ and $\gamma_2(\xi) = 0$ a.e. $d\eta_1$. On the other hand, we have (S. Kotani [3])

$$\int_{\mathbb{C}_+} (w_1'(\lambda) - w_2'(\lambda))^2 d\lambda = - \pi \int_R \{\gamma_1(\xi) - \gamma_2(\xi)\}\{d\eta_1(\xi) - d\eta_2(\xi)\}$$

for any $w_1, w_2 \in W$ satisfying $- i\sqrt{\lambda}w(\lambda) = \lambda + m + O\left(\frac{1}{|\lambda|}\right)$.

Therefore we easily see $w_1 = w_2$. So the first problem is

Prob. 1. To characterize a closed set $\Sigma$ of R coming from $w \in W_0$.

As a necessary condition we know already that $\Sigma$ should fulfill $\overline{\Sigma}^{cap} = \Sigma$. And in a discrete version of random Schrödinger operators we know this is necessary and sufficient if $\Sigma$ is a compact set.

Prob. 2. To characterize $w \in W_0$ or $\Sigma \subset R$ (closed set) such that any $P \in \mathcal{P}_e$ with $w = w_P$ has only absolutely continuous spectrum.

Prob. 3. For a given $w \in W_0$ solving Prob. 2, to investigate the structure of $\{P \in \mathcal{P}_e;\ w_P = w\}$. Although a part of the above problems will be solved in a forthcoming paper, a complete solution to the Prob. 1 ~ 3 will generalize the results obtained in the study of periodic inverse spectral problems.

## References


[1] K. Ishii: Localization of eigenstates and transport phenomena in one-dimensional disordered systems, Supp. Prog. Theor. Phys., 53(1973) 77-138.

[2] S. Kotani: Ljapunov indices determine absolutely continuous spectra of stationary random one-dimensional Schrödinger operators, Proc. Taniguchi Symp. S.A. Katata (1982) 225-247.

[3] __________: One-dimensional random Schrödinger operators and Herglotz functions, Proc. Taniguchi Symp. Prob. method of Math. Phys., Katata (1985), to appear.

[4] L.A. Pastur: Spectral properties of disordered systems in one-body approximation, Comm. Math. Phys., 75 (1980) 167-196.

[5] R. Johnson - J. Moser: The rotation number of almost periodic potentials, Comm. Math. Phys., 84 (1982) 403-438.


**Correction:** The author has noticed that $\Sigma$ does not determine $w \subset W_0$ but $N = \{\xi \subset R\ ;\ \gamma(\xi) = 0\}$. Therefore in Prob. 1, 2 we should replace $\Sigma$ by $N$.

# Numerical Simulation of Itô Stochastic Differential Equations on Supercomputers

W. P. Petersen
9C-01, BCS, 565 Andover Park W
Tukwila, WA 98188

ABSTRACT. Second order Monte-Carlo simulations of Itô stochastic differential equations are described. Stability and the importance of sample size in approximating the Brownian motion are discussed along with implications for implementation on vector computers.

**1. INTRODUCTION.** In the following we consider second order accurate algorithms for numerical simulation of Itô stochastic differential equations. There is some review of recent work [1,2] done in collaboration with J. R. Klauder. Further, we consider long time stability of solutions in implicit and explicit rules.

1.1 ITÔ RULES. In the Itô formulation for the interpretation of the stochastic differential equation (SDE)

$$dx = b(x)\,dt + \sigma(x)\,dw(t) \tag{1.1}$$

as a stochastic integral

$$x_t = x_0 + \int_0^t b(x(\tau))\,d\tau + \int_0^t \sigma(x(\tau))dw(\tau) \tag{1.2}$$

the integration over the Brownian motion w(t) is defined to be non-anticipating. That is,

$$\int_0^t \sigma(x(t))dw(t) = \lim_{\Delta t \to 0} \sum_k \sigma(x(t_k))\{w(t_{k+1}) - w(t_k)\}$$

where $\Delta t = \max(t_{k+1}-t_k)$ on the partition $0 = t_0<t_1<\ldots<t_N = t$. For the case that the drift $b(x) \equiv 0$ in (1.1), the resulting process

$$x_t = x_0 + \int_0^t \sigma(x(\tau))dw(\tau)$$

is a martingale, namely the expectation

$$\langle x_t \rangle = \mathbf{E}x(t) = x_0.$$

It is desireable for a numerical simulation that the discretized process should have this property.

**2. DISCRETE APPROXIMATIONS.** First, we need a discrete model for dw(t) with the Itô rules [4],

$$\langle dw(t) \rangle = 0$$

and

$$dw(t)^2 = dt.$$

Our model is

$$\Delta w = \sqrt{h}\, z \qquad (1.3)$$

where h is the time step, and z is a gaussian random variable of mean zero and unit variance. A sample of 64 Wiener paths approximated by this model ( h = 1/10) is shown in **Figure 1**.

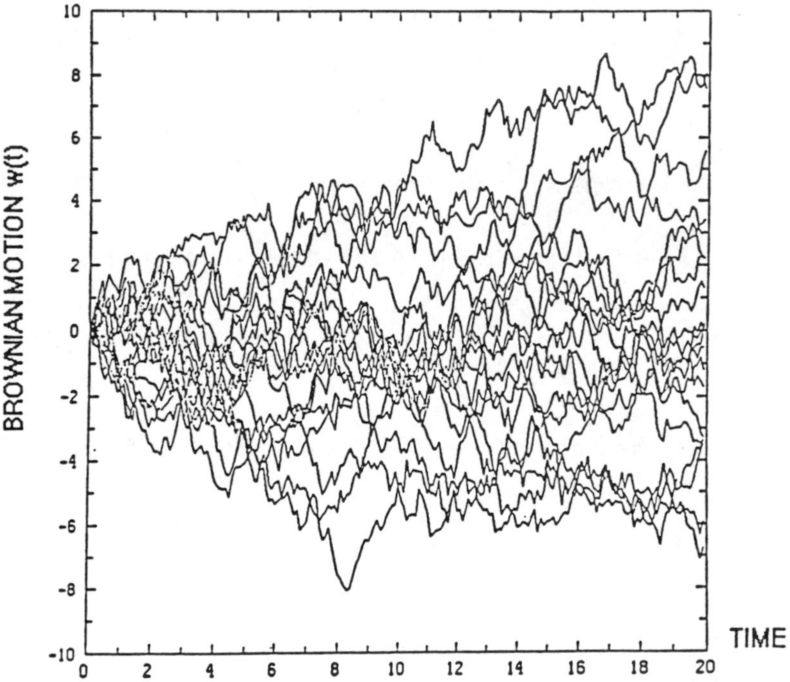


Figure 1

At each new time step $t \geq s + h$, $\Delta w(t)$ will be uncorrelated with any process $f(s)$. Namely,

$$\langle f(s)\, \Delta w(t) \rangle = \langle f(s) \rangle \langle \Delta w(t) \rangle = 0$$

and

$$\langle f(s)\, \Delta w(t)^2 \rangle = \langle f(s) \rangle \langle \Delta w(t)^2 \rangle = h \langle f(s) \rangle.$$

It follows that in a distribution sense

$$\Delta w(t)^2 = h \approx dt.$$

Generation of the z's will be discussed in section (3).

2.1 FORWARD EULER APPROXIMATION. As a first approximation to (1.2) the forward Euler rule is

$$x_h = x_0 + h\, b(x_0) + \sqrt{h}\, \sigma(x_0)\, z,$$

where a new z is generated at each time step. For $b(x) \equiv 0$, the Euler approximation yields a martingale for the discrete process

$$\{x_k\} = \{ x(t_k) \mid t_k = kh,\ k = 0, 1, \ldots \}.$$

On the other hand, the Euler method is only $O(h)$ accurate in $\langle x^2 \rangle$, easily seen from the example

$$x(t) = e^{w - t/2} \tag{2.1}$$

where the SDE is

$$dx = x\, dw.$$

Here x is a martingale

$$\langle x_t \rangle = \exp((\langle w^2 \rangle - t)/2) = 1$$

and

$$\langle x_t^2 \rangle = e^t.$$

The forward Euler approximation for $\langle x^2 \rangle$ at $t = h$ is

$$\langle x^2(h) \rangle = \langle (1 + \sqrt{h}\ z)^2 \rangle$$
$$= 1 + h$$

which is $e^h$ only to $O(h)$, motivating a more accurate procedure.

2.2 A RUNGE-KUTTA METHOD. Again consider the martingale case

$$dx = \sigma(x)\, dw$$

for which the Euler approximation is

$$x_h = x_0 + \sqrt{h}\,\sigma(x_0)\, z.$$

Instead of of the Euler 'predictor' $x_0$ argument of $\sigma(x)$, let's try

$$x_h = x_0 + \sqrt{h}\,\sigma(x_0 + \eta\sqrt{h}\,\sigma(x_0)\, z)\, z$$

where the predictor, $x_0 + \eta\sqrt{h}\,\sigma(x_0)z$ has a parameter $\eta$ to be determined. The expected value is

$$\langle x_h \rangle = x_0 + \eta h \langle \sigma_0 \sigma_0' \rangle + O(h^2).$$

To this order, $O(h^2)$, this predictor is an Euler approximation of step $\Delta w(h\eta) = \eta\,\Delta w(h)$ as an $\eta h$ step linear interpolant [3] to $\Delta w$. We notice that although $\langle \Delta w(h\eta)\,\Delta w(h) \rangle = \eta h$, $\langle \Delta w(h\eta)^2 \rangle = \eta^2 h$ for this interpolant. If $\{x(kh), k = 0, 1, \ldots\}$ is a martingale, $\eta \equiv 0$, so we still get forward Euler.

Modifying the above predictor to $x_0 + \eta\sqrt{h}\,\sigma(x_0)\, z_1$ with another **independent** gaussian random variable $z_1$, on the other hand, yields

$$x_h = x_0 + \sqrt{h}\,\sigma(x_0 + \eta\sqrt{h}\,\sigma(x_0)\, z_1)\, z_0 \qquad (2.2)$$

for which $\langle x_h \rangle = x_0$, independent of h. Now parameter $\eta$ is easily determined from our example (2.1), $x = \exp(w - t/2)$ by

$$\langle (1 + \sqrt{h}(1 + \eta\sqrt{h}\, z_1)\, z_0)^2 \rangle$$

$$= 1 + h + \eta^2 h^2$$

$$= e^h + O(h^3)$$

for the choice $\eta = \pm 1/\sqrt{2}$ . The distribution for $z_1$ is symmetric so the sign is unimportant. This predictor is an Euler approximation in the sense that it uses a $\Delta w(h/2)$ which is **uncorrelated** with $\Delta w(h)$, namely $\langle \Delta w(h/2)\, \Delta w(h) \rangle = 0$, but $\langle \Delta w(h/2)^2 \rangle = h/2$ and is consistent with the model (1.3). More generally [1], an explicit Runge Kutta rule

$$x_h = x_0 + \frac{h}{2}\left( b(x_0) + b(x_0 + hb(x_0) + \sqrt{h}\,\sigma(x_0)\, z_0)\right) \qquad (2.3)$$

$$+ \frac{\sqrt{h}}{2}\left( \sigma(x_0 + \sqrt{h/2}\,\sigma(x_0)\, z_1) + \sigma(x_0 + h\, b(x_0) + \sqrt{h/2}\,\sigma(x_0)\, z_1)\right) z_0$$

for problem (1.1) is $O(h^2)$ accurate in mean and variance. No such claims are made for higher moments, however. Again, using our martingale example (2.1), $\langle x(h)^3 \rangle = e^{3h}$, while (2.2) gives $\langle x(h)^3 \rangle = 1 + 3h + 3h^2/2$, which is only $O(h)$ accurate. But, $\langle x(h)^m \rangle$ **is however**, $O(h)$ accurate, $\forall\, m \geq 1$.

2.3 STABILITY AND LONG TIME BEHAVIOR OF THE DISCRETE PROCESS. Are samples of (1.1) solution paths stable? In particular, is **zero** a stable solution of the linearization

$$b(x) = \beta x + O(x^2)$$
$$\sigma(x) = \alpha x + O(x^2)$$

of (1.1)? Zero is not a stable solution if $\langle |x|^m \rangle$ is growing for some $m \geq 1$, given some initial perturbation $x_0$. The Itô solution of

$$dx = \beta x\, dt + \alpha x\, dw \qquad (2.4)$$

is

$$x = x_0\, e^{\alpha w + (\beta - \alpha^2/2)\, t} \qquad (2.5)$$

and x is a martingale if $\langle x(t) \rangle = x_0$, or $\beta = 0$. So, moments $\langle |x|^m \rangle$ won't grow if

$$m \cdot \mathbf{Re}\,( m\, \alpha^2/2 + \beta - \alpha^2/2) < 0 \qquad (2.6)$$

which for arbitrary $m \geq 1$ requires $\mathrm{Re}\,\alpha^2 < 0$; that is, for example $\alpha$ **purely imaginary**. **Figures 2**[a], **2**[b] show results using (2.3) for $\beta = 1$ and $\alpha = 1/4$. Monte Carlo results clearly show stable accurate averages of 10240 paths. Conversely, our martingale example (2.1), for which parameters $\alpha = 1$, and $\beta = -1/2$, exhibits relative instability since (2.6) is not satisfied for $m \geq 1$. **Figure 3** illustrates some results for 20480 paths using (2.3).

REAL(MOMENTS) vs. TIME STEP

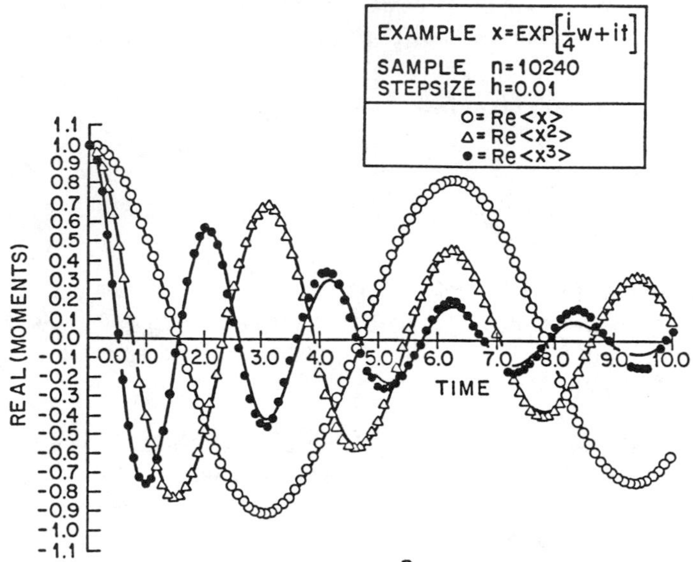


Figure 2[a]

IMAG(MOMENTS) vs. TIME STEP

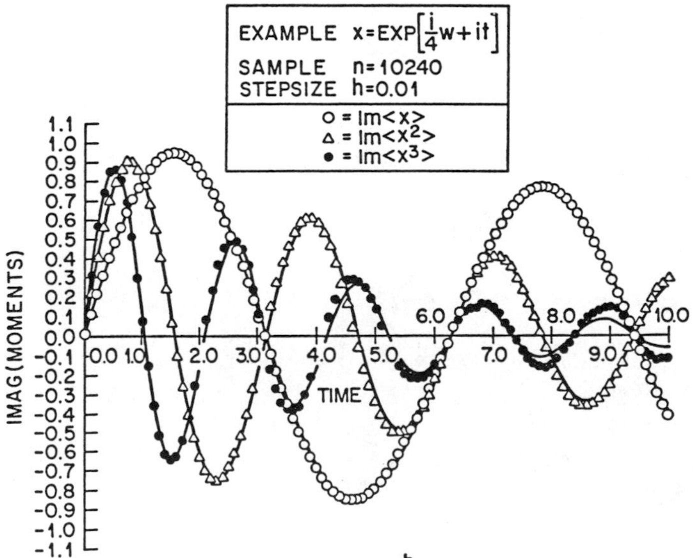


Figure 2[b]

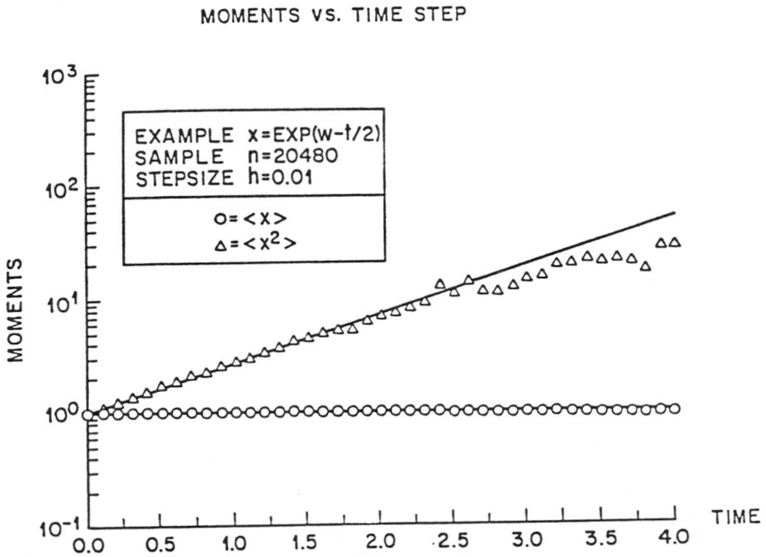


Figure 3

2.4 STABILITY TO ERROR IN THE APPROXIMATION OF THE BROWNIAN MOTION. The discrete approximation (1.3) of w depends in part on the accuracy of representations for $z_0$ and $z_1$. Given a sample (say n = 10240) of simulated paths we can imagine a variance of $\langle z_0 \rangle$ from zero to be proportional to 1/n. Likewise $\langle z_0^2 \rangle$ will have variance from unity by this same law of large numbers.

Bounds on errors for general approximations are given in Ikeda and Watanabe (ref. [4]), but for our purposes we illustrate the effect of variance due to sampling by **Figures** $\mathbf{4^a}$ and $\mathbf{4^b}$. The sample $\mathbf{4^a}$ of n =10240 yields a clearly more accurate representation than $\mathbf{4^b}$ with n = 512.

2.5 STIFFNESS. Our explicit Runge Kutta algorithm (2.3) for the linearized problem (2.4) is

$$x_h = [\,(1+\beta h+(h\beta)^2/2) + \{\sqrt{h}\,\alpha z_0 + h\,\alpha^2 z_0 z_1/\sqrt{2} + h^{3/2}\,\beta\alpha\, z_0\}\,]\,x_0$$

$$= A(h)\,x_0,$$

where the stochastic matrix A is contracting if

$$\| A(h) \|^2 = \langle | A |^2 \rangle \qquad (2.7)$$

$$= |\,|1+\beta h+(h\beta)^2/2|^2 + h\,(1 + h\,|\alpha|^2/2 + h^2(|\beta|^2+\mathbf{Re}\beta))\,|\alpha|^2 < 1.$$

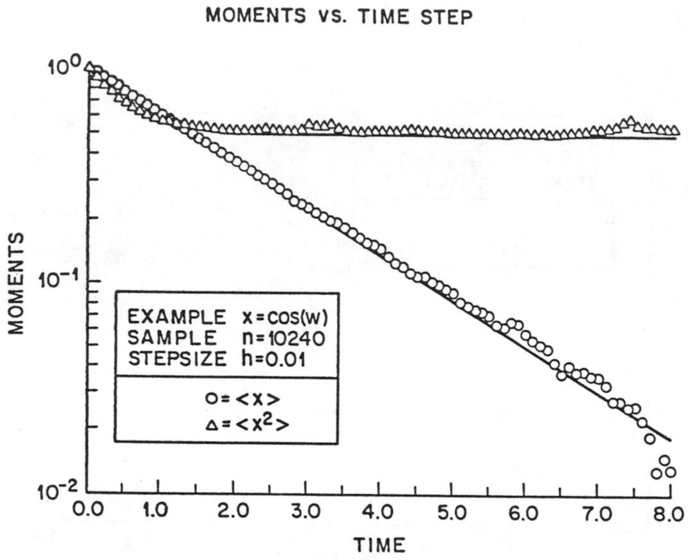


Figure 4[a]

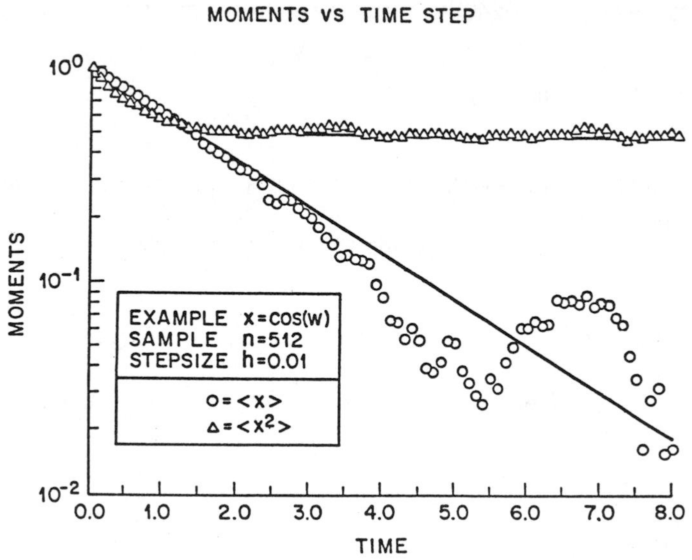


Figure 4[b]

To lowest order in $|\beta h|^2$, this is $|(1+\beta h)|^2 + h|\alpha|^2 < 1$ or

$$2\,\mathbf{Re}\,\beta + |\alpha|^2 < 0 \tag{2.8}$$

similar to (2.6). If, however, $|\beta|$ is comparable in size to $1/h$, $A(h)$ is **not contracting** even though the **solution should be qualitatively decreasing.** This type of behavior is known to give frightful numerical instabilities in fixed step size Ordinary Differential Equation (ODE) integrators (see section 1.3 of ref. [5]). For variable step size integrators ($h$ allowed to change) forcing $h << 1/|\beta|$ can lead to very small step sizes even though according to (2.5), time scales where $(-\,\mathbf{Re}\,\beta)t$ is small will dominate.

2.4 AN IMPLICIT FORMULATION. From the theory of Ordinary Differential Equations (see section 2.3 of Gear, ref. [5]), it is known that implicit algorithms are generally more stable. Since we are working in the Itô rule, however, the diffusion term must be a martingale. Let the discretized process be

$$x_h = x_0 + h\,B(x_h, x_0) + \sqrt{h}\;S(x_h, x_0)\,z_0. \tag{2.9}$$

The following easy theorem shows that S is an explicit formula.

**Theorem 1.** In the Itô rule for (2.9), S depends only on $x_0$ and contains no dependence on $z_0$.

**Proof.** Substituting the right hand side of (2.9) for the occurence of $x_h$ in the martingale part of (2.9) gives

$$\sqrt{h}\,S(x_h,x_0)\,z_0 = \sqrt{h}\,S(x_0,x_0)\,z_0 + h\,\frac{\partial S(x0,x0)}{\partial x_h}\,S(x_0,x_0)\,z_0^2 + O(h^{3/2}).$$

The expectation value of this must vanish, so to orders $h^{1/2}$, $h$ respectively:

(i) $\quad \langle S(x_0,x_0)\,z_0\rangle = 0$

(ii) $\quad \langle \dfrac{\partial S(x0,x0)}{\partial x_h}\,S(x_0,x_0)\,z_0^2\rangle = 0$

hence S cannot contain any $z_0$'s by (i) and therefore by (ii) $\partial S/\partial x_h = 0$.□

By the following bit of trickery we can derive an implicit second order method. We'll use the linearized form (2.4) and solution (2.5) to resolve a solution for the weights $0 \le \xi, \gamma, \eta \le 1$ in

$$x_h = x_0 + h\,\{\xi_1 b(x_h) + \xi_2 b(x_0)\}$$
$$+ \sqrt{h}\,\{\gamma_1 \sigma(x_0 + \eta_1\sqrt{h}\,\sigma_0 z_1)$$
$$+ \gamma_2\sigma(x_0 + \eta_2\sqrt{h}\,\sigma_0 z_1 + \eta_3 h\, b_0)\}\, z_0. \qquad (2.10)$$

**Theorem 2.** The second order accurate (in mean and variance) implicit Itô rule for (2.10) is the trapezoidal rule.

**Proof**. We get a unique result from the linear case (2.4), namely $\sigma_0 = \alpha x_0$ and $b_0 = \beta x_0$. By using solution (2.5), to $O(h^2)$ accuracy the expression for $\langle x(h) \rangle$ is

$$(1 - \xi_1 h\beta)\, e^{\beta h} = 1 + \xi_2 h\beta.$$

The $O(h)$ term of this produces $\xi_1 + \xi_2 = 1$, and the $O(h^2)$ term requires $\xi_1 = 1/2$. The drift part of (2.10) will then be the **trapezoidal rule** [5]. We can get $\gamma_1 + \gamma_2 = 1$ by the same device as in (2.2) for $\beta = 0$. The remaining coefficients must satisfy (to $O(h^2)$ accuracy)

$$(1 - h\beta/2)^2\, e^{(\alpha^2 + 2\beta)h} = (1 + h\beta/2)^2 + h\alpha^2$$
$$+ h^2(\gamma_2\eta_3\,\alpha^2\beta + (\gamma_1\eta_1 + \gamma_2\eta_2)^2\,\alpha^4)$$

which comes from the expected values $\langle x(h)^2 \rangle$ from (2.10) and solution (2.5). Using the independence of $\alpha^2$ and $\beta$ we get

$$(\gamma_1\eta_1 + \gamma_2\eta_2)^2 = 1/2 \qquad (\alpha^4 \text{ term})$$

$$\gamma_2\eta_3 = 1 \qquad (\alpha^2\beta \text{ term}).$$

Since $\gamma_2$, $\eta_3$ are **weights**, $\gamma_2 = \eta_3 = 1$; therefore $\gamma_1 = 0$, $\eta_2^2 = 1/2$. Thus, the trapezoidal method for (1.1) in the Itô rule is

$$x_h = x_0 + \frac{h}{2}(b(x_h) + b(x_0))$$

$$+ \sqrt{h}\,\sigma(x_0 + h\,b(x_0) + \sqrt{h/2}\,\sigma(x_0)\,z_1)z_0. \; \square \qquad (2.11)$$

Algorithm (2.11) agrees with the explicit method (2.3) when $b \equiv 0$, and the trapezoidal rule when $\sigma \equiv 0$.

2.5 STABILITY FOR THE IMPLICIT FORM. The linearized problem (2.4), (2.11) has the algorithm

$$(1 - h\beta/2)\, x_h = [1 + h\beta/2 + \sqrt{h}\,\alpha(1 + h\beta + \sqrt{h/2}\,\alpha z_1)z_0\,]\, x_0$$

$$= [\,1 + h\beta/2 + S\,]\, x_0$$

where S is the stochastic part of the matrix on the right hand side. Stability will be determined by the norm of the fundamental matrix

$$T(h) = (1 - h\beta/2\,)^{-1}\,[\,1 + h\beta/2 + S\,].$$

At time t = kh, the discrete approximation to the fundamental matrix is

$$T(t) = \prod_k [\,(1 - h\beta/2\,)^{-1}\,[\,1 + h\beta/2 + S_k\,]]$$

Small solutions of (2.4) by (2.11) will be stable if

$$\|\,T(h)\,\|^2 = |1 - h\beta/2|^{-2}\,(\,|1 + h\beta/2|^2 + \|S\|^2) \leq 1 \qquad (2.12)$$

Which means the stochastic part S must have bounded norm

$$\|\,S\,\|^2 = h \langle\, |\,\alpha(1 + h\beta + \sqrt{h/2}\,\alpha z_1)z_0|^{\,2} \rangle$$

$$= h\,|\,\alpha\,|^2\,|\,(1 + h\beta\,)|^2 + h^2|\alpha\,|^4$$

$$\leq |1 - h\beta/2|^2 - |1 + h\beta/2|^2$$

$$= -\,h\,\mathbf{Re}\,\beta \qquad (2.13)$$

which for $\alpha = 0$ says that the trapezoidal rule is unconditionally stable for decreasing solutions, namely **Re** $\beta < 0$. For $\alpha \neq 0$, however, the situation is more complicated. By analogy with the situation in the ODE case, we expect that the case where $|h\beta| \geq 1$ will give us problems. For small $|\alpha|^2$ and large $|\,\beta\,|$ the stability condition (2.13) is roughly

$$h^2 \leq \frac{-\,\mathbf{Re}\,\beta}{|\alpha|^2|\,\beta\,|^2}\,.$$

Notice in this case that when $-$ **Re** $\beta$ (therefore $|\,\beta\,|$) is large, stability of the zero solution demands $h^2$ decrease in proportion to $1/|\beta|$ unlike h in the explicit formula (2.7). We expect (2.11) to be substantially more stable than the explicit Runge Kutta method (2.3) when the Jacobian of $b(x_\mu)$ has eigenvalues with widely differing negative real parts for some **particular sample path** $x_\mu(t)$.

**3. IMPLEMENTATION ON VECTOR COMPUTERS.** At this writing we have conducted no experiments on variable step size nor implicit methods. **Figures** 2-4 were done using the Runge Kutta method (2.3) and h = 1/100. In other experiments [2] we have found instabilities leading to the discussion about implicit rules in section 2. This section contains remarks about the implementation of (2.3) on a Cray-1 vector processor. Finally there are some speculations about the implicit formula (2.11).

3.1 SAMPLING. As we noted in section 2.4 and Figures $4^a$, $4^b$, the accuracy of our results depend on familiar Monte Carlo considerations about the sample size. Namely, for small solutions of (1.1), we expect errors (in the mean and variance) which are typically proportional to 1/n, where n is the number of independent sample paths [1]. We compute the expectation values from

$$\langle f(x(t)) \rangle = (1/n) \sum_{\mu=1}^{n} f(x_\mu(t))$$

given a sample of paths $\{ x_\mu \mid \mu = 1, 2, \ldots n \}$. On vector computers, the approach is to follow several thousand sample paths simultaneously. Integration rule (2.2), for example looks like

$$\begin{bmatrix} x_1(h) \\ x_2(h) \\ \cdot \\ \cdot \\ x_n(h) \end{bmatrix} = \begin{bmatrix} x_1(0) \\ x_2(0) \\ \cdot \\ \cdot \\ x_n(0) \end{bmatrix} + \begin{bmatrix} \sigma(x_1(0) + \sqrt{h/2}\,\sigma(x_1(0))\, z_{11})\, z_{10} \\ \sigma(x_2(0) + \sqrt{h/2}\,\sigma(x_2(0))\, z_{21})\, z_{20} \\ \cdot \\ \cdot \\ \sigma(x_n(0) + \sqrt{h/2}\,\sigma(x_n(0))\, z_{n1})\, z_{n0} \end{bmatrix}$$

given a step size h. Similarly for (2.3). As long as $b(x)$ , $\sigma(x)$, $z_0$ and $z_1$ can be **vectorized**, all the paths in the sample can be integrated by (2.3) simultaneously. This is easy when b an $\sigma$ only contain familiar FORTRAN library functions: for example, b and $\sigma$ may be polynomials in x, contain functions like sin(x), exp(x), or powers of x [6]. All that is left is the vector of gaussian random variables $z_0$, $z_1$. We used a variant of the Box Muller method [7]. **Figure 5** shows the Cray Fortran (**CFT**) code to generate n-vectors $z_0$ and $z_1$. Function **ranf()** is a library random number generator (0.0<ranf()<1.0) and is a standard linear congruential method.

```
real z0(n), z1(n)
do 1 i = 1,n
   t1    = twopi*ranf()
   t2    = sqrt(-2.*log(ranf()))
   z0(i) = cos(t1)*t2
   z1(i) = sin(t1)*t2
1 continue
```

**Figure 5**

In **Figure 5**, twopi = $2\pi$, and log, sqrt are standard library routines.

3.2 SPECULATIONS ABOUT IMPLICIT RULES. Implict rules require more work than described in 3.1. Usually b(x) is a non-linear function, so (2.11) will likely have to be solved by an iterative method. Newton methods are the usual procedure for implicit equations. In the vectorized recipe described in 3.1, however, all of the solutions $\{x_\mu\}$ must be computed simultaneously. This will likely cause some difficulties in a Newton method. Indeed, if the Newton step

$$x^{(m+1)} = x^{(m)} - (1 - h\beta/2)^{-1}[(x^{(m)} - x_0) - h\,(b(x^{(m)}) + b(x_0))/2 - S]$$

is executed for iterations m until $|x^{(m+1)} - x^{(m)}|$ reaches a certain tolerance, the number of iterations m will likely depend on the initial data $x_0$ and the stochastic term $S(x_0)$ on the right hand side of (2.11) **for each path** $x_\mu$. Strategies for solving several uncoupled non-linear equations by iterative methods need to be developed. A fixed number of steps rather than convergence by some tolerance is the easiest. This fixed number must be chosen large enough such that any errors introduced will be permissible. The alternative is to process the whole n-vector set until the slowest converging $x_\mu$ satisfies the tolerance criteria, or else introducing an additional step of compressing the computed vector such that only non-converging elements are processed.

4. CONCLUSIONS. We find that for the limited class of test cases we have studied, explicit Runge Kutta formula (2.3) gives stable and accurate results when the Jacobian of b(x) has only negative real parts and the martingale part of (2.3) is small enough such that inequality (2.7), or roughly (2.8) holds. When the magnitude of one of these eigenvalues is large enough, however, the explicit algorithm (2.3) will be unstable. In this case an implicit form like the trapezoidal rule (2.11) will likely be necessary.

## REFERENCES

[1] J. R. Klauder and W. P. Petersen, SIAM Journal on Numerical Analysis, vol. 22, no. 6, Dec. 1985.

[2] J. R. Klauder and W. P. Petersen, Journal of Statistical Physics, Vol 39, p53, April 1985.

[3] W. Rumelin, SIAM Journal on Numerical Analysis, 19, No. 3, June 1982.

[4] N. Ikeda and S. Watanabe, Stochastic Differential Equations and Diffusion Processes, ( North Holland/Kodansha, Tokyo 1981).

[5] C. W. Gear, Numerical Initial Value Problems in Ordinary Differential Equations, ( Prentice Hall, Inc., Englewood Cliffs, NJ, 1971).

[6] Cray Research Inc. FORTRAN (CFT) reference manual, SR-0009, version 1.14, 1985, (Mendota Heights, MN).

[7] D. Knuth, The Art of Computer Programming, vol. 2, Addison Wesley Publ. Co., Menlo Park, CA, 1969.

# LYAPUNOV EXPONENT AND ROTATION NUMBER OF THE STOCHASTIC HARMONIC OSCILLATOR

Mark A. Pinsky*
Northwestern University

## 1a. Introduction

Many talks at this workshop have dealt with the spectral properties of random operators of the form $-\Delta + q(\cdot,\omega)$ where $\Delta$ is the Laplacian of a suitable space and $q(\cdot,\omega)$ is a random family of multiplication operators. In this report we shall deal with one of the simplest such problems, the random harmonic oscillator. This is defined as the solution of the following initial-value problem:

$$\ddot{y} + [\gamma + \sigma F(\xi(t,\omega))]y = 0$$

$$y(0) = y_0, \ \dot{y}(0) = y_1 \ .$$

Here $\gamma$ is a fixed positive number, $F(\xi(t,\omega))$ is a random process with mean-value zero and $\sigma$ is a positive parameter which will be studied in the limit $\sigma \downarrow 0$.

This problem has a long history, beginning with work of Pastur and Feldman [12] and most recently the work of Arnold, Papanicolaou and Wihstutz [3]. We represent the solution in terms of the solution of the following first order system of equations:

$$\dot{\phi}(t) = h(\phi(t),\xi(t)) \doteq -\sqrt{\gamma} + \frac{\sigma F(\xi(t))}{\sqrt{\gamma}} \cos^2\phi(t), \ \phi(0) = \phi_0$$

$$\dot{r}(t)/r(t) = q(\phi(t),\xi(t)) \doteq \frac{\sigma F(\xi(t))}{2\sqrt{\gamma}} \sin 2\phi(t), \ r(0) = r_0 \ .$$

Setting $y(t) = r(t)\cos\phi(t)/\sqrt{\gamma}$, then $\dot{y}(t) = r(t)\sin\phi(t)$ and $\ddot{y}(t) + [\gamma + \sigma F(\xi(t))]y(t) = 0$. The Lyapunov exponent and rotation number are defined respectively by the following limits:

$$\lambda(\sigma) = \lim_{t\uparrow\infty} t^{-1}\log r(t)$$

$$\alpha(\sigma) = \lim_{t\uparrow\infty} t^{-1}\phi(t) \ .$$

* Research supported by the National Science Foundation DMS 8403154-01.

Arnold, Papanicolaou and Wihstutz [3] have obtained asymptotic expansions for $\lambda(\sigma)$, $\alpha(\sigma)$ in case $\xi(t)$ is a reversible diffusion process on a compact manifold. In this paper we assume that $\xi(t)$ is a reversible Markov process with N ($<\infty$) states. Then we have the following results.

Theorem 1: The Lyapunov exponent and rotation number have the following asymptotic expansions when $\sigma \downarrow 0$:

$$\lambda(\sigma) = C_1\sigma^2 + O(\sigma^3)$$

$$\alpha(\sigma) = -\sqrt{\gamma} + D_1\sigma^2 + O(\sigma^3)$$

where the constants $C_1, D_1$ are strictly positive unless the noise process is identically zero (in which case $C_1 = D_1 = 0$). In detail we have

$$C_1 = (1/4\gamma)\sum_{k=2}^{N} \frac{\lambda_k <F,\psi_k>^2}{\lambda_k^2 + 4\gamma}$$

$$D_1 = (1/2\sqrt{\gamma})\sum_{k=2}^{N} \frac{<F,\psi_k>^2}{\lambda_k^2 + 4\gamma}$$

where $(\lambda_k, \psi_k)$ is the eigensystem for the generator G of the Markov process: $G\psi_k = -\lambda_k\psi_k$ for $k = 2,\ldots,N$ and the inner product is defined by $<F,\psi> = \Sigma F_i\psi(i)\nu_i$ where $\nu = (\nu_i)$ is the invariant measure satisfying $\nu G = 0$.

We may paraphrase this result by stating that the small noise term induces instability and retards rotation in the stochastic harmonic oscillator process. It is an open problem to prove that $\alpha(\sigma) > -\sqrt{\gamma}$ for all $\sigma > 0$ (it is known that $\lambda(\sigma) > 0$ for all $\sigma > 0$, cf. Arnold [1]).

In order to compare the first order asymptotics with the true values of $\alpha(\sigma)$, $\lambda(\sigma)$ we may compute additional terms in the asymptotic expansions. This has been done for the case $N = 2$ according to the following:

Theorem 2: In the case of a 2-state Markov process, we have the expansions

$$\lambda(\sigma) = C_1\sigma^2 + C_2\sigma^4 + O(\sigma^5)$$

$$\alpha(\sigma) = -\sqrt{\gamma} + D_1\sigma^2 + D_2\sigma^4 + O(\sigma^5)$$

<u>where</u> $C_2 > 0$ <u>unless the noise process is identically zero</u>.

This settles a long-standing conjecture as to the exact form of the exponential growth rate namely that $C_1\sigma^2$ is merely the first term in the asymptotic expansion, rather than the exact form of the exponential growth rate.

## 1b. Notations and definitions

Let $\{\xi(t): t \geqslant 0\}$ be a temporally homogeneous Markov process on the state space $M = \{1,...,N\}$. This means that the conditional probability $P\{\xi(t+s) = j \mid \xi(u): u \leqslant t\} = (e^{sG})_{ij}$ where $i = \xi(t)$ and G is a finite $N \times N$ matrix whose properties are detailed in section 2, equations (2.1) - (2.3). In particular, the invariant measure $\nu = (\nu_i)$ satisfies the matrix equation

$$\sum_{i=1}^{N} \nu_i G_{ij} = 0 \qquad (1 \leqslant j \leqslant N). \tag{1.1}$$

Let $F = F(\xi)$ be a real-valued function on M with mean value zero: $\sum_{i=1}^{N} F(i)\nu_i = 0$, and let $\gamma, \sigma$ be positive constants. The stochastic harmonic oscillator process is the unique solution of the equation (1.1) with the initial conditions $y(0) = y_0$, $\dot{y}(0) = y_1$. We take polar coordinates in the form $y\sqrt{\gamma} = r\cos\phi$, $\dot{y} = r\sin\phi$. With this transformation, equation (1.1) can be solved by first solving the following equations:

$$\dot{\phi} = h(\phi(t),\xi(t)) \qquad \phi(0) = \phi_0 \tag{1.2}$$

$$\dot{r}/r = q(\phi(t),\xi(t)) \qquad r(0) = r_0 \tag{1.3}$$

where

$$h(\phi,\xi) = -\sqrt{\gamma} + \frac{\sigma F(\xi)}{\sqrt{\gamma}} \cos^2\phi \tag{1.4}$$

$$q(\phi,\xi) = \frac{\sigma F(\xi)}{2\sqrt{\gamma}} \sin 2\phi \, . \tag{1.5}$$

The pair $(\phi(t),\xi(t))$ is a temporally homogeneous Markov process on the space $S^1 \times M$, where $S^1$ denotes the circumference of the unit circle. The infinitesimal generator of the Markov process is defined by

$$Lf(\phi,\xi) = \lim_{t\downarrow 0} t^{-1} E_{\phi,\xi}\{f(\phi(t),\xi(t)) - f(\phi,\xi)\}$$

where $E_{\phi,\xi}\{\ldots\}$ denotes the mathematical expectation for the condition $\phi(0)=\phi$, $\xi(0)=\xi$. The explicit form of the infinitesimal generator is given by

$$Lf(\phi,\xi) = (Gf)(\phi,\xi) + h(\phi,\xi)\frac{\partial f}{\partial \phi}(\phi,\xi) .$$

Let $\nu(d\xi)$ be the invariant measure of the Markov process $\xi(t)$ and let $d\phi$ be Lebesgue measure on $S^1$. Then the process $(\phi(t),\xi(t))$ has an invariant measure of the form $p_\sigma(\phi,\xi)d\phi\,\nu(d\xi)$ which can be obtained as the solution of the equations

$$0 = L^* p_\sigma = Gp_\sigma - \frac{\partial}{\partial\phi}\{h(\phi,\xi)\,p_\sigma(\phi,\xi)\}$$

$$= \sum_{j=1}^{N} G_{ij}\,p_\sigma(\phi,j) - \frac{\partial}{\partial\phi}\{h(\phi,i)\,p_\sigma(\phi,i)\} \qquad (\phi,i)\in S^1\times M$$

$$1 = \int_{S^1\times M} p_\sigma(\phi,\xi)d\phi\,\nu(d\xi) .$$

According to a theorem of Crauel [6], the long-time behavior of $r(t)$ can be computed in terms of the "top Lyapunov exponent," in the form

$$\lambda(\sigma) = \lim_{t\uparrow\infty} t^{-1}\log r(t) = \int_{S^1\times M} q(\phi,\xi)\,p_\sigma(\phi,\xi)d\phi\,\nu(d\xi) .$$

The spiraling properties of the solution are described by the "rotation number," which is computed as

$$\alpha(\sigma) = \lim_{t\uparrow\infty} t^{-1}\phi(t) = \int_{S^1\times M} h(\phi,\xi)\,p_\sigma(\phi,\xi)d\phi\,\nu(d\xi) .$$

The paper is organized as follows. In section 2 we collect some preliminary facts on finite Markov chains and related equations. In section 3 we develop the corresponding properties for the pair process $(\phi(t),\xi(t))$. Then in Section 4 we use these to develop a perturbation scheme to obtain an asymptotic formula for $p_\sigma(\phi,\xi)$ when $\sigma\downarrow 0$. This is then applied to obtain the expansion of the Lyapunov exponent and rotation number. In section 5 we specialize to the case $N=2$ to obtain an additional term in the asymptotic expansions.

## 2. Finite Markov Chains And Related Equations

Let $G = (q_i \pi_{ij})$ be an $N \times N$ matrix which generates a continuous-parameter irreducible, time-reversible Markov chain with invariant measure $\nu = (\nu_i)$. The state space of the Markov chain is the finite set $M = \{1,\ldots,N\}$. The matrix elements satisfy

$$q_i > 0, \ \pi_{ii} = -1, \ \sum_{j=1}^{N} \pi_{ij} = 0 \qquad (1 \leq i \leq N) \tag{2.1}$$

$$0 < \pi_{ij} < 1 \qquad (1 \leq i \neq j \leq N) \tag{2.2}$$

$$\nu_i q_i \pi_{ij} = \nu_j q_j \pi_{ji} \qquad (1 \leq i,j \leq N) \ . \tag{2.3}$$

The inner product of two vectors $p = (p_i)$ and $q = (q_i)$ is defined by

$$<p,q> = \sum_{i=1}^{N} p_i q_i \nu_i = \int_M pq\nu(d\xi) \ . \tag{2.4}$$

The matrix $G$ is self-adjoint in the sense that $<Gp,q> = <p,Gq>$ for any $p = (p_i)$, $q = (q_i)$.

Let $\psi_1,\ldots,\psi_N$ be the normalized eigenfunctions of $G$. These satisfy

$$\psi_1 = 1, \ G\psi_1 = 0 \tag{2.5}$$

$$<\psi_i,\psi_1> = 0, \ <G\psi_i,\psi_i> = -\lambda_i < 0 \qquad (2 \leq i \leq N) \ . \tag{2.6}$$

The spectral theorem for finite matrices yields the decomposition

$$e^{tG}p = <p,\psi_1>\psi_1 + \sum_{i=2}^{N} e^{-\lambda_i t} <p,\psi_i>\psi_i \ . \tag{2.7}$$

The equation $(c\frac{\partial}{\partial\phi} + G)p = -f$ is solved by the following proposition.

<u>Proposition 2.1</u>: <u>Suppose that</u> $\int_{S^1 \times M} f(\emptyset,\xi)d\emptyset\nu(d\xi) = 0$. <u>Define</u>

$$p(\emptyset,\xi) = \lim_{\alpha\downarrow 0} \int_0^\infty e^{-\alpha t}(e^{tG}f)(\emptyset + ct,\cdot)dt \tag{2.8}$$

Then

$$c\,\frac{\partial p}{\partial \phi} + Gp = -f\ . \tag{2.9}$$

In terms of Fourier components we may equivalently write

$$f(\emptyset,\xi) = \sum_{(n,k)\neq(0,1)} a_{nk}e^{in\emptyset}\psi_k(\xi) \tag{2.10}$$

$$p(\phi,\xi) = \sum_{(n,k)\neq(0,1)} \frac{a_{nk}}{\lambda_k - inc}\, e^{in\emptyset}\psi_k(\xi)\ . \tag{2.11}$$

The proof is by direct computation. The uniqueness of the solution is given by the following proposition.

Proposition 2.2: Let $p = p(\emptyset,\xi)$ be a solution of the equation $c\frac{\partial p}{\partial \phi} + Gp = 0$ where $c \neq 0$. Then $p(\emptyset,\xi)$ is a constant, independent of $(\emptyset,\xi)$.

Proof. Let $\emptyset(t) = \emptyset + ct$ and let $\xi(t)$ be the N-state Markov chain generated by G. Then the real-valued stochastic process $p(\emptyset(t),\xi(t))$ is a bounded martingale [7] and we can write

$$\begin{aligned} p(\emptyset,\xi) &= E\{p(\emptyset(0),\xi(0))\} \\ &= E\{p(\emptyset(t),\xi(t))\} \\ &= (e^{tG}p)(\emptyset + ct,\cdot) \\ &= \langle p(\emptyset + ct,\cdot),1\rangle + \sum_{i=2}^{N} e^{-\lambda_i t}\langle p(\emptyset + ct,\cdot),\psi_i\rangle\,\psi_i\ . \end{aligned}$$

The second term tends to zero when $t \to \infty$, whereas the first term does not depend on $\xi$. Hence the first term has a limit $p(\emptyset)$ and $p(\emptyset,\xi) = p(\emptyset)$ independent of $\xi$. Referring back to the equation $c\frac{\partial p}{\partial \phi} + Gp = 0$, we see that $Gp = 0$ and hence $\frac{\partial p}{\partial \phi} = 0$, thus p is constant in both $\emptyset$ and $\xi$.

Corollary 2.3: Let $p = p(\emptyset,\xi)$ be a solution of the equation $c\frac{\partial p}{\partial \phi} + Gp = 0$ where $c \neq 0$ and such that $\int_{S^1\times M} p(\phi,\xi)d\phi\nu(d\xi) = 0$. Then $p(\emptyset,\xi) = 0$ for all $\emptyset,\xi$.

## 3. The Angular Process And Related Equations

We now consider equation (1.2) for the angular process, written in detail as

$$\dot{\phi} = h(\phi(t),\xi(t)) \tag{3.1}$$
$$= -\sqrt{\gamma} + \frac{\sigma F(\xi(t))}{\sqrt{\gamma}} \cos^2\phi(t) \; .$$

The joint process $(\phi(t),\xi(t))$ is a Markov process on the state space $S^1 \times M$ with infinitesimal generator $G + h(\phi,\xi)\frac{\partial}{\partial\phi}$. To obtain the necessary estimates we shall need to estimate the mean hitting time of points by this process. For this purpose let $\Phi_{\phi,\xi}(t)$ be the solution of the ordinary differential equation $\dot{\phi}(t) = h(\phi(t),\xi)$ with $\phi(0) = \phi$ and the second variable is frozen at the state $\xi \in M$. Consider the deterministic hitting times

$$T_{\phi,\xi}(\psi) = \inf\left\{t > 0 \colon e^{i\Phi_{\phi,\xi}(t)} = e^{i\psi}\right\} .$$

For $|\sigma| \leq \gamma/(2 \max |F|)$ we have $h(\phi,\xi) \leq -\frac{1}{2}\sqrt{\gamma}$ and thus $\Phi_{\phi,\xi}(t) \leq \phi - \frac{1}{2}t\sqrt{\gamma}$ for all t. This immediately yields

<u>Proposition 3.1</u>: <u>For</u> $|\sigma| \leq \gamma/(2 \max |F|)$ <u>we have</u>

$$T_{\phi,\xi}(\psi) \leq 4\pi/\sqrt{\gamma} \; .$$

We now consider the hitting times of the random process $(\phi(t),\xi(t))$. Let

$$\tau(\psi,\eta) \doteq \inf\left\{t > 0 \colon e^{i\phi(t)} = e^{i\psi}, \xi(t) = \eta\right\}$$
$$u(\phi,\xi;\psi,\eta) \doteq E\left\{\tau(\psi,\eta) \mid \phi(0) = \phi, \xi(0) = \xi\right\} .$$

Using the Markov property we obtain

Lemma 3.2: We have the following integral equations

$$u(\emptyset,\xi;\psi,\xi) = T_{\emptyset,\xi}(\psi)e^{-q_\xi T_{\emptyset,\xi}(\psi)} \tag{3.2}$$

$$+ \sum_{\xi'\neq\xi} \pi_{\xi\xi'} \int_0^{T_{\phi,\xi}(\psi)} \left[q_\xi^{-1} + u(\Phi_{\emptyset,\xi}(s),\xi';\psi,\xi)\right] q_\xi e^{-sq_\xi}\, ds$$

$$u(\emptyset,\xi;\psi,\eta) = q_\xi^{-1} + \pi_{\xi\eta} \int_0^\infty u(\Phi_{\phi,\xi}(s),\eta;\psi,\eta) q_\xi e^{-sq_\xi}\, ds \tag{3.3}$$

$$+ \sum_{\xi'\notin\{\xi,\eta\}} \pi_{\xi\xi'} \int_0^\infty u(\Phi_{\phi,\xi}(s),\xi;\psi,\eta) q_\xi e^{-sq_\xi}\, ds \qquad (\eta\neq\xi)\ .$$

Furthermore we have the bounds

$$u(\phi,\xi;\psi,\xi) \leqslant \frac{1}{q_{min}}\left(e^{-1} + \frac{1}{1-s}\right)$$

$$u(\phi,\xi;\psi,\eta) \leqslant \frac{1}{q_{min}}\left(\frac{1}{1-r} + \frac{pe^{-1}}{(1-p)^2}\right) \qquad (\eta\neq\xi)$$

$$r \doteq \max_{\xi,\eta} \pi_{\xi\eta}\ , \quad s = 1-e^{-(4\pi/\sqrt{\gamma})q_{max}}\ , \quad p = \max(r,s)\ .$$

Corollary 3.3: For any $\alpha > 0$ we have

$$E\left\{e^{\alpha\tau(\psi,\eta)} \mid \emptyset(0) = \emptyset, \xi(0) = \xi\right\} \leqslant C_\alpha\ .$$

Proof. From Lemma 3.2 we have $E\{\tau(\psi,\eta)\mid \emptyset(0) = \phi, \xi(0) = \xi\} \leqslant C$ where the constant $C$ is independent of $\emptyset,\xi,\psi,\eta$. By Chebyshev's inequality $P\{\tau(\psi,\eta) \geqslant t_0\} \leqslant t_0^{-1}E\{\tau(\psi,\eta)\} \leqslant Ct_0^{-1}$. Choosing $t_0 = 2C$ this probability is less than $\frac{1}{2}$; now by the simple Markov property, $P(\tau(\psi,\eta) \geqslant 2t_0) = P(\tau(\psi,\eta) \geqslant t_0, \tau(\psi,\eta)\cdot\theta_{t_0} \geqslant t_0) \leqslant \frac{1}{2}P(\tau(\psi,\eta) \geqslant t_0) \leqslant (\frac{1}{2})^2$. Continuing inductively we have $P(\tau(\psi,\eta) \geqslant nt_0) \leqslant (\frac{1}{2})^n$. Thus $E(e^{\alpha\tau(\psi,\eta)}) < \infty$ with a bound which only depends on $\alpha$.

<u>Proposition 3.4</u>: <u>Let</u> $p = p(\emptyset,\xi)$ <u>be the unique solution of the equation</u>

$$(L_0 + \sigma L_1)^* p = -f$$

$$\int_{S^1 \times M} p(\emptyset,\xi)d\emptyset\nu(d\xi) = 0 .$$

<u>Then</u>

$$|p(\emptyset,\xi)| \leq C \max_{\phi,\xi} |f(\emptyset,\xi)| .$$

<u>Proof</u>. Since the integral of p is zero, there must be a zero of $p(\emptyset,\xi)$ somewhere on the state space $(-\pi,\pi) \times \{1,\dots,N\} = S^1 \times M$. Let $T = \inf\{t > 0 : p(\emptyset(t),\xi(t)) = 0\}$. The equation for p is written in detail as

$$Gp + \sqrt{\gamma}\,\frac{\partial p}{\partial \emptyset} - \cos^2\emptyset\,\frac{\sigma F(\xi)}{\sqrt{\gamma}}\,\frac{\partial p}{\partial \emptyset} + 2\sin\emptyset\cos\emptyset\,\frac{\sigma F(\xi)}{\sqrt{\gamma}}\,p = -f .$$

The left member is the generator of a Markov process which is killed according to the multiplicative functional

$$M_t = \exp\left[\int_0^t (2\sigma/\sqrt{\gamma}) \sin\phi(s) \cos\phi(s) F(\xi(s))\,ds\right] .$$

By Dynkin's formula for killed processes [8], we have

$$E\left\{p(\emptyset(T),\xi(T))M_T\right\} - p(\emptyset,\xi) = E\left\{\int_0^T f(\emptyset(s),\xi(s))M_s\,ds\right\} .$$

The first term is zero by definition of T whereas the integral on the right is estimated by $|fM| \leq e^{\alpha T}\max|f|$. Using the inequality $x \leq e^x$ in the form $Te^{\alpha T} \leq (1/\alpha)e^{2\alpha T}$ and applying Corollary 3.3 gives the desired result.

## 4. <u>Expansion Of The Invariant Density And Applications</u>

We can now apply the preceding estimates to give the expansion for the invariant density $p(\emptyset,\xi)$. By definition it satisfies

$$(L_0 + \sigma L_1)^* p = 0$$

$$\int_{S' \times M} p(\emptyset,\xi)d\emptyset\nu(d\xi) = 1$$

where

$$L_0 = G - \sqrt{\gamma}\,\frac{\partial}{\partial\phi}$$

$$L_1 = \frac{F(\xi)}{\sqrt{\gamma}}\cos^2\phi\,\frac{\partial}{\partial\phi} \quad .$$

We construct an approximate invariant density by the finite expansion

$$p^{(N)} = p_0 + \sigma p_1 + \cdots + \sigma^N p_N \tag{4.1}$$

where

$$p_0 = (1/2\pi) \tag{4.2}$$

$$L_0^* p_N + L_1^* p_{N-1} = 0 \qquad (N \geqslant 1) \tag{4.3}$$

$$\int_{S^1 \times M} p_N(\phi,\xi)\,d\phi\,\nu(d\xi) = 0 \qquad (N \geqslant 1) \tag{4.4}$$

These may be constructed using the results of section 2. For each $N \geqslant 1$, the second term of (4.3) satisfies the compatability condition, since

$$\int_{S^1 \times M} (L_1^* p_{N-1})(\phi,\xi)\,d\phi\,\nu(d\xi)$$

$$= -\int_M F(\xi)\,\nu(d\xi) \int_{-\pi}^{\pi} (\partial/\partial\phi)\left\{\cos^2\phi\; p_{N-1}(\phi,\xi)\right\} d\phi$$

$$= 0 \ .$$

The remainder $R \doteq p - p^{(N)}$ satisfies

$$(L_0^* + \sigma L_1^*)R = -\sigma^{N+1} L_1^* p_N$$

$$\int_{S^1 \times M} R(\phi,\xi)\,d\phi\,\nu(d\xi) = 0 \ .$$

Therefore we may apply proposition 3.4 with the result

$$|R(\phi,\xi)| \leqslant C\sigma^{N+1} \max_{\phi,\xi} |L_1^* p_N(\phi,\xi)| \ .$$

We now specialize to the case $N = 1$ with the following result

Proposition 4.1:

$$p_1(\phi,\xi) = \left(\frac{1}{2\pi\sqrt{\gamma}}\right) \mathrm{Im}\left\{e^{2i\phi}\sum_{k=2}^{N} \frac{\langle F,\psi_k\rangle}{\lambda_k - 2i\sqrt{\gamma}}\,\psi_k(\xi)\right\}$$

$$\int_{S^1\times M} q(\phi,\xi)p_1(\phi,\xi)\,d\phi\,\nu(d\xi) = (\sigma/4\gamma)\sum_{k=2}^{N} \frac{\lambda_k\langle F,\psi_k\rangle^2}{\lambda_k^2 + 4\gamma} > 0$$

$$\int_{S^1\times M} h(\phi,\xi)p_1(\phi,\xi)\,d\phi\,\nu(d\xi) = (\sigma/2\sqrt{\gamma})\sum_{k=2}^{N} \frac{\langle F,\psi_k\rangle^2}{\lambda_k^2 + 4\gamma} \quad .$$

In particular no higher terms are necessary to verify the instability so long as $F$ is not identically zero.

Proof. $p_1(\phi,\xi)$ is determined by solving the equation $L_0^* p_1 + L_1^* p_0 = 0$, or in detail

$$\left(G + \sqrt{\gamma}\,\frac{\partial}{\partial\phi}\right)p_1 = \frac{F(\xi)}{\sqrt{\gamma}}\,\frac{\partial}{\partial\phi}\left(\frac{\cos^2\phi}{2\pi}\right)$$

$$= -\,\frac{F(\xi)}{2\pi\sqrt{\gamma}}\,\sin 2\phi$$

$$= -\,\frac{1}{2\pi\sqrt{\gamma}}\sum_{k=2}^{N} \langle F,\psi_k\rangle\,\psi_k(\xi)\,\mathrm{Im}(e^{2i\phi})$$

where we have used the fact that $0 = \langle F,\psi_1\rangle = \int F(\xi)\nu(d\xi)$. According to Proposition 2.1, the unique solution of this equation may be obtained in the form

$$p_1(\phi,\xi) = \frac{1}{2\pi\sqrt{\gamma}}\sum_{k=2}^{N} \langle F,\psi_k\rangle\psi_k(\xi)\,\mathrm{Im}\left(\frac{e^{2i\phi}}{\lambda_k - 2i\sqrt{\gamma}}\right)$$

which is the required form. To compute the necessary integrals, we write

$$q(\phi,\xi) = (\sigma/2\sqrt{\gamma})\,\sin 2\phi \sum_{\ell=2}^{N} \langle F,\psi_\ell\rangle\psi_\ell(\xi) \quad .$$

Thus

$$\int q(\phi,\xi)p_1(\phi,\xi)d\phi\nu(d\xi) = (\sigma/2\sqrt{\gamma})\sum_{\ell=2}^{N}\langle F,\psi_\ell\rangle \int_{S^1\times M}\sin 2\phi\, \psi_\ell(\xi)p_1(\phi,\xi)d\phi\nu(d\xi)$$

$$= (\sigma/2\sqrt{\gamma})(1/2\pi\sqrt{\gamma}) \sum_{k,\ell=2}^{N} \langle F,\psi_k\rangle\langle F,\psi_\ell\rangle \int_{S^1\times M} \sin 2\phi\, \psi_k(\xi)\psi_\ell(\xi)\,\mathrm{Im}\left(\frac{e^{2i\phi}}{\lambda_k-2i\sqrt{\gamma}}\right)d\phi\nu(d\xi)$$

$$= (\sigma/2\sqrt{\gamma})(1/2\pi\sqrt{\gamma}) \sum_{k,\ell=2}^{N} \langle F,\psi_k\rangle\langle F,\psi_\ell\rangle\,\mathrm{Im}\left(\frac{i\pi}{\lambda_k-2i\sqrt{\gamma}}\right)\langle \psi_k,\psi_\ell\rangle$$

$$= (\sigma/4\gamma)\sum_{k=2}^{N} \frac{\lambda_k}{\lambda_k^2+4\gamma}\langle F,\psi_k\rangle^2 .$$

Similarly, to compute the second integral we first note that the constant term in $h(\phi,\xi)$ integrates to zero with respect to $p_1(\phi,\xi)$ and thus

$$\int h(\phi,\xi)p_1(\phi,\xi)d\phi\nu(d\xi) = (\sigma/\sqrt{\gamma})\sum_{\ell=2}^{N}\langle F,\psi_\ell\rangle \int_{S^1\times M}\psi_\ell(\xi)\cos^2\phi\, p_1(\phi,\xi)d\phi\nu(d\xi)$$

$$= (\sigma/2\pi\gamma) \sum_{k,\ell=2}^{N} \langle F,\psi_k\rangle\langle F,\psi_\ell\rangle \int_{S^1\times M} \cos^2\phi\,\psi_k(\xi)\psi_\ell(\xi)\,\mathrm{Im}\left(\frac{e^{2i\phi}}{\lambda_k-2i\sqrt{\gamma}}\right)d\phi\nu(d\xi)$$

$$= (\sigma/2\pi\gamma) \sum_{k,\ell=2}^{N} \langle F,\psi_k\rangle\langle F,\psi_\ell\rangle\,\mathrm{Im}\left(\frac{\frac{1}{2}\pi}{\lambda_k-2i\sqrt{\gamma}}\right)\langle \psi_k,\psi_\ell\rangle$$

$$= (\sigma/2\sqrt{\gamma})\sum_{k=2}^{N} \frac{\langle F,\psi_k\rangle^2}{\lambda_k^2+4\gamma}$$

<u>Corollary 4.2</u>: <u>The Lyapunov exponent and the rotation number have the following asymptotic developments when</u> $\sigma \to 0$:

$$\lambda = (\sigma^2/4\gamma) \sum_{k=2}^{N} \frac{\lambda_k\langle F,\psi_k\rangle^2}{\lambda_k^2+4\gamma} + O(\sigma^3)$$

$$\alpha = -\sqrt{\gamma} + (\sigma^2/2\sqrt{\gamma}) \sum_{k=2}^{N} \frac{<F,\psi_k>^2}{\lambda_k^2 + 4\gamma} + O(\sigma^3) .$$

In conclusion, we see that the small noise term induces instability and retards rotation in the stochastic harmonic oscillator process.

In other works on this problem, the basic equation (1.1) is written in the form $\ddot{y} + k^2(1 + b\xi(t))y = 0$, where $\xi(t) = \pm 1$ is a two-state Markov chain with infinitesimal generator $G = \begin{pmatrix} -a & a \\ a & -a \end{pmatrix}$. Therefore we make the substitutions $\gamma = k^2$, $\sigma < F,\psi_2 > = k^2 b$, $\lambda_2 = 2a$ with the resulting asymptotic formula

$$\lambda = \frac{k^2 ab^2}{8(a^2 + k^2)} + O(b^3) ,$$

$b \downarrow 0$. In the next section we compute the correction term which is nonzero at the level $O(b^4)$. Compare with formula (3.34) in [5].

## 5. Refined Computation Of The Lyapunov Exponent and rotation Number

We now assume that G is the generator of the standard 2-state Markov chain with

$$G = \begin{pmatrix} -a & a \\ a & -a \end{pmatrix} .$$

Thus $\nu = (1/2,1/2)$, $\psi_1(\xi) \equiv 1$, and $\psi_2(\xi) = (-1)^\xi$ for $\xi = 1,2$. We let $F = < F,\psi_2 >$.

Proposition 5.1:

$$p_2(\phi,\xi) = \frac{F^2}{8\pi\gamma(a^2+\gamma)} \left\{ \frac{a}{\sqrt{\gamma}}(\sin 2\phi + \tfrac{1}{2}\sin 4\phi) + (\cos 2\phi + \tfrac{1}{2}\cos 4\phi) \right\} \tag{5.1}$$

$$p_3(\phi,\xi) = -\frac{F^3\psi_2(\xi)}{8\pi\gamma^{3/2}(a^2+\gamma)} \left\{ \frac{a}{\sqrt{\gamma}} \operatorname{Re}\left( \frac{5}{4}\frac{e^{2i\phi}}{2a-2i\sqrt{\gamma}} + 2\frac{e^{4i\phi}}{2a-4i\sqrt{\gamma}} + \frac{3}{4}\frac{e^{6i\phi}}{2a-6i\sqrt{\gamma}} \right) \right. \tag{5.2}$$

$$\left. - \operatorname{Im}\left( \frac{5}{4}\frac{e^{2i\phi}}{2a-2i\sqrt{\gamma}} + 2\frac{e^{4i\phi}}{2a-4i\sqrt{\gamma}} + \frac{3}{4}\frac{e^{6i\phi}}{2a-6i\sqrt{\gamma}} \right) \right\}$$

$$= \frac{5F^3\psi_2(\xi)}{32\pi\gamma^{3/2}(a^2+\gamma)^2} \left\{ \frac{\gamma^2 - a}{2\sqrt{\gamma}} \cos 2\phi + a\sin 2\phi + \sum_{n\neq\pm 2} \beta_n e^{in\phi} \right\}$$

$$\int_{S^1\times M} q(\phi,\xi)p_2(\phi,\xi)d\phi\nu(d\xi) = 0 \tag{5.3}$$

$$\int_{S^1\times M} h(\phi,\xi)p_2(\phi,\xi)d\phi\nu(d\xi) = 0 \tag{5.4}$$

$$\int_{S^1\times M} q(\phi,\xi)p_3(\phi,\xi)d\phi\nu(d\xi) = \frac{5\sigma F^4 a}{64\gamma^2(a^2+\gamma)^2} \tag{5.5}$$

$$\int_{S^1\times M} h(\phi,\xi)p_3(\phi,\xi)d\phi\nu(d\xi) = \frac{5\sigma F^4(\gamma-a^2)}{128\gamma^{5/2}(a^2+\gamma)^2} \tag{5.6}$$

Proof. Direct computation, following the method of the previous section.

Corollary 5.2: *For the two-state Markov chain we have the following asymptotic formulas for the Lyapunov exponent and rotation number when* $\sigma \to 0$:

$$\lambda = \frac{\sigma^2 aF^2}{8\gamma(a^2+\gamma)} + \frac{5\sigma^4F^4a}{64\gamma^2(a^2+\gamma)^2} + O(\sigma^5)$$

$$\alpha = -\sqrt{\gamma} + \frac{\sigma^2F^2}{8\sqrt{\gamma}(a^2+\gamma)} + \frac{5\sigma^4F^4(\gamma-a^2)}{128\gamma^{5/2}(a^2+\gamma)^2} + O(\sigma^5) \ .$$

More detailed computations are contained in [13].

References

1. L. Arnold, A formula connecting sample and moment stability of linear stochastic systems, SIAM J. Appl. Math 44 (1984), 793-802.

2. L. Arnold and W. Kliemann, Qualitative theory of stochastic systems, in A. T. Bharucha-Reid (ed) Probabilistic Analysis and Related Topics, Vol. 3, Academic Press, New York, 1983.

3. L. Arnold, G. Papanicolaou, V. Wihstutz, Asymptotic analysis of the Lyapunov exponent and rotation number of the random oscillator and applications, SIAM J. Appl. Math (to appear).

4. E. I. Auslender and G. N. Milshtein, Asymptotic expansion of the Lyapunov index for linear stochastic systems with small noise, PMM USSR 46 (1983), 277-283.

5. G. L. Blankenship and K. A. Loparo, Almost sure instability of a class of linear stochastic systems with jump process coefficients, preprint, January 1985.

6. H. Crauel, Lyapunov numbers of Markov solutions of linear stochastic systems, Stochastics 14 (1984), 11-28.

7. J. L. Doob, Stochastic Processes, John Wiley and Sons, New York, 1953.

8. E. B. Dynkin, Markov Processes, 2 vols., Springer Verlag, New York, 1965.

9. George Hermann (ed), Dynamic Stability of Structures, Proceedings of an International Conference held at Northwestern University, Pergamon Press Ltd., Oxford, 1967.

10. R. Z. Khasminskii, Stochastic Stability of Differential Equations, Sijthoff and Nordhoff, Alpen aan den Rijn, 1980 (translation and revision of the Russian edition, Nauka, Moskow, 1969).

11. F. Kozin and S. K. Prodmorou, Necessary and sufficient conditions for almost sure sample stability of linear Itô equations, SIAM J. Appl. Math 21 (1971), 413-424.

12. L. A. Pastur and E. P. Feldman, Wave transmittance for a thick layer of a randomly inhomogeneous medium, Soviet Physics JETP, 30 (1975), 158-162.

13. M. Pinsky, Instability of the harmonic oscillator with small noise, SIAM J. Appl. Math (to appear).

# REGULARITY OF THE DENSITY OF STATES FOR STOCHASTIC JACOBI MATRICES: A REVIEW

Barry Simon
Division of Physics, Mathematics and Astronomy
California Institute of Technology
Pasadena, CA 91125

## 1. The Density of States, the Lyaponov Exponent and Their Relation

In this paper, we will discuss stochastic Jacobi matrices which are operators on $\ell^2(\mathbb{Z}^\nu)$. Indicate elements of this Hilbert space by u(n) with $n \in \mathbb{Z}^\nu$. The free (kinetic) energy operator is given by:

$$(H_0 u)(n) = \sum_{|j|=1} u(n+j) \tag{1}$$

and we will consider operators, $H = H_0 + V_\omega$, where $\omega$ is a label in a probability measure space. The potential, V, is a family of random variables forming an ergodic process. To be explicit, we let $(\Omega, \mu)$ be a probability measure space with a family, $T_1, \ldots, T_\nu$ of commuting, measure preserving transformations which generate an ergodic action. Pick f, a measurable function on $\Omega$, and define

$$V_\omega(n) = f(T_1^{n_1} \cdots T_\nu^{n_\nu} \omega) \tag{2}$$

For simplicity, we will normally suppose that f is bounded, although many results only require the minimal regularity property

$$\int \ell n[|f(\omega)|+1] d\mu(\omega) < \infty \tag{3}$$

We will occasionally discuss unbounded f's, in which case we will freely use those results which hold in the more general setting.

Two special subclasses of stochastic Jacobi matrices have received especial attention: The situation where V is an almost periodic function and $\Omega$ is just the hull of V (see, for example, the appendix of Avron-Simon [1] for background on almost periodic functions), and the situation where V is a family of independent, identically distributed random variables, a setup known as the Anderson model. We will let $d\kappa$ denote the probability density of V in this case.

These families of operators have received considerable attention in the theoretical physics literature. The Anderson model is supposed to be a caricature of the effect of impurities on electron motion in solids, and of electron motion in amorphous materials, like glass. The almost periodic models may describe certain alloys and the recently discovered quasicrystals. The mathematical physics literature has discussed both these models and their continuum analogs where $\ell^2(\mathbb{Z}^\nu)$ is replaced by $L^2(\mathbb{R}^\nu)$ and H becomes a differential operator. We will occasionally mention results that are not known to extend to the continuum case, but in the interests of simplicity, we will restrict our discussion to the discrete Jacobi matrix case. This way, one can avoid getting bogged down in technical subtleties; indeed, these technical problems can often be nontrivial, so that much more is known currently about the discrete case than about the continuum case.

The deepest and most interesting feature of these models concerns the spectral properties of the operators. There are recent reviews of these things in the book of Cycon et al. [14], and the lecture notes of Carmona [8] and Spencer [43]. The density of states, which we discuss here, is a less interesting object, but one which has evoked a considerable literature because it is a basic object of some use in the deeper analysis, and simply related to directly measurable quantities in physical systems.

Let $H_{\omega,L}$ denote the restriction of H, thought of as an infinite matrix, to indices with $|i_1|,\dots,|i_\nu| \le L$. Thus H is a matrix of dimension $(2L+1)^\nu$. The *integrated density of states* (ids) is defined by

(4) $$k(E) = \lim_{L\to\infty}(2L+1)^{-\nu}\#(\text{of e.v. of } H_{\omega,L} \le E)$$

That the limit exists is a result going back to Benderskii-Pastur [6]. There have been numerous refinements of this existence theorem which is essentially a consequence of the ergodic theorem. The following result is proven in Avron-Simon [1]. It discusses the "typical spectrum" using another consequence of the ergodic theorem, namely, that there is a subset, W, of $\Omega$ of full measure and a subset, $\Sigma$, of $\mathbb{R}$ so that the spectrum of $H_\omega$ is $\Sigma$ whenever $\omega \in W$. The definition (4) is not so convenient as an initial definition since the ergodic theorem allows a set of measure zero where the limit fails to exist, and because the set of E is uncountable, this can cause problems. This explains why we deal with vague convergence: The separability of the continuous functions allows one to deal with one set of full measure. One thus defines a measure $dk_{\omega,L}$ to be the point measure giving weight $(2L+1)^{-\nu}$ to each eigenvalue of $H_{\omega,L}$. Degenerate eigenvalues are given multiple weight, so that $dk_{\omega,L}$ is a probability measure. We will also define $\chi_L$ to be the projection onto those vectors in $\ell^2$ supported in the region where

$|i_1| \le L, \ldots, |i_\nu| \le L.$

<u>Theorem 1.1</u> There is a subset, $W_1$, of $\Omega$ of full measure, and an $\omega$-independent measure, $dk$, on $\mathbb{R}$ so that, for all $\omega$ in $W_1$

$$dk_{\omega,L} \to dk \tag{5}$$

in the vague topology. The typical spectrum of H is precisely the support of dk. Moreover, for any continuous function, f:

$$\int f(E)dk(E) = \text{Exp}(f(H)(0,0))$$

where f(H)(i,j) are the matrix elements of the operator f(H), and thus, by translation invariance, for such f,

$$\int f(E)dk(E) = \lim_{L\to\infty}(2L+1)^{-\nu}\ \text{Exp}(\text{Tr}(\chi_L f(H_{\omega,L}))$$

$$= \lim_{L\to\infty}(2L+1)^{-\nu}\text{Tr}(\chi_L f(H_{\omega,L}))$$

where the final equality holds for a.e. $\omega$. If k(E) is defined to be the weight dk assigns to $(-\infty,E)$, then (Note: $H_\omega$, not $H_{\omega,L}$)

$$k(E+0) - k(E-0) = \text{Exp}(P_{\{E\}}(H_\omega)(0,0)) \tag{6}$$

$$= \lim_{L\to\infty}(2L+1)^{-\nu}\text{Tr}(\chi_L P_{\{E\}}(H_\omega))$$

where $P_S(A)$ represents the spectral projections of A. If this last quantity is zero (i.e. if k is continuous at E), then (4) holds and E is an eigenvalue of $H_\omega$ with probability zero.

One can also show that k can be defined for cutoff H's with a wide variety of boundary conditions.

The last statement in the theorem shows that continuity properties of k, which we discuss in sections 2 and 3, are especially important. Essentially, we will show that k is always continuous. We caution the reader that, because the set of reals is uncountable, it can happen that $H_\omega$ has point spectrum with probability one even though any fixed energy is an eigenvalue with probability zero.

In the case $\nu = 1$, there is a second quantity of interest which is related to k. Let $T_\omega(E,n)$ be the two by two matrix which maps data at (0,1) for solutions of the Schrödinger equation:

$$(7) \qquad u(n+1) + u(n-1) + V_\omega(n)u(n) = Eu(n)$$

(thought of as a numerical equation and not as an $\ell^2$-equation, i.e. u need not be $\ell^2$) to data at (n,n+1). Then the *Lyaponov exponent* is defined by

$$(8) \qquad \gamma(E) = \lim_{n\to\infty} \frac{1}{|n|} \ln \left|\left| T_\omega(n,E) \right|\right|$$

By the subadditive ergodic theorem, for each fixed complex E, the limit exists for a.e. $\omega$ and is independent of $\omega$. Unlike the case of the ids, one cannot be sure that one can find a set of full measure where the limiting relation (7) holds for all real E, although one can arrange this for all complex E with non-zero imaginary part and a.e. real E. The following result is known as the *Thouless formula*:

<u>Theorem 1.2</u> For all E, one has that

$$(9) \qquad \gamma(E) = \int \ln \left| E-E' \right| dk(E')$$

This result first appeared in the physics literature in papers of Thouless and Herbert-Jones. Their proof works directly for E with Im E $\neq 0$, but there are technical difficulties for real E. The first rigorous proof for real E was obtained by Avron-Simon [2]; see Craig-Simon [12] for a different proof along the same lines. The idea of these proofs is to relate the eigenvalues of H (or more properly, the restrictions to [0,L-1] and [1,L]) to the vanishing of matrix elements of T(E), to note that these matrix elements are monic polynomials in E, and so write them in terms of eigenvalues. [2] handles real E using the theory of Hilbert transforms, while [12] uses subharmonic functions. There is a second approach to the Thouless formula using Weyl m-functions: See Johnson-Moser [20], Kotani [23] and Simon [36].

One should emphasize that k(E) is a bad indication of the spectral properties of H. As we will see in Section 7, there exist two distinct families of stochastic Jacobi matrices which have identical ids's, but so that one family has pure point spectrum with probability one, and the other has singular continuous spectrum with probability one. The ids does determine the absolutely continuous spectrum because of the Thouless formula and the following theorem of Kotani [23]:

Theorem 1.3 For any one-dimensional stochastic Jacobi matrix, the set of real E for which the Lyaponov exponent vanishes is the essential support of the absolutely continuous spectrum for a typical H.

Even here, the determination from k(E) is a global one, and doesn't really have much to do with the density of states. If the conjecture that the higher dimensional Lloyd model has some extended states is correct, then in higher dimensions, one would know that the absolutely continuous spectrum is not determined by the ids.

## 2. Continuity of k

In this section, we will discuss the idea beyond the very simplest proof of the following fundamental result:

Theorem 2.1 The ids, k(E), is continuous for any stochastic Jacobi matrix.

This result was proven in the one-dimensional case by Pastur [34]. The higher dimensional result was proven by Craig-Simon [13], who proved a stronger result which we will discuss in the next section. An elementary proof in the higher dimensional case was subsequently found by Delyon-Souillard [16].

The key idea in both the Pastur and Delyon-Souillard proofs is to exploit formula (6). To prove that k is continuous, one needs only show that

$$(2L+1)^{-\nu}\mathrm{Tr}(\chi_L P_{\{E\}}(H_\omega)) \to 0 \tag{10}$$

In the one-dimensional case, this is trivial since $P_{\{E\}}(H_\omega)$ is at most two-dimensional. In the higher dimensional case, Delyon-Souillard make use of the fact that, for projections P and Q

$$\mathrm{Tr}(PQ) \leq \dim \ Q[\mathrm{Ran} \ P] \tag{11}$$

so that one needs only show that the restriction of the set of $\ell^2$-eigenfunctions to a finite box forms a space whose dimension grows at a rate small compared to the volume of the box. In fact, Delyon-Souillard obtain a bound which only grows as the surface area of the box.

It is an annoying and unfortunate fact that there is still no proof of continuity of the ids in the continuum case except in one-dimension, where the Pastur argument goes through. This is the most important open question in the study of the ids for stochastic Schrödinger operators.

### 3. Log-Hölder Continuity of k

Craig-Simon [13] proved the following extension of Theorem 2.1:

<u>Theorem 3.1</u> The ids, k(E), is log-Hölder continuous for any stochastic Jacobi matrix.

A function, f, is called log-Hölder continuous if and only if there is a constant, C, so that

$$|f(x)-f(y)| \le C(\ln|x-y|^{-1})^{-1}$$

for all x,y with $|x-y| \le \frac{1}{2}$.

This theorem depends on the following elementary lemma:

<u>Lemma 3.2</u> Let dq be a measure of compact support with distribution function q. If

$$\int \ln|x-y|\,dq(y) \ge 0$$

then q is log-Hölder continuous.

The idea is that one cannot lose log-Hölder continuity at a point E' without dq being so concentrated that the integral diverges to $-\infty$ at E'. Given the lemma, the Thouless formula immediately implies Theorem 3.1 in the one-dimensional case; this was already noted in [12]. This is because the Lyaponov exponent, as a limit of positive quantities (the matrix T has determinant 1, and thus norm at least 1) is positive. The general case is proven by showing that the integral is positive also in the multi-dimensional case, for the integral can be shown to be the limit of the average of the non-negative Lyaponov exponents for strips.

There is a sense in which Theorem 3.1 is essentially optimal, for given $\varepsilon$, Craig [11] has constructed examples of almost periodic functions (actually, only in a weak sense; see Pöschel [35] for strictly almost periodic examples) for which there are points, E, with

$$\lim_{\delta \downarrow 0} \left|k(E+\delta)-k(E)\right| / [\ell n\ \delta^{-1}]^{1+\varepsilon} = \infty$$

Moreover, we will see that there exist random potentials which yield a k which is not Hölder continuous of any prescribed strictly positive order.

## 4. The One-Dimensional Anderson Model: Positive Results

To go beyond Theorem 3.1 and find smoothness properties of k, one must be prepared to make some special hypotheses, as the discussion at the end of the last section makes clear. It is clear that one should not look for much smoothness in the case of almost periodic potentials, for it is a general phenomenon (see the discussion in section 9) that the spectra of such operators tend to be Cantor sets, that is, closed, nowhere dense sets. The corresponding k cannot be $C^1$ because its derivative is zero on the complement of the spectrum, which is dense. It is therefore natural to look at the Anderson model. In general, LePage [25,26] has proven the following result:

<u>Theorem 4.1</u> The density of states, k(E), associated to any one-dimensional Anderson model, is Hölder continuous of some strictly positive order.

In the next section, we will mention examples of Anderson models whose ids fails to be Hölder continuous of any given prescribed order. Thus, one must make some additional assumptions on the input measure, $d\kappa$, in order to be certain that k has greater regularity properties. Given that the k associated to V = 0 is not $C^1$ but has a divergent first derivative at E = 2 and -2, one might naively expect that k cannot be too smooth but, in fact, the randomness is smoothing. Not only is k $C^\infty$ if $d\kappa$ is $C^\infty$, but under some minimal regularity assumptions on $d\kappa$, k is already $C^\infty$. This phenomenon was first proven to occur by Simon-Taylor [41], whose results applied to the case originally studied by Anderson, where

$$d\kappa(x) = \frac{1}{b-a} \chi_{[a,b]}(x)dx$$

Subsequently, Campinino-Klein [7] and March-Snitzman [30] proved results which complement and/or extend the results of [41]. The following is proven by Campinino-Klein:

<u>Theorem 4.2</u> Consider the one-dimensional Anderson model with input distribution $d\kappa$. Suppose that $d\kappa$ has moments of all orders, and its Fourier transform $m(p) = \int e^{-ipx} d\kappa(x)$ obeys

$$|m(p)| \leq C(1+|p|)^{-\alpha}$$

for some $C, \alpha > 0$. Then the ids, k(E), is $C^\infty$.

The smoothness of k associated to random operators viz-a-viz the singularities of the free case is illuminated by the fact that the singularities in the free case are at the edge of the spectrum, where the random case has the Lifschitz tail behavior to be discussed in

Section 8.

## 5. The One-Dimensional Anderson Model: Negative Results

There is an Anderson-type model of especial interest in providing counterexamples for regularity results that one might conjecture. This is what might be called the Bernoulli-Anderson model, where the input measure, $d\kappa$, is a pure point measure with two point support, i.e.

$$d\kappa = \theta\delta_a + (1-\theta)\delta_b$$

We will call this the Bernoulli-Anderson model with parameters $a,b,\theta$. Halperin [19] studied a closely related continuum model and showed nonregularity of k. His argument can easily be made rigorous, and this was done by Simon-Taylor. The result is:

<u>Theorem 5.1</u> The Bernoulli-Anderson model with parameters $a,b,\theta$ has a $k(E)$ which is not Hölder continuous of any order greater than

$$\alpha_0 = 2\left|\log(1-\theta)\right|/\text{Arc cosh}(1 + \tfrac{1}{2}\left|a-b\right|)$$

Note that $\alpha_0$ goes to zero as $\left|a-b\right| \to \infty$, or as $\theta \downarrow 0$, showing that one cannot improve on Theorem 4.1 without making a restriction on $d\kappa$, which will eliminate the Bernoulli-Anderson model. The idea behind the proof is quite simple. One finds certain energies about which the finite volume eigenvalues are clumped. These are eigenvalues for the operator obtained by surrounding a finite array of a's and b's by a sea of b's. Since the corresponding eigenfunctions decay exponentially, one can show that the system in an enormous box will have one eigenvalue exponentially near the infinite volume eigenvalue for each large sub-box of the big box containing the finite array surrounded by b's.

The Bernoulli-Anderson model has evoked considerable interest in the physics and chemical physics literature. This is partly because it models a binary alloy and partly because, before the advent of high speed computers, it was about the only model where one could reasonably compute the ids numerically. There are a number of features of the ids of the model which are hinted at by numerical and theoretical studies, but not yet rigorously proven:

(1) It is likely that, for suitable values of the parameters, the Bernoulli-Anderson model has an ids, k, for which dk has a singular continuous component; see Simon-Taylor [41].

(2) Luck-Nieuwenhuizen [29] have an analysis of the structure of k(E) at the energies described above (eigenvalues for the operator obtained by surrounding a finite array of a's and b's by a sea of b's), which suggests, but does not rigorously prove, the precise nature of the singularities at these energies.

(3) There are certain "special energies" at which the density of states is supposed to vanish; see Endrullis and Englisch [17].

## 6. The Higher Dimensional Case

Much less is known about the Anderson model in dimension greater than one. All indications are that the ids gets better behaved as dimension increases, so it is not an unreasonable conjecture that, for any Anderson model in dimension greater than one, the ids is $C_\infty$. Unfortunately, all the results proven thus far have involved showing that dk is about as well behaved as the input measure $d\kappa$, and there are no $C^\infty$ results for cases where $d\kappa$ has compact support. One of the nicest results is the following one proven by Wegner [45]:

<u>Theorem 6.1</u> Let k be the ids of an Anderson model whose input measure, $d\kappa$, is absolutely continuous with bounded Radon-Nikodym derivative. Then k is Lipschitz continuous.

In addition to the ideas of Wegner, there is an alternate proof using ideas of Simon-Wolff [42] on averages of the spectral measures under rank one perturbations.

There is also a result of Constantinescu, Fröhlich and Spencer [10] which says that if $d\kappa$ is absolutely continuous with a Radon-Nikodym derivative which is analytic in a sufficiently wide strip, then k is real analytic either in the region where $|E|$ is large or for large coupling constant.

## 7. Cauchy Models

There is one class of stochastic Jacobi matrices which is useful because one can compute the ids precisely. These are models where V has a Cauchy distribution with restrictions to be made precise on the correlations between V's at distinct sites. The first model of this type for which the ids was computed is the Anderson model with a Cauchy density for $d\kappa$, i.e.

$$d\kappa(x) = \frac{\lambda}{\pi} \frac{dx}{x^2+\lambda^2}$$

This model is known as the Lloyd model [28]. Much more recently, Grempel et al. [18] computed the ids in the almost periodic potential with

(12) $$V(n) = \lambda \tan(\pi\alpha n+\theta)$$

a model which has come to be called the Maryland model, after the place where Grempel et al. worked. They found the remarkable fact that the ids was the same in the two models for the same value of $\lambda$, and in particular, the ids in the Maryland model is independent of the frequency $\alpha$ so long as it is irrational. Motivated by this discovery, Simon [37] proved the following:

Theorem 7.1 Let k be the ids for a stochastic Jacobi matrix whose potential has the form:

$$V(n) = \sum_{j=1}^{J} \alpha_j \tan(\beta_{j,n}+\theta)$$

where $\alpha_j$, $\beta_{j,n}$, $\theta$ are random variables with the two restrictions that $\theta$ is uniformly distributed and $\alpha_j \geq 0$, $\Sigma\alpha_j = \lambda$. Then

$$k(E) = \int \frac{\lambda}{\pi} \frac{1}{\lambda^2+(E-E')^2} k_0(E')dE'$$

where $k_0(E)$ is the ids for the free model in the corresponding dimension.

The Lloyd model corresponds to the case where J = 1 and the $\beta_j$ are uniformly and identically distributed, independent random variables.

Simon [38] has proven that there are some values of $\alpha$ for which the Hamiltonian corresponding to (12) has point spectrum, and other values for which the Hamiltonian has singular continuous spectrum. Since the ids in the two cases are the same, one sees that the ids cannot distinguish between point and singular continuous spectrum.

## 8. Lifschitz Tails

It is not hard to show that (see Kunz-Souillard [24]) the spectrum of a typical H for the $\nu$-dimensional Anderson model is given by

$$\text{spec}(H) = \text{spec}(H_0) + \text{supp}(d\kappa) \tag{13}$$

Suppose that $\text{supp}(d\kappa) = [a,b]$ so that $a - 2\nu$ is the bottom of the spectrum of H and thus, by Theorem 1.1, $k(E) = 0$ for $E < a - 2\nu$. In cases where k is smooth, it must go to zero as E approaches $a - 2\nu$ from above faster than any polynomial. The rate at which it goes to zero was first determined by E.M. Lifschitz [27], so that this region is known as the Lifschitz tail. The leading behavior is given by the formula

$$k(E) \sim \exp[-(E-a-2\nu)^{-\nu/2}] \tag{14}$$

Lifschitz provided a simple intuition about why this formula holds: For a state to have energy only $\varepsilon$ above the minimum value, both its kinetic and potential energies must be small. Since the kinetic energy of a state of extent L is of order $L^{-2}$, we must have that

$$\varepsilon \sim L^{-½}$$

For the potential energy to be of order $\varepsilon$, most of the sites in this box must have a potential value very close to the minimum value a, and this will have a small probability of order exp(-S) where S is the number of sites in the box, i.e. $S \sim L^\nu = \varepsilon^{-\nu/2}$.

The earliest proofs of Lifschitz's result tended to use rather sophisticated arguments from the theory of large deviations. More recently, proofs have been given closer to the spirit of Lifschitz's original arguments; these proofs exploit Dirichlet-Neumann bracketing;

see, for example, Kirsch-Martinelli [21], Simon [39] and Mezincescu [31]. (14) is proven in the sense that

(14')
$$\lim_{E \downarrow a+2\nu} [\ell n(E-a-2\nu)]^{-1} \ell n[\ell nk(E)^{-1}] = -\nu/2$$

under the requirement that $[a,a+\delta)$ doesn't vanish faster than polynomially as $\delta \downarrow 0$.

These Lifschitz tails which occur at the outer edges of the spectrum are occasionally called "external Lifschitz tails". There has also been study of the situation where $\text{supp}(d\kappa) = [a,b] \cup [c,d]$ with $c - b > 4\nu$. In that case, there is a gap in the spectrum of H, and one expects that the approach of k(E) to its value in the gap as E approaches the gap from within the spectrum has Lifschitz behavior. These "internal Lifschitz tails" have been proven to occur by Mezincescu [32], and subsequently by Simon [40].

This isn't the end of the story concerning Lifschitz tails: For random plus periodic potentials, Kirsch-Simon [22] have proven that there are external Lifschitz tails, but no proof of internal Lifschitz tails has been found.

**9. Gap Labeling**

There is a final aspect of the density of states special to the almost periodic case that we should mention, especially since it suggests that "normally" the ids will not be $C^1$ in these cases. The *frequency module* of an almost periodic function on $\mathbb{Z}$ is defined as follows: Any almost periodic function has an average:

$$Av(f) = \lim_{L\to\infty} \frac{1}{2L+1} \sum_{|n|\le L} f(n)$$

The frequency module of an almost periodic function is the additive subgroup of $\mathbb{R}$ generated by 1 and those frequencies, $\alpha$, for which $Av(e^{-2\pi i\alpha n}f) \neq 0$. There is a more elegant and illuminating definition which is also longer; it is discussed, for example, in [1]. In the continuum case, the 1 is not included; it is a reflection of the periodicity of the lattice. There are definitions around which differ by factors of $\pi, 2\pi$, and 2 from the one we give here, and the fact that the gap labeling theorem we give and the one in Johnson-Moser [20] differ by a factor of 2 is resolved by differing definitions of the frequency module. Our definition is such that, if f is a periodic function of period L, its frequency module is $\{n/L \mid n\in\mathbb{Z}\}$.

If f is quasiperiodic, i.e. if

$$f(n) = F(2\pi\alpha_1 n, \ldots, 2\pi\alpha_\nu n)$$

for a continuous function, F, on the $\nu$-dimensional torus, then the frequency module is always contained in the set $\{\sum_{j=0}^{\nu} n_j\alpha_j \mid n_j\in\mathbb{Z}; \alpha_0=1\}$ and will equal that set if F has enough non-zero Fourier coefficients. f is quasiperiodic if and only if its frequency module is finitely generated, and it is limit periodic (i.e. a uniform limit of periodic functions) if and only if its frequency module has the property that any two elements in it have a common divisor in it.

The basic gap labeling theorem is:

<u>Theorem 9.1</u> Let H be an almost periodic one-dimensional Jacobi matrix. Then the value of the ids in any gap in the spectrum of H lies in the

frequency module.

Of course, the value also always lies in [0,1]. This phenomenon was first found by Claro and Wannier [44], and first rigorously proven in continuum models by Johnson-Moser [20], and then for discrete models by Bellissard, Lima and Testard [4]. Johnson-Moser use a homotopy argument which was carried over to the discrete case by Delyon-Souillard [15]. Bellissard, Lima and Testard use some C*-algebra techniques, and their argument has been extended to the higher dimensional case by them [3].

The relevance of the gap labeling theorem to regularity of the ids comes from the following meta-theorem:

Meta-theorem 9.2 A "generic" almost periodic one-dimensional Jacobi matrix has gaps in its spectrum where the ids takes each possible allowed value (i.e. all numbers in the frequency module which lie in (0,1)).

This result has been proven in the limit periodic case [9.33,1] and (in a weakened form) for the case where $V(n) = \lambda \cos(2\pi\alpha n+\theta)$ [5] for suitable notions of generic. If the almost periodic function is not strictly periodic, then the frequency module is dense in $\mathbb{R}$. Gaps in the spectrum are open sets on which k is constant, so on which k has a zero derivative. If every allowed value occurs in a gap, then the spectrum is a Cantor set (nowhere dense, but not necessarily of zero measure), and k is a Cantor function, which means it cannot be $C^1$ in the neighborhood of any point of the spectrum.

Acknowledgment

This research was partially supported by the United States National Science Foundation under grant DMS84-16049.

References

1. Avron, J. and Simon, B., Almost periodic Schrödinger operators, I. Limit periodic potentials, Comm. Math. Phys. 82 (1982), 101-120

2. Avron, J. and Simon, B., Almost periodic Schrödinger operators, II. The integrated density of states, Duke Math. J. 50 (1983), 369-391

3. Bellissard, J., Lima, R. and Testard, D., Almost periodic Schrödinger operators, ZIF preprint

4. Bellissard, J., Lima, R. and Testard, D., A metal-insulator transition for the almost Mathieu model, Comm. Math. Phys. 88 (1983), 207-234

5. Bellissard, J. and Simon, B., Cantor spectrum for the almost Mathieu equation, J. Func. Anal. 48 (1982), 408-419

6. Benderskii, M. and Pastur, L., On the spectrum of the one-dimensional Schrödinger equation with a random potential, Math. USSR Sb. 11 (1970), 245

7. Campanino, M. and Klein, A., A supersymmetric transfer matrix and differentiability of the density of states in the one-dimensional Anderson model, Comm. Math. Phys., to appear

8. Carmona, R., Lectures on random Schrödinger operators, to appear in Proc. 14th St. Flour Probability Summer School (1985), Springer

9. Chulaevsky, V., On perturbations of a Schrödinger operator with periodic potential, Russian Math. Surveys 36 (1981), 143

10. Constantinescu, F., Fröhlich, J. and Spencer, T., Analyticity of the density of states and replica method for random Schrödinger operators on a lattice, J. Stat. Phys. 34 (1984), 571-596

11. Craig, W., Pure point spectrum for discrete almost periodic Schrödinger operators, Comm. Math. Phys. 88 (1983), 113-131

12. Craig, W. and Simon, B., Subharmonicity of the Lyaponov index,

Duke Math. J. 50 (1983), 551-560

13. Craig, W. and Simon, B., Log Hölder continuity of the integrated density of states for stochastic Jacobi matrices, Comm. Math. Phys. 90 (1983), 207-218

14. Cycon, H., Froese, R., Kirsch, W. and Simon, B., *Topics in the theory of Schrödinger operators*, to appear

15. Delyon, F. and Souillard, B., The rotation number for finite difference operators and its properties, Comm. Math. Phys. 89 (1983), 415

16. Delyon, F. and Souillard, B., Remark on the continuity of the density of states of ergodic finite difference operators, Comm. Math. Phys. 94 (1984), 289

17. Endrullis, M. and Englisch, H., Special energies and special frequencies, preprint

18. Grempel, D., Fishman, S. and Prange, R., Localization in an incommensurate potential: An exactly solvable model, Phys. Rev. Lett. 49 (1982), 833

19. Halperin, B., Properties of a particle in a one-dimensional random potential, Adv. Chem. Phys. 31 (1967), 123-177

20. Johnson, R. and Moser, J., The rotation number for almost periodic potentials, Comm. Math. Phys. 84 (1982), 403

21. Kirsch, W. and Martinelli, F., Large deviations and Lifshitz singularity of the integrated density of states of random Hamiltonians, Comm. Math. Phys. 89 (1983), 27-40

22. Kirsch, W. and Simon, B., Lifshitz tails for periodic plus random potentials, J. Stat. Phys., to appear

23. Kotani, S., Ljaponov indices determine absolutely continuous spectra of stationary random one-dimensional Schrödinger operators, Stochastic analysis (1984), K. Ito (ed.), North Holland, pp. 225-248

24. Kunz, H. and Souillard, B., Sur le spectre des operateurs aux differences finies aleatoires, Comm. Math. Phys. 78 (1980), 201-246

25. LePage, E., Empirical distribution of the eigenvalues of a Jacobi matrix, Probability measures on groups, VII, Springer Lecture Notes Series 1064, 1983, pp. 309-367

26. LePage, E., in prep.

27. Lifschitz, I., Energy spectrum structure and quantum states of disordered condensed systems, Sov. Phys. Usp. 7 (1965), 549

28. Lloyd, P., Exactly solvable model of electronic states in a three-dimensional disordered Hamiltonian: Non-existence of localized states, J. Phys. C2 (1969), 1717-1725

29. Luck, J. and Nieuwenhuizen, T., Singular behavior of the density of states and the Lyaponov coefficient in binary random harmonic chains, J. Stat. Phys., to appear

30. March, P. and Snitzman, A., Some connections between excursion theory and the discrete random Schrödinger equation, with applications to analyticity and smoothness properties of the density of states in one dimension, Courant preprint

31. Mezincescu, G., Bounds on the integrated density of electronic states for disordered Hamiltonians, Phys. Rev., to appear

32. Mezincescu, G., Internal Lifschitz singularities of disordered finite-difference Schrödinger operators, Comm. Math. Phys., to appear

33. Moser, J., An example of a Schrödinger equation with an almost periodic potential and nowhere dense spectrum, Comm. Math. Helv. 56 (1981), 198

34. Pastur, L., Spectral properties of disordered systems in one-body approximation, Comm. Math. Phys. 75 (1980), 179

35. Poschel, J., Examples of discrete Schrödinger operators with pure point spectrum, Comm. Math. Phys. 88 (1983), 447-463

36. Simon, B., Kotani theory for one-dimensional stochastic Jacobi matrices, Comm. Math. Phys. 89 (1983), 227

37. Simon, B., The equality of the density of states in a wide class of tight binding Lorentzian models, Phys. Rev. B27 (1983), 3859-3860

38. Simon, B., Almost periodic Schrödinger operators, IV. The Maryland model, Ann. Phys. 159 (1985), 157-183

39. Simon, B., Lifschitz tails for the Anderson model, J. Stat. Phys. 38 (1985), 65-76

40. Simon, B., Internal Lifschitz singularities in the Anderson model, in prep.

41. Simon, B. and Taylor, M., Harmonic analysis on SL(2,R) and smoothness of the density of states in the one dimensional Anderson model, Comm. Math. Phys., to appear

42. Simon, B. and Wolff, T., Singular continuous spectrum under rank one perturbations and localization for random Hamiltonians, Comm. Pure Appl. Math., to appear

43. Spencer, T., The Schrödinger equation with a random potential - a mathematical review, to appear in Proc. 1984 Les Houches Summer School

44. Wannier, G. and Claro, F., Magnetic subband structure of electrons in hexagonal lattices, Phys. Rev. B19 (1979), 6068

45. Wegner, F., Bounds on the density of states in disordered systems, Z. Phys. B44 (1981), 9-15.

# LIMIT THEOREMS FOR MEASURE-VALUED DIFFUSION PROCESSES THAT ARISE IN WAVE PROPAGATION IN A RANDOMLY INHOMOGENOUS OCEAN *

Halil Mete Soner

Department of Mathematics
Carnegie-Mellon University
Schenley Park
Pittsburgh, P.A. 15213

## I. Introduction

In this paper we investigate the limit behaviour of a certain class of diffusion processes $Y^N(t) \in \Sigma^N$, (N-1)-dimensional simplex, as N approaches to infinity. We obtain an analog of the law of large numbers for this system. That is, the weak convergence of $Y^N(\cdot)$ to a deterministic process in C ([0,∞); M([0,1])), the space of probability measure valued continuous functions on [0,∞). Then, we prove the convergence of appropriately normalized fluctuations of $Y^N(t)$ around its mean, to a distribution-valued Gaussian process.

The diffusion process we study is obtained by W. Kohler and G.C. Papanicolau [3] as the limit behavious of the trapped modes of a sound pressure field propagating in a randomly inhomegenous ocean. This limit is valid for small noise intensities and large distances from the source. Also, see the paper of F.D. Tappert [4] and the references therein for more information about the physical problem. First, we briefly describe the underlying physical model and some results of W. Kohler and G.C. Papanicolau. Let $e^{iwt}$ $p(r,\theta,z)$ be the sound pressure field in cylindirical coordinates with z measured ownward from the surface of the ocean. Then, $p(r,\theta,z)$ satisfies the following equation

$$\frac{\partial^2}{\partial r^2} p + \frac{1}{r}\frac{\partial}{\partial r}p + \frac{1}{r^2}\frac{\partial^2}{\partial \theta^2} p + \frac{\partial^2}{\partial z^2} p + k^2[n^2(z) + \varepsilon\mu(r,z)]p = \frac{\delta(r)}{2\pi r}\,\delta(z-z_0)$$

---

* This research was supported in part by the Institute for Mathematics and its Applications with funds provided by the National Science Foundation and the Office of Naval Research

with boundary condition $p(r,\theta,0)=0$. Here $\varepsilon$ is a small parameter, $n(z)$ denotes the mean index of refraction and $\varepsilon\mu(r,z)$ is its random fluctuations.
Define $\bar{p}(r,z) = (\sqrt{r})^{-1} \int_0^{2\pi} p(r,\theta,z)d\theta$. Neglecting the terms of the order $r^{-2}$, we obtain

$$\frac{\partial^2}{\partial r^2}\bar{p} + \frac{\partial^2}{\partial z^2}\bar{p} + k^2[n^2(z)+\varepsilon\mu(r,z)]\,\bar{p} = 0 \tag{1.1}$$

with boundary condition $p(r,0) = 0$ and a deterministic initial condition $p(0,z)= =\phi(z)$. The function $\phi$ is to be obtained by the matching-to-the-source procedure and assumed to be deterministic because for small $r$ random fluctuations do not accumulate.

If one assumes that, there is $d > 0$ such that $n^2(z)=1$ for $z > d$ and $n^2(z) > 1$ for $z < d$, the differential operator $L= \frac{d^2}{dz^2} + k^2n^2(z)$ with zero boundary condition at $z=0$ has finitely many points and a semi-infinite line in its spectrum. Let $\{\nu_j(\cdot) : j=1,\dots,N\}$ and $\{\nu(\cdot,x) : x < k^2\}$ be the normalized eigenfunctions (the modes) of $L$ with eigenvalues $\{\lambda_j : j=1\dots,N\}$ and $\{\lambda(x) : x < k^2\}$. Note that $N=N(k)$, the number of trapped modes, is an increasing function of the wave number $k$. Now, expand the solution $p(r,z)$ of (1.1) in terms of the modes.

$$\bar{p}(r,z) = \sum_{j=1}^{N} c_j(r)\nu_j(z) + \int_{-\infty}^{k^2} c(r,x)\,\nu(z,x)dx \tag{1.2}$$

One obtains a system of differential equations satisfied by the mode amplitudes $c_j(r)$ and $c(r,x)$ by substituting (1.2) into (1.1). Rewrite the mode amplitues as $c_j(r) = [c_j^+(r)\, e^{i\lambda_j r} + c_j^-(r)\, e^{-i\lambda_j r}]/\sqrt{\lambda_j}$ with the additional relation $e^{i\lambda_j r}\frac{d}{dr}c_j^+(r) + e^{-i\lambda_j r}\frac{d}{dr}c_j^-(r) = 0$. Define $c^+(r,x)$ and $c^-(r,x)$ similarly.

The complex random functions $c_j^+(r)$ and $c_j^-(r)$ are called forward and backward mode amplitudes. Finally, neglecting $c_j^-(r)$ and $c^-(r,x)$ (this constitutes the forward scattering approximation) and $c(r,x)$ for $x < 0$, one obtains the following stochastic differential equation.

$$\frac{d}{dr}c_j^+(r) = \varepsilon i \sum_{\ell=1}^{N} \mu_{ji}\; e^{i(\lambda_\ell-\lambda_j)r}\; c_\ell^+(r) + \varepsilon i \int_0^{k^2} \mu_j(r,x)e^{i(\sqrt{x}-\lambda_j)r}\; c^+(r,x)$$

where $\mu_{j\ell}(r) = 1/2\; k^2(\lambda_j\lambda_\ell)^{-1}\int_0^\infty \mu(r,z)\nu_j(z)\nu_\ell(z)dz$ and $\mu_j(r,x)$ is given by a similar expression. Also $c^+(r,x)$ solves a similar equation.

Suppose that in r fluctuations are stationary and satisfy on exponential mixing condition. Also, that they do not vary rapidly as a function of z and localized in the region $0 < z < d$. Under these assumptions, the energies of the trapped modes $(|c_1^+(r\varepsilon^{-2})|^2,\ldots,|c_N^+(r\varepsilon^{-2})|^2)$ with the distance r rescaled by $\varepsilon^{-2}$ have a limit as $\varepsilon$ tends to zero. The limit process for the trapped modes is decoupled from the one for the radiation modes. Moreover, if the eigenvalues $\lambda_j$'s are distinct along with their sums and differences, the infinitesimal generator $\mathcal{L}^N$ of the limit process $u^N(t) = (u^N(1,t),\ldots,u^N(N,t))$ has the following form:

$$\mathcal{L}^N f(u) = \sum_{i=1}^{N} \frac{\partial}{\partial u_i} f(u)\,[-b^N(i)u_i + \sum_{j=1}^{N} a_{ij}^N\,(u_j-u_i)] \; + \tag{1.3}$$

$$+ \sum_{i,j=1}^{N} \frac{\partial^2}{\partial u_i\,\partial u_j} f(u)\, u_i\,[\delta_{ij} \sum_{\ell=1}^{N} a_{i\ell}^N u_\ell - a_{ij}^N\, u_j] \quad , \quad \forall f \in C^2(R^N)$$

where $\delta_{i,j} = 0$ if $i \neq j$ and $\delta_{i,i} = 1$. The coefficients $a_{ij}^N$ and $b^N(i)$ are given in terms of time averages of the correlation functions of $\mu_{ij}$ and $\mu_i(\cdot,x)$ and for large N (equivalently for large k) they have the following form:

$$\begin{aligned}
&\text{(i)} \quad a_{ij}^N = N^2\, a^N\left(\tfrac{i+j}{2N}\right)\, \delta_{|i-j|,1} && ;i,j=1\ldots,N \\
&\text{(ii)} \quad \lim_{N\to\infty} |b^N(i)| = 0 && ;i<N \\
&\text{(iii)} \quad \lim_{N\to\infty} b^N(N)\,N^{-2} = +\infty &&
\end{aligned} \tag{1.4}$$

where $a^N(\cdot)$ is a bounded function of $[0,1]$. In addition to (1.4) we assume that there is a $C^2([0,1])$ function $a(\cdot)$ such that

$$\text{(i)} \quad \lim_{N\to\infty} \|a^N - a\|_{C^2} = 0$$

$$\text{(ii)} \quad a(x) > 0 \quad , \quad \forall\, x \in [0,1]$$

Also we will take $b^N(i)=0$ for $i < N$.

## 2. Measure-Valued Diffusion Processes

Let $M_c([0,1]$ denotes the set of positive measures on $[0,1]$ with total mass less than c. We construct a measure-valued process $Y^N(t)$ from the diffusion process $u^N(t)$ by $\langle Y^N(t),f\rangle = \sum u^N(i,t)f(i/N)$ for all $f \in C([0,1])$. The form of the generator $^N$ yields that if $Y^N(0) \in M_c([0,1]$ then $Y^N(t) \in M_c([0,1])$ for all $t > 0$. This simply means that the total energy of the trapped modes is not increasing in time. For simplicity, we assume that $Y^N(o) \in M_1([0,1])$ and drop the subscript c in the notation.

Now, let $\phi(\mu) = F(\langle\mu,f_1,\rangle\ldots,\langle\mu,f_k\rangle)$ for $\mu \in M(0,1])$, $F \in C^2(R^k)$ and $f_i \in C^2([0,1])$. Then, the infinitesimal generator of the Markov process $Y^N(t)$, denoted by $\mathcal{L}^N$ again, has the following form

$$\mathcal{L}^N \phi(\mu) = \sum_{i=1}^{N} \frac{\partial}{\partial z_i} F(\langle\mu,f_1\rangle,\ldots)\langle G^N\mu,\Delta^N f_i\rangle$$

$$+ \sum_{i,j=1}^{N} \frac{\partial^2}{\partial z_i \partial z_j} F(\langle\mu,f_1\rangle,\ldots)\langle G^N\mu \otimes G^N\mu, D^N(f_i \otimes f_j)\rangle$$

where $\otimes$ denotes the tensor product ,ie for all $h \in C([0,1]^2)$ $\langle \mu\otimes\mu,h\rangle = \int_0^1\int_0^1 h(x,y)\mu(dx)\mu(dy)$, and $G^N$ is a mapping from $M([0,1]$ into $M(\{\frac{1}{N}\ldots,\frac{N}{N}\})$ given by $\langle G^N\mu,f\rangle = \sum f(\frac{i}{N})\langle\mu,\zeta_i^N\rangle$. Here $\{\zeta_i^N(\cdot) : i=1..N\}$ is a partition of unity for $[0,1]$ with the property $\zeta_i^N(\frac{i}{N}) = 1$. Since $G^N\mu \in M(\{\frac{1}{N},\ldots,\frac{N}{N}\})$, we have to define $\Delta^N f(x)$ for $x=\frac{i}{N}$ and $D^N(f\otimes g)(x,y)$ for $x=\frac{i}{N}$, $y=\frac{j}{N}$.

$$\Delta^N f(\tfrac{i}{N}) = \begin{cases} N^2([f(\frac{i+1}{N}) - f(\frac{i}{N})]a^N(\frac{2i+1}{2N}) + [f(\frac{i-1}{N}) - f(\frac{i}{N})]a^N(\frac{2i-1}{2N})) & \text{if } i\neq 1,N \\ N^2([f(\frac{2}{N}) - f(\frac{1}{N})]\, a^N(\frac{3}{2N})) & \text{if } i=1 \\ N^2([f(\frac{N-1}{N}) - f(1)]\, a^N(1-\frac{1}{2N}) - b^N(N)f(i)) & \text{if } i=N \end{cases} \quad (2.2)$$

$$D^N(f\otimes g)(\tfrac{i}{N},\tfrac{j}{N}) = \begin{cases} N^2 a^N(\frac{2i+1}{2N})[f(\frac{i+1}{N}) - f(\frac{i}{N})][g(\frac{i+1}{N}) - g(\frac{i}{N})] & \text{if } i+1=j \\ 0 & \text{otherwise} \end{cases}$$

Remark 2.1
(i) Take $F(z)=z$ in (2.1) to obtain that $E\langle Y^N(t),f\rangle=\langle Y^N(0)f\rangle+\int_0^t E\langle Y^N(s),\Delta^N f\rangle ds$. Formally, passing to the limit as N tends to infinity we obtain the following equation for any limit point $Y(t)$ of $Y^N(t)$.

$$E\langle Y(t),f\rangle = \langle Y(0),f\rangle + \int_0^t E \langle Y (s), \frac{d}{dx}[a(\cdot) \frac{d}{dx}f(\cdot)]\rangle ds \tag{2.3}$$

And (2.3) should hold for all $f \in C^2([0,1])$ satisfying $\frac{d}{dx}f(0)=f(1)=0$.

(ii) Again use (2.1) to obtain,

$$\begin{aligned} \operatorname{var} \langle Y^N(t),f\rangle &= 2 \int_0^t E \langle Y^N(s) \times Y^N(s), D^N(f \times f)\rangle ds \\ &\leq 2 \|f'\|_\infty^2 \int_0^t \sum_{i=1}^{N-1} U^N(i,i+1,s)ds. \end{aligned} \tag{2.4}$$

where

$$U^N(i,j,t) = E[u^N(i,t)u^N(j,t)] \tag{2.5}$$

We now state a result which is the analog of the law of large numbers.

Theorem 2.2: Suppose that the initial conditions $Y^N(0)$ almost surely converge to a deterministic measure $\mu$ in weak$^*$. Then $Y^N(\cdot)$ converges weakly to a deterministic process $Y(\cdot)$ in $C([0,\infty);M([0,1]))$ and $Y(t,dx)=v(t,x)dx$ where $v(t,x)$ is the only solution of the following equation

$$\begin{aligned} &\frac{\partial}{\partial t}v(t,x) = \frac{\partial}{\partial x}[a(x) \frac{\partial}{\partial x}v(t,x)] && ,t>0 \ , \ x\in(0,1) \\ &\frac{\partial}{\partial x}v(t,0) = v(t,1)=0 && ,t>0 \\ &\lim_{t\downarrow 0} \int_0^1 f(x)v(t,x)dx = \langle f,\mu\rangle && ,f\in C([0,1]) . \end{aligned} \tag{2.6}$$

In order to prove the above theorem we need estimates of $U^N(i,j,t)$ defined by (2.5). So we will give the proof of Theorem 2.2 at the end of the next section.

## 3. Coupled Fluctuation Equations

Using (1.3) one derives the following equation for $U^N(i,j,t)$.

$$\frac{d}{dt}U^N(i,j,t) = [F\ U^N(\cdot,\cdot,t)](i,j) \quad , \ t>0, \ i,j=1..N$$

with initial conditions $U^N(i,j,0) = E[u^N(i,0)u^N(j,0)]$. Here F is a linear mapping from $R^{N^2}$ into itself and for $\{V(i,j):i,j=1...N\}$ , $[F\,V](i,j)$ is given by

$$[FV](i,j) = \begin{cases} 2 \sum_{k\neq i} a^N_{ki}\,[V(k,i)+V(i,k)-2V(i,i)] - 2\,b^N(i)V(i,i) & ,\text{if } i=j \\ \sum_{k=1}^{N} [a^N_{ik}\,[V(k,j)-V(i,j)]+a^N_{kj}\,[V(i,k)-V(i,j)]] - (b^N(i)+b^N(j)+2a^N_{ij}) & ,\text{if } i\neq j \end{cases} \quad (3.1)$$

We will obtain pointwise estimates of $U^N(i,j,t)$ by studying the spectral properties of F . But F is not self-adjoint and we introduce the following transformation to overcome this difficulty.

$$[SV](i,j) = [V(i,j)\ \delta_{i\neq j} + V(i,i)\ \delta_{i=j}\ \tfrac{1}{\sqrt{2}}\,] \quad (3.2)$$

$$[\bar{F}\ V](i,j) = [S[F[S^{-1}V]]](i,j) \quad (3.3)$$

It is elementary to check that $\bar{F}$ is self-adjoint and non-positive. Let $\{V_m(i,j) :i,j=1...N\}$ be normalized eigenvectors of $\bar{F}$ corresponding to eigenvalues $\lambda_m$ for $m=1...,N^2$ ,ie

(i) $[\bar{F}V_m](i,j)=\lambda_m V_m(i,j)$

(ii) $(V_m,V_n):= \sum V_n(i,j)V_m(i,j) = \delta_{m,n}$

(iii) $\lambda_{N^2} \leq \lambda_{N^2-1} \leq .... \leq\lambda_1 \leq 0$

Proposition 3.1: There is a constant $c>1$ ,independent of N and m, such that

$$-c\ (m^2\ ) \leq \lambda_m \leq -\ c^{-1}\ (m^2\ ) \quad (3.4)$$

$$\|V_m\|_\infty \leq c\ (m+1)N^{-1} \quad (3.5)$$

Proof: For any $V\varepsilon R^{N^2}$ define $Q(V)$ by

$$Q(V)=(V,\bar{F}V) = -\sum_{i\neq j} [\tfrac{1}{2} \sum_{k\neq i} a^N_{kj}(V(i,k)-V(i,j))^2 + \tfrac{1}{2} \sum_{k\neq j} a^N_{ik}(V(k,j)-V(i,j))^2 \quad (3.6)$$

$$+ a^N_{ij}((\sqrt{2}\ V(i,j)-V(i,i))^2 + (\sqrt{2}\ V(i,j)-V(j,j))^2)]$$

$$- \sum_{i,j} (V(i,j))^2\ (b^N(i)+b^N(j)).$$

Now let $\overline{a}^N_{ij} = N^2 \, \delta_{|i-j|=1}$ and $\overline{\Delta}^N$ be defined by

$$[\overline{\Delta}^N V](i,j) = \sum_{k=1} \overline{a}^N_{i,k}(V(k,j)-V(i,j)) + \overline{a}^N_{k,j}(V(i,k)-V(i,j))] - (b^N(i)+b^N(j))V(i,j)$$

Then, the quadratic form $\overline{Q}(V)=(V,\overline{\Delta}^N V)$ of $\overline{\Delta}^N$ has the same form as $Q(V)$. In fact, $\overline{Q}(V)$ is equal to the summation obtained by replacing $a^N$ by $\overline{a}^N$ and $\sqrt{2}$ by 1 in (3.6). By using the uniform positivity of $a^N$ we obtain the following inequalities.

$$c\,\overline{Q}(SV) \leq Q(V) \leq c^{-1}\overline{Q}(SV) \quad , \quad \forall V \in R^{N^2} \tag{3.7}$$

Now let $\overline{\lambda}_{N^2} \leq \ldots \leq \overline{\lambda}_1 \leq 0$ be the eigenvalues of $\overline{\Delta}^N$ with eigenvectors $\overline{V}_m$ and

$R_m = \{\sum_{i=1}^{m} \alpha_i V_i : \alpha_i \in R\}$ , $\overline{R}^c_m = \{\sum_{i=m}^{N^2} \alpha_i V_i : \alpha_i \in R\}$. Invoke the self adjointness of $\overline{F}$ and $\overline{\Delta}^N$ together with (3.7) to obtain:

$$\lambda_m = \inf\{Q(V)\|V\|_2^{-2} : V \in R_m\}$$

$$\leq C^{-1} \inf \{\overline{Q}(SV)\ \|V\|_2^{-2} \quad : \quad SV \in S(R_m) \quad R_m^{-c}\}$$

$$\leq (2C)^{-1} \inf \{\overline{Q}(U)\ \|U\|_2^{-2} \quad : \quad U \in S(R_m) \quad R_m^{c}\}$$

Since $\overline{R}_m$ is m-dimensional and $\overline{R}^c_m$ is $(N^2-m+1)$-dimensional linear subspaces of $R^{N^2}$, $S(R_m) \quad \overline{R}^c_m$ is non-empty. Moreover, $\overline{Q}(u)\|U\|_2^{-2} \leq \overline{\lambda}_m$ for $U \in \overline{R}^c_m$. Therefore,

$\lambda_m \leq (2c)^{-1}\,\overline{\lambda}_m$ . Similarly, one obtains the inequality $\lambda_m \geq (2c)\overline{\lambda}_m$. A simple calculation yields that $\overline{\lambda}_m$ and $\overline{V}_m$ are given by

$$\overline{\lambda}_m = 4N^2\,[\mathrm{Cos}\left(\frac{\pi(2k-1)}{2N-1} + \frac{\nu(k,N)}{N-1}\right) - 1][\mathrm{Cos}\left(\frac{\pi(2\ell-1)}{2N-1}\right) + \frac{\nu(\ell,N)}{N-1})-1]$$

$$V_m(i,j) = C_m \,\mathrm{Cos}([\frac{\pi(2k-1)}{2N-1} + \nu(k,N)\,](\frac{i-1}{N-1}) + \frac{(2k-1)\pi}{2(2N-1)} + \frac{\nu(k,N)}{2(N-1)}) \;\times$$

$$\mathrm{Cos}([\frac{\pi(2\ell-1)}{2N-1} + \nu(\ell,N)]\,(\frac{j-1}{N-1}) + \frac{(2\ell-1)\pi}{2(2N-1)} + \frac{\nu(\ell,N)}{2(N-1)})$$

for some $k,\ell \in \{1,..,N\}$ and a normalizing constant $C_m$. (Also $\nu(k,N)$ solves the equation $(1+b(N)N^{-2})\ \mathrm{Sin}(\nu(k,N)(2N-1)(2N-2)^{-1}) = \mathrm{Sin}\ (\nu(k,N)(2N-2)^{-1}+(2k-1)\pi(2N-1)^{-1} + \nu(k,N)\ (N-1)^{-1})$. Since $b(N)N^{-2}$ is large (see 1.4)) , $\nu(\ell,N)$ is of the order of $N^2(b(N))^{-1}$. Also $\overline{\lambda}_m$'s are ordered, therefore there is $c > 1$, independent of N, such that $-cm^2 \leq \lambda_m \leq -c^{-1}m^2$. This proves (3.4).

To prove (3.5) observe that,

$$\begin{aligned} \|V_m\|_\infty &\leq \max\ \{\|V\|_\infty \quad : \quad Q(V)=\lambda_m \quad \text{and} \quad \|V\|_2^2 = 1\} \\ &\leq \max\ \{\|V\|_\infty \quad : \quad Q(V) \geq -Cm^2 \quad \text{and} \quad \|V\|_2^2 = 1\} \\ &\leq \sqrt{2}\ \max\ \{\|U\|_\infty : \ \overline{Q}(U) \geq -C^2m^2 \quad \text{and} \quad \|S^{-1}U\|_2^2 = 1\}\ . \end{aligned} \tag{3.8}$$

Any $U \in R^{N^2}$ has the representation $U(i,j) = \sum_m \alpha_m\ \overline{V}_m(i,j)$ . Also, the explicit form of $\overline{V}_m$ yields that $\|\overline{V}_m\|_\infty \leq cN^{-1}$. Therefore, (3.8) implies the following inequality

$$\|V_m\|_\infty \leq cN^{-1}\ \max\ \{\sum_n |\alpha_n| \quad : \quad \sum_n \alpha_n^2\ |\overline{\lambda}_n| \ \leq cm^2\}$$

But for any $\{\alpha_n\}$ , $\sum_n |\alpha_n| \leq C(\sum_n \alpha_n^2 n^2)^{1/2} \leq c^2\ (\sum_n \alpha_n^2\ |\lambda_n|)^{1/2}$ . □

<u>Proposition 3.2</u>: <u>There is a positive constant c, independent of N and t, such that</u>

$$\text{(i)} \quad \sum_{i,j} [U^N(i,j,t)]^2 \ \leq C(N^2t\sqrt{t})^{-1}$$

$$\text{(ii)} \quad \sum \ U^N(i,i+1,t) \ \leq C(Nt)^{-1}$$

$$\text{(iii)} \quad -\ Q(SU^N(;;t)) \quad \leq C(N^2t^2\sqrt{t})^{-1}$$

$$\text{(iv)} \quad \lim_{N\to\infty}\ NE[u^N(N,t)] = 0 \qquad \forall\ t > 0\ .$$

<u>Proof</u>: $U^N(i,j,t) = \sum_m e^{\lambda_m t}\ V_m(i,j)\ (V_m,U^N(.,.,0))$ . Since $\sum_{i,j} U^N(i,j,0)=1$ , (3.5) yields that $|(V_m,U^N(.,.0))| \leq c(m+1)N^{-1}$. Therefore, (3.4) together with this observation yields (i) and (ii). The continuous analog of $-N^2Q(SU^N(;;t))$ is $\int_0^1 \int_0^1 \ |\nabla\nu(x,y,t)|^2\ dxdy$ where $\nu$ is the solution of the heat equation on $[0,1]^2$ with proper boundary conditons and diffusion coefficient. Therefore, same

type of "energy estimates" devised for the heat equation yields (iii).

$$NE[u^N(N,t)] \leq [N^2U^N(N,N,t)]^{1/2} \leq c \left[ \frac{N^4}{b^N(N)} \quad (-Q(SU^N(.,.,t))\right]^{1/4}$$

$$\leq c(t) \; [N^2/b^N(N)]^{1/4}$$

Use the assumption (1.4) (iii) to complete the proof of the proposition. □

Remark 3.3: Suppose that $E(u^N(i,0))^2 \leq CN^{-2}$ for all i. Then, the estimates in Proposition 3.2 would be time-independent.

Proof of Theorem 2.2

Tightness of the family $Y^N(\cdot)$ in $C([0,\infty) \; ; \; M \; ([0,1]))$ can be proved by using the technique developed by W.H. Fleming and M.Viot ([1], Lemma 4. Also see [6] and the estimates obtained in the previous section. In view of (2.4) and Proposition 3.2, any limit point of $Y^N(\cdot)$ is deterministic. So, it suffices to verify (2.3) to complete the proof of the theorem. Rewrite $E\langle Y^N(t),\Delta^N f\rangle$ as
verify (2.3) to complete the proof of the theorem. Rewrite $E\langle Y^N(t),\Delta^N f\rangle$ as

$$E\langle Y^N(t),\tilde{\Delta}^N f\rangle + E[u^N(N,t)].[a^N(1-\tfrac{1}{2N})(f(1-\tfrac{1}{N}) - f(1)) - b^N(N)f(1)]\, N^2 \text{ where } \tilde{\Delta}f(i/N) = \Delta^N f(i/N)$$

for $i<N$ and $\tilde{\Delta}f(N)=0$. Therefore, $E\langle Y^N(t),\tilde{\Delta}^N f\rangle$ converges to $E\langle Y(t), \frac{d}{dx}[a(\cdot)\frac{d}{dx}f(\cdot)]\rangle$. Since $f(1)=0$, $|E\langle Y^N(t),\Delta^N f-\Delta^N\tilde{f}\rangle| \leq C\,N\,E\,u^N(N,t)$. Now use Proposition 3.2 to complete the proof of the theorem. □

## 4. Fluctuations of $Y^N(t)$

In this section we study the fluctuations of $Y^N(t)$ around its means normalized by $\sqrt{N}$. Let $(Z^N(t),f)[\langle Y^N(t),f\rangle - E\langle Y^N(t),f\rangle]$. It is clear that any limit point of $Z^N(t)$ lies in a space larger than $M([0,1]$. So, following G. Kallianpur [2] we introduce the following spaces.

Let $\theta_n$ and $\lambda_n$ be the $L^2([0,1])$-normalized eigenfunctions and eigenvectors of the differential operator $Lf(x)=\frac{d}{dx}\,[a(x)\frac{d}{dx}\,f(x)]$ with boundary conditions

$\frac{d}{dx} f(0)=f(1)=0$. Define H to be the $|||\cdot|||_H$ - closure of the linear space spanned by $\{\theta_n : n=1,..\}$ where $|||\theta|||_H^2 = \sum \lambda_n^4 <\theta,\theta_n>_{L^2}$ . Finally, $H_*$ is the dual of H with $L^2([0,1])$ pairing, endowed with weak$^*$-topology. Then the strong norm on $H_*$ is, $|||z|||_*^2 = \sum \lambda_n^{-4} (z,\theta_n)^2$ on $H_*$, where $(z,\theta)$ is the action of $z\varepsilon H_*$ at $\theta\varepsilon H$.

Lemma 4.1: Suppose that $E[u^N(i,0)]^4 \leq CN^{-4}$ for all $i=1..,N$ . Then for any $T>0$, there is $c(T) > 0$ such that

$$\sup_N E ( \sup_{0\leq t\leq T} |||z^N(t)|||_*^2 ) \leq c(T)$$

Proof: Set $M_N^n(t) = < Z^N(t),\theta_n) - \int_0^t (Z^N(s),\Delta^N\theta_n)ds$. Then, $M_N^n(t)$ is a martingale and the supermartingale inequality yields,

$$E(\sup_{0\leq t\leq T} (M_N^n(t))^2 \leq 4 E (M_N^n(T))^2 \tag{4.1}$$

$$= 4N\int_0^T E[<Y^N(s) \times Y^N(s),D^N(\theta_n \times \theta_n) >]ds$$

$$\leq 4 \|\theta_n'\|_\infty^2 \int_0^T N \sum_i U^N(i,i+1,s)ds.$$

$$\leq c(T) \|\theta_n'\|_\infty^2$$

In the last inequality, we have used Remark 3.3. Since $\Delta^N$ is the discretization of L, $|\Delta^N\theta_n( \frac{i}{N} )| \leq |\lambda_n| \|\theta_n^1\|_\infty \frac{1}{N}$ for all i. ( In fact, in order to obtain this estimate at i=N, one has to work with $\overline{\theta}_n(\cdot) = \theta_n(\cdot+\partial_N)$, where $\alpha_N$'s are properly chosen constants converging to zero as N tends to infinity. But this detail only complicates the proof, so we shall omit it). Therefore, $|(Z^N(t),\theta_n) - M_N^n(t) - \int_0^t \lambda_n(Z^N(s),\theta_n)ds| \leq |\lambda_n|\|\theta_n'\|_\infty(\overline{N})^{-1/2}$ . Use the variation of paramaters formula to obtain

$$|(Z^N(t),\theta_n) - M_N^n(t) - \int_0^t \lambda_n e^{\lambda_n(t-s)} M_N^n(s)ds| \leq |\lambda_n| \|\theta_n'\|_\infty N^{-1/2} \tag{4.2}$$

The form of $|||\cdot|||_*$, (4.1) and (4.2) yields

$$E( \sup_{0\leq\tau\leq T} |||Z^N(t)|||_*^2 ) \leq c(T)[ \sum_n \lambda_n^{-4} (\|\theta_n'\|_\infty^2 + |\lambda n|\|\theta_n'\|_\infty N^{-1/2})]$$

Since $a(\cdot)$ is smooth and uniformly positive, the sum in the above inequality is convergent.

□

Lemma 4.2 : Let $U^N(i,j,k,\ell,t)=E[u^N(i,t)u^N(j,t)u^N(k,t)u^N(\ell,t)]$ for $i,j,k,\ell=1\ldots,N$. Then, under the hypothesis of Lemma 4.1, $U^N(i,j,k,\ell,t) \leq cN^{-4}$ for some $c>0$.

Proof: $U^N$ satisfies the equation $\frac{d}{dt}U^N(i,j,k,\ell,t) = [F^{N,4}U^N(.,.,.,.,t)]$ where for $\{V(ijk\ell) : i,j,k,\ell=1\ldots N\}$ , $F^{N,4}V$ is given by

$$(F^{N4}V)(ijk\ell) = [F^N V(.,j,.,\ell)](i,k) + \ldots\ldots + [F\ V(i,j,.,.)(k,\ell)$$

and F is as in (3.1). Therefore, the same analysis as in Section 3 can be applied to this equation. Also, in view of the hypothesis the conclusion of the lemma holds at t=0. Now, combine these two observations to prove the estimate for all t>0.

□

Theorem 4.3: Under the hypothesis of Lemma 4.1, $Z^N(\cdot)$converges to $Z(\cdot)$ in $C([0,\infty); H_*$, weak$^*$). Moreover, for any $F\in C^2(R^k)$ and $Lf_i \in H$ the following expression is a martingale

$$F((Z(t),f_1),\ldots,(Z(t),f_k)) - \int_0^t \{\sum_{i=1}^{k} \frac{\partial}{\partial z_i} F(Z(s),f_1)\ldots)\ (Z(s),Lf_i) +$$

$$+ \sum_{i,j=1} \frac{\partial^2}{\partial z_i\,\partial z_j} F((Z(s),f_1)\ldots)\ [\int_0^1 v^2(x,s)a(x)\ f_i'(x)f'_j(x)dx]\}\ ds$$

where $v(x,s)$ is given by (2.6).

Proof: Let $M_N^n(t)$ be as in the proof of Lemma 4.1 and $\langle M_N^n\rangle(t)$ be its quadratic variation process, ie $\langle M_N^n\rangle(t) = N\int_0^t \langle Y^N(s)\times Y^N(s),D^N(\theta_n\times\theta_n)\rangle ds$. So, the previous lemma impies that $E\,|\langle M_N^n\rangle(t) - \langle M_N^n\rangle(s)|^2 \leq C_n\,(t-s)^2$ . Therefore,

using Lemma 4.1 and the technique used by W.H. Fleming and M. Viot one proves the tightness of $M_N^n(\cdot)$ for every n . In view of Lemma 4.1, this implies the tightness of $Z^N(\cdot)$ in $C([0,\infty)$, $H_*$ weak$_*$). Also, (2.1) yields that for every N the following expression is a martingale.

$$F((Z^N(t),f_1),\ldots.) - \int_0^t \{ \sum_{i=1}^{k} \frac{\partial}{\partial z_i} F((Z^N(s),f_1)\ldots) (Z^N(s),\Delta^N f_i) +$$

$$+ \sum_{i,j=1}^{k} \frac{\partial^2}{\partial z_i \partial z_j} F((Z^N(s),f_1)\ldots) \; N < Y^N(s) \times Y^N(s), D^N(f_i \times f_j) > \} ds.$$

The estimates we have proved enable us to pass to the limit as N tends to infinity. But note that $\Delta^N f_i(N)$ is of order N and to offset this term one has to shift $f_i$. Again we omit this technical argument.

□

**Acknowledgements**

I would like to thank W.H. Fleming for suggesting the problem and valuable comments and G.C. Papanicolaou for helpful discussions.

**References**

[1] W.H. Fleming and M. Viot, Some measure-valued Markov processes in population genetic theory, Indiana University Math. J. 28(5), 1979, 817-843.

[2] G. Kallianpur, Infinite dimensional stochastic differential equations, IMA preprint 1986.

[3] W. Kohler and G.C. Papanicolaou, Wave propagation in a randomly inhomegenous ocean, Springer Lecture Notes in Physics, No. 70, 1977.

[4] F.D. Tappert, The Parabolic approximation method, Springer Lecture Notes in Physics, No. 70, 1977.

[5] M. Viot, A stochastic partial differential equation arising in population genetics theory, Brown University LCDS Technical Report 75-3, 1975.

[6] M. Viot, Methodes de composite et de monotinie compacite pour les equations aux derivees partielles stochastique, These, Univ. de Paris, 1975.

# SPECTRAL PROPERTIES OF DISCRETE AND CONTINUOUS RANDOM SCHRÖDINGER OPERATORS: A REVIEW*

Bernard SOUILLARD

Centre de Physique Théorique**
Ecole Polytechnique
F-91128 Palaiseau
France

Abstract:
We review the mathematical results concerning the spectral properties of discrete and continuous Schrödinger equations with a random potential and of other random wave equations.

**Introduction**

The mathematical properties of *random* discrete and continuous Schrödinger equations have been studied now for more than ten years and become a very fashionable subject in the last five years. We are thus now in possession of many interesting results produced by many groups. It is the purpose of the present review paper to serve as a guide for the reader to most of the results in this domain; but we emphazise that we restrict here to review the spectral properties of random Hamiltonians, other types of properties such as the properties of the density of states being reviewed in [1]. In any case, no proofs are reproduced here, the reader being send back to the original papers.

Most of the motivations for this field and the interesting conjectures come from the physics of disordered solids and from the subject of wave propagation in inhomogeneous media. However we omit here references to the physics litterature, being concerned in this review only with mathematical results and we refer the interested reader to [2],[3],[4],[5] for the connection with physics.

The review is organized as follows:

1 - the discrete Schrödinger equation with a random potential.

2 - the one-dimensional ergodic discrete random Schrödinger operator.

3 - the one-dimensional non ergodic case: an example of a transition in the nature of the spectrum.

4 - the quasi-one dimensional case: systems in a strip.

5 - the multidimensional case: localization for large disorder.

6 - the Bethe lattice: the Anderson transition and the mobility edge.

7 - other discrete equations.

8 - non independant random potentials.

9 - continuous Schrödinger equations and other continuous wave equations.

## 1 - The Discrete Schrödinger Equation with a Random Potential

In most of this review we will consider the discrete Schrödinger equation with a random potential: it is a typical model and the most studied by physicists under the name "Anderson model", but we will also describe later the results for other classes of discrete random equations in Section 7, as well as the results for the continuous Schrödinger equation and other continuous wave equations in Section 9.

The discrete Schrödinger operator, also known as a Jacobi matrix is defined for $x,y \in \mathbb{Z}^d$ (or a subset of $\mathbb{Z}^d$) by

$$[H\Psi](x) = \Sigma_{|x-y|=1}\Psi(y) + V(x)\Psi(x) = [(H_0 + V)\Psi](x) \tag{1.1}$$

These Schrödinger hamiltonians define self adjoint operators on $L^2(\mathbb{Z}^d)$ and the set of $\Psi$ with finite support form a core for them. If the potential V is bounded these operators are bounded operators. Note that (1.1) is, up to an irrelevant constant diagonal part, the discretization of a continuous Schrödinger operator.

We want to study the case of random potentials; what does this mean? it means that the V(x) will be given according to some random process. The simplest case is the one where the V(x) will be choosen at each site x as independant random variables; actually for most of this review we are going to restrict to this case but in Section8 below, we will indicate the results known in the case of non independant random variables.

So we consider, for the time beeing, the V(x) to be choosen at each site as independant random variables with a probability distribution $\mu_x$ and thus for each realisation of V we have an operator H: we can thus consider from now the operator H as a random operator. The independant random variables V(x) can be identically distributed or not and we will be interested in both cases: as we will see below each of them leads to very interesting situations and results.

If the random variables are identically distributed, the random process associate to V is ergodic in the sense that if we consider the probability space $(\Omega,\mathrm{B},\mathbb{P})$ where $\Omega$ is the set of potentials on $\mathbb{Z}^d$, with the appropriate $\sigma$-algebra B and $\mathbb{P}$ is the probability on $\Omega$, then the probability $\mathbb{P}$ is ergodic with respect to the translations of $\mathbb{Z}^d$. Since as a general fact the spectral properties of H are measurable with respect to V [6] (see also [7]), in the case of an ergodic process any spectral property will hold with probability zero or one. For example, the spectrum $\sigma$ of H is then almost-surely a constant set [8], the pure point, singular continuous and absolutely continuous spectrum $\sigma_{pp}$, $\sigma_{sc}$, $\sigma_{ac}$ are also almost surely constant sets [6],[7].

These results do not depend on the independance assumption made above but as indicated in the quoted refences they do depend only on the ergodicity properties and thus apply to non independant potentials, to other discrete equations and to continuous equations.

In the identically distributed random case, the spectrum of H (i.e. the set which is almost-surely the spectrum of H) is exactly known [6]:

$$\sigma = [-2d,+2d] + \mathrm{Supp}\,\mu \tag{1.2}$$

where Supp $\mu$ denotes the support of the probability $\mu$, and A+B denotes the set $\{a+b;\ a\in A,\ b\in B\}$; note that $[-2d,+2d]$ is simply the spectrum of the operator $H_0$ and Supp $\mu$ almost surely the spectrum of V.

## 2 - The One-Dimensional Ergodic Discrete Random Schrödinger Operator

In this section we study the case of the operator (1.1) with the V(x) independant and identically distributed random variables according to a probability measure $\mu$. We postpone to other Sections below the discussion of the results for other situations.

For one-dimensional ergodic systems, it is useful to introduce the Lyapunov exponent $\gamma(E)$: it is defined as the following almost-sure limit

$$\gamma(E) = \lim_{L\to\infty} (2L)^{-1} \mathrm{Ln}\, \| M_L \| \tag{2.1}$$

where for a given potential V and a given energy E, $M_L$ is the 2×2 matrix given by $M_L\phi_0=\phi_L$ with $\phi_L=\{\Psi(L),\Psi(L+1)\}^t$, $\Psi$ solving the equation $H\Psi=E\Psi$ with given initial datas $\phi_0=\{\Psi(0),\Psi(1)\}^t$. The limit is known to be almost sure because of the subadditive ergodic theorem and $\gamma(E)$ does not depend on the initial condition choosen although the set of measure 1 for which the convergence holds do depend on the initial datas.

It follows from a well known result of Furstenberg [9] that:

**Theorem 2.1:** *Suppose the measure $\mu$ to be non trivial (i.e. its support has more than one point). Then for any* E,

$$\gamma(E) > 0 \tag{2.2}$$

This result concerns the behaviour of the solutions of the Cauchy problem (fixed datas at 0 and 1) and thus apparently does not tell us anything on the spectral properties of H. In fact the connection with the spectral properties arises in the following theorem discovered in specific situations by Casher and Lebowitz [10] and Ishii [11] and put in a general and simple form by Pastur [8]

**Theorem 2.2:** *Suppose that the measure $\mu$ is non trivial. Then almost-surely the Hamiltonian* H *has no absolutely continuous spectrum, i.e. for almost every* V, $\sigma_{ac} = \varnothing$.

This result can be in fact reexpressed in the following way: if the Lyapunov exponent in non zero for almost every E in some interval I, then the operator H has almost surely no absolutely continuous spectrum in I. The converse can be proven [12][13], yielding that the absolutely continuous spectrum is exactly the essential closure of the set of E for which the Lyapunov exponent vanishes.

Under some additional assumptions one can go farther and prove that the spectrum of H is in fact almost surely pure point with exponentially decaying eigenfunctions. In the present situation the strongest result is the following one, proven in [14],[15],[16],[17]:

**Theorem 2.3:** *Suppose that the measure* $\mu$ *has some absolutely continuous component. Then almost surely* H *has only pure point spectrum, the corresponding eigenfunctions decay exponentially at infinity and the rate of exponential decay is given by the Lyapunov exponent.*

Historically, pure point spectrum was first proven for a class of continuous random Schrödinger equations by Golds'heid, Molchanov and Pastur [18],[19]. The first proof for discrete equations was obtained in [6]. These results as well as several further results obtained later, [20],[21],[23],[22],[24] supposed that the distribution $\mu$ of the potential was absolutely continuous and also used some other technical conditions on the distribution of the potential. The above result 2.3 assuming only the much weaker hypothesis of existence of an absolutely continuous component for $\mu$, was obtained through two proofs which are not only much more general but also much simpler than the previous ones. These proofs take roots in works of Carmona [24] and of Kotani [25] and particularly use a very clarifying idea of the later one discussing the influence of the boundary conditions on an operator of type (1.2) but on the semi-axis (more precisely the type of results obtained in [25] is that if $\gamma(E) > 0$, a.e. E, then for a.e. boundary conditions, (1.1) on the half-line $\mathbb{N}$ has only pure point spectrum. These proofs, and particularly the one of [14], extend to a much wider class of situations than the previous ones as we will see in Sections 7,8,9.

As far as the decay of the eigenfunctions is concerned all the proofs, in the limit of their respective conditions of application, yielded exponential decay of the eigenfunctions. The first result on a connection between the rate of decay of the eigenfunctions and the Liapunov exponent, was the proof in some class of continuous models that the first one should be larger than or equal to the second [20]; on the other hand it was proven under quite general conditions that for ergodic systems the rate of decay should not be larger than the Lyapunov exponent [26]. In the new approaches, the fact that the rate of decay of the wave functions is exactly given by the Lyapunov exponent is not only proven in a more general situation but stems out directly and in an intuitive way from the proof.

## 3 - The One-Dimensional Non-Ergodic Case; an Example of a Transition in the Nature of the Spectrum

In this Section, we still consider equation (1.1) in dimension 1 and we still assume the potential V(x) to consist of independant random variables at each site x, but now in contrast to the previous Section, we do not asssume them any more to be identically distributed. We thus do not have anymore an ergodic situation and the results of the previous Section fail now to apply.

In such a situation we may have several cases. Suppose that we know that the solutions of the corresponding Cauchy problem still behave exponentially as in the previous Section, or at least grow sufficiently fast at infinity. Then the proofs of Theorem 2.3 from [14],[15],[16],[17] can be applied to yield almost surely pure point spectrum under the condition that the distribution of the potential at 0 and 1 is absolutely continuous. Hovever such a growth property is not anymore proven from general theorems. In fact the only result in this direction is the one of [22] which proves in the same time the growth property and pure point spectrum: basically this result says that if the disorder varies from site to site but remains of the same order of magnitude (e.g. the variance of the distribution at each site does not go to zero with distance), then the spectrum of H is almost surely pure point and the eigenfunctions decay exponentially. One of the applications of this result is the following [22]:

**Theorem 3.1:** *Suppose that the potential* V(x) *in (1.1) is given by*

$$V(x) = V_0(x) + V_1(x)$$

*where* $V_0(x)$ *is a random potential given from identically distributed random variables with an absolutely continuous and sufficiently regular distribution and where* $V_1(x)$ *is any fixed potential. Then* H *has almost surely only pure point spectrum and the corresponding eigenfunctions decay exponentially.*

It is interesting to note that this result is not true in the continuous case as we will see in Section 9 below. But one can go farther using the proof of [22] and discuss the case where the disorder is not homogeneous and for example decays in space. And in fact quite interesting results arise from this situation.

We thus consider equation (1.1) where we suppose that

$$V(x) = \lambda\, |x|^{-\alpha} \vartheta(x)$$

with the $\vartheta(x)$ being given by independant and identically random variables which we suppose to have a density with some regularity in particular at infinity (for other distributions in particular for a Cauchy distribution we can prove some different results [28]). Then we have

**Theorem 3.2:** *The following properties of* H *hold for almost all sequences* $\{\vartheta(x)\}_{x\in \mathbb{Z}}$ *:*

*i) if* $0\leq\alpha<1/2$, *the spectrum of* H *is pure point and the corresponding eigenfunctions* $\Psi(x)$ *satisfy*

$$|\Psi(x)| \leq C \exp\{-C' |x|^{1-2\alpha}\}$$

*and*

$$\underline{C} \exp\{-\underline{C}' |x|^{1-2\alpha}\} \leq (\Psi(x)^2 + \Psi(x+1)^2)^{1/2}$$

*ii) if* $\alpha=1/2$ *and* $\lambda > \lambda_1(r)$, *the spectrum of* H *is pure point and the corresponding eigenfunctions* $\Psi_E(x)$ *satisfy*

$$|\Psi(x)| \leq C(E)\ |x|^{3/2+\varepsilon-c\lambda^2}$$

*and for energies* E *within the interval* ]-2,2[ *satisfy*

$$C'(E)\ |x|^{-a\lambda^2(4-E^2)^{-1}} \leq \Psi(x)^2 + \Psi(x+1)^2$$

*iii) if* $\alpha=1/2$, K *any compact of* ]-2,2[, *and* $\lambda< \lambda_2(r,K)$, *the spectrum of* H *is purely singular continuous.*

*iv) if* $\alpha>1/2$, *the spectrum of* H *within the interval* ]-2,2[ *is purely absolutely continuous.*

This Theorem is essentially a result of [27],[28]. However the first half of i) was proven previously in [29] where it was first realized that the proof of [22] could be used for studying random potentials decaying in space. Also, [27],[28] proved only in iii) that the spectrum is purely continuous within ]-2,2[ and the fact that it is also singular is a result of [30], and in iv) only continuity whereas Kotani [31] noted how to extend the proof in order to obtain absolute continuity.

This model offers us an example of a transition from pure point to continuous spectrum when varying the intensity of the disorder in some class of random Schrödinger operator, a phenomenon which otherwise is only known on the Bethe lattice (see Section 6) but is of a different nature since here we have a transition to singular continuous and not absolutely continuous spectrum; the present example is also related to the behaviour of Schrödinger equations with some class of random potentials and in an electric field (see Section 9).

Finally let us mention that one can produce various classes of coexistence of continuous and point spectrum for one dimensional inhomogeneous systems as was done in [24] (the proof is worked out there for continuous systems but does adapt to discrete ones): for example if the potential is a random one on the left hand side and an appropriate periodic one on the right hand side, then in the bands of the periodic one, there will be absolutely continuous spectrum and outside of them pure point spectrum.

## 4 - Quasi-Onedimensional Systems: Random Discrete Schrödinger Equations in a Strip

In this Section we are concerned with equations of the type (1.1) in a strip, that is in a subset of $\mathbb{Z}^d$ of the form $A \times \mathbb{Z}$ where A is a finite connected subset of $\mathbb{Z}^{d-1}$. We will suppose that the potential V(x) is constituted by independant and identically distributed random variables with a common probability distribution $\mu$ and we are thus in an ergodic situation. For such a system, it is natural to introduce the transfer matrices associate in the natural way to our operator and which are now $2|A|\times 2|A|$ random matrices where $|A|$ is the cardinality of the set A; it is also natural to try to study the various Lyapunov exponents associate to the corresponding product of random matrices. One can check that these matrices are symplectic matrices and this fact has the consequence that the Lyapunov exponents come by opposite pairs. We denote and order them as

$$\gamma_1 \geq \gamma_2 \geq \ldots \geq \gamma_{|A|} \geq -\gamma_{|A|} \geq \ldots \geq -\gamma_2 \geq -\gamma_1 \tag{4.1}$$

It is a non trivial extension of Furstenberg type of results to prove that the Lyapunov exponents are all non zero [32],[33]:

**Theorem 4.1:** *Suppose that the distribution* $\mu$ *of the potential is absolutely continuous. Then all the Lyapunov exponents are non zero for every* E.

Pastur's result 2.2 can be easily extended yielding:

**Theorem 4.2:** *Suppose that all the Lyapunov exponents are non zero for almost every* E. *Then almost surely, the Hamiltonian* H *has no absolutely continuous spectrum.*

As in the one-dimensional case it is possible to go farther and to prove under some hypothesis that we have only pure point spectrum. From the results of [14],[15],[16],[34] one gets that

**Theorem 4.3:** *Suppose that the distribution* $\mu$ *of the potential has some absolutely continuous part. Then almost-surely the operators* H *has only pure point spectrum.*

In fact one can go farther and prove that [14],[15]

**Theorem 4.4:** *Under the same hypothesis as 4.3, for almost every realisations of the potential, the eigenvectors are exponentially decaying with the smallest Lyapunov exponent* $\gamma_{|A|}$ *as a rate and furthermore the spectrum of* H *is non degenerate (i.e. has multiplicity one).*

The pure point character of the spectrum for a discrete Schrödinger equation in a strip was first shown in [35] for a situation with a purely absolutely continuous $\mu$ and a complete proof of such a result can be found in [36]. The more recent proof quoted above not only applies to a much wider class of situations but is also much simpler and transparent.

When studying problems in a strip a natural and interesting question to be asked is the following one: let us suppose that we choose a sequence of subsets A of $\mathbb{Z}^{d-1}$ growing to $\mathbb{Z}^{d-1}$. We can then ask what is the behaviour of $\gamma_{|A|}$ in this limit and essentially we would like to know whether it goes to zero or not. This is a very hard problem to tackle with directly; one of the main reason being that there does not exist, at least up to now, a good expression for the smallest positive Lyapunov exponent $\gamma_{|A|}$; but using indirect methods connected with random discrete Schrödinger equations it was possible [14],[15] to prove the following result:

**Theorem 4.5:** *Let us suppose that the probability measure* $\mu$ *is absolutely continuous and that its density is* $L_\infty$ *with a sufficiently small* $L_\infty$*-norm. Then the "smallest" Lyapunov exponent* $\gamma_{|A|}$ *remains strictly positive in the limit* $A\uparrow \mathbb{Z}^{d-1}$.

This result combines methods connected to discrete Schrödinger equations in [15] together with a result of [37] that we will see later.

## 5 - The Multidimensional Case: Localization for Large Disorder

In this Section, we consider the discrete equation (1.1) on $\mathbb{Z}^d$, and with the random variables V(x) supposed to be independant and identically distributed with $\mu$ as a common probability distribution. We first present a result on the Green's function which will be useful later: the Green's function $G_E(x,y)$ for $x\in \mathbb{Z}^d$, $y\in \mathbb{Z}^d$, is defined as the solution of the equation

$$(H-E)\, G_E(x,.) = \delta_x(.) \qquad (5.1)$$

which has solutions when the imaginary part of E is positive and the Green's function is defined for E real as the limit for E going to the real line if it exists. The result obtained by Fröhlich and Spencer [37], [38] is the following one:

**Theorem 5.1:** *Let us suppose* $\mu$ *to be absolutely continuous. Then there exists some* $\lambda_c$, *and for every* $\lambda$ *there exists in the almost sure spectrum of* H *some* $E_0(\lambda), E_0'(\lambda)$, $(-2d+\inf \operatorname{supp} \mu< E_0(\lambda)\le E_0'(\lambda)<2d+\max \operatorname{supp} \mu)$ *such that for* $\lambda>\lambda_c$ *and every* E, *and for any* $\lambda$ *and* $E<E_0(\lambda)$ *or* $E>E_0'(\lambda)$ *the Green's function*

$G_E(x,y)$ *exists for almost every realisation of the potential* $V(x)$ *and satisfies*

$$|G_E(x,y)| \leq C\, e^{-\gamma|x-y|} \tag{5.2}$$

*for some* $\gamma$ *and for some* C *depending on* x *and on the realisation of the potential.*

In other words, for "large enough disorder" or for "energies near enough to the edges of the spectrum", the Green's function decays exponentially with distance.

In fact as can be seen in the references there is also some uniformity of this result with respect to the approximation of the infinite system by finite boxes and approximating real energies by complex ones. From this result follows [39] by an argument which represents the multi-dimensional analogue of Pastur's argument of Section 2:

**Theorem 5.2:** *Let us assume that for some given finite difference Schrödinger equation* H *of the type (1.1), the Green's functions decays exponentially for almost every* E *in some interval* I. *Then* H *has no absolutely continuous spectrum.*

In fact under the hypothesis of Theorem 5.1, one can go farther and prove that the spectrum of H is pure point:

**Theorem 5.3:** *Let us assume that* $\mu$ *is absolutely continuous. Then for* $\lambda > \lambda_c$, *and for any* $\lambda$ *and* $E < E_0(\lambda)$ *or* $E > E_0'(\lambda)$ *the spectrum of* H *is almost surely pure point and the corresponding eigenfunctions are exponentially decaying.*

This result was proven in [14],[40] and in [16],[17] using the exponential decay of Theorem 5.1, the idea of Theorem 5.2 connecting the Green's function decay to the nature of the spectrum, and ideas from [25] used by these authors in the works on one dimensional systems mentionned in Section 2 above. It was also proven in [41] by a different method, essentially extending the method of [37] (in this respect it is interesting to note the proof of pure point spectrum in a class of hierarchical models [42] which served as a laboratory for the proof of these authors). The same result was also announced in [43]. The main consequence of the clarification due to Ref.[14],[40],[16], is the direct and simple connection between the decay of the Green's function and the pure point caracter of the spectrum which is exhibited there. A consequence of their approach is the fact that the good analogue of the smallest Lyapunov exponent of a system in a strip for a multidimensional system is the rate of decay of the Green's function. This understanding was used in the proof [15] mentionned in Section 4 above that in some circumstances the smallest Lyapunov exponent for a problem in a strip does not go to zero as the strip becomes larger and larger. Another application of the same type of ideas yields a solution to a natural but otherwise unsolved question [40]:

**Theorem 5.4:** *Under the same conditions as Theorem 5.3 and in the same regions of the parameters, for almost every realization of the potential the spectrum of* H *is non degenerate (i.e. has multiplicity one).*

We have seen that in dimension 1 and in a strip (which is in some sense of the one dimensional type), the spectrum is pure point for any disorder and any energy whereas it is proven to be pure point in higher dimension only for large enough disorder or near enough to the band edges. Is this a weakness of the proofs or do we expect new phenomenon to occur in dimensions larger than one? The following conjectures indicates our belief based on various physical methods, especially renormalization group treatments.

**Conjectures 5.5:** *Let us suppose for example that* $\mu$ *is absolutely continuous and has a bounded density possessing two moments. Then*

*- in dimension* $d=2$*, for almost every realisation of the potential the spectrum is pure point with exponentially decaying eigenfunctions.*

*- in dimension* $d\geq 3$*, for almost every realisation of the potential, the spectrum is pure point with exponentially decaying eigenfunctions for large enough disorder (i.e.* $\lambda>\lambda_c$ *for some* $\lambda_c$*) whereas for smaller disorder (*$\lambda<\lambda_c$*) some part of the spectrum is again pure point with exponentially decaying eigenfunctions and the rest of the spectrum is absolutely continuous and one should have* $\sigma_{pp}\cap\sigma_{ac}=\emptyset$.

*In particular in any* d *almost surely* H *does not possess singular continuous spectrum.*

These results are far from being proven but there is an example where partial results in these directions can be proven, and this is the topic of the next Section.

## 6 - The Bethe Lattice: the Anderson Transition and the Mobility Edge

In this Section, we consider equation (1.1) on a Bethe lattice. Let us first define what is a Bethe latttice which is also called a Cayley tree: it is a connected graph with no cycles but with a constant coordination number K+1 so that any point of the tree has exactly K+1 nearest neighbours. The Bethe lattice is widely used in the theory of critical phenomena, and it is known on physical grounds that the results on the Bethe lattice (interpreted in the appropiate way) give results analogous to those of a $\mathbb{Z}^d$ lattice in dimension d sufficiently large.

Thus in this Section we consider equation (1.1) on a Bethe lattice $\mathbb{T}$ with K>1, and we suppose the potential variables $\{V(x)\}_{x\in\mathbb{T}}$ to be independant and identically distributed random variables with a density that we assume sufficiently regular and decaying at infinity. The Mott-Anderson transition has been proven in this model [44],[45] for a large class of distributions:

**Theorem 6.1:** *There exist some* $\lambda_1>0$, $\lambda_2>0$ *and some* $E_1(\lambda)>0$, $E_2(\lambda)>0$ *such that*

*(i) for* $\lambda>\lambda_1$ *or for* $|E|>E_1(\lambda)$ *the spectrum of* H *is almost-surely pure point.*

*(ii) for* $\lambda<\lambda_2$ *and* $|E|<E_2(\lambda)$ *the spectrum of* H *is almost-surely absolutely continuous.*

It is also of great interest to understand what happens when the transition from pure point to absolutely continuous spectrum takes place. In this respect it is interesting to note the following result obtained in the same work which yields the exact value of a critical exponent on the Bethe lattice:

**Theorem 6.2:** *Under some additional assumptions on the density of the random potential, the localization length which governs the rate of decay of the eigenfunctions in the pure point spectrum diverges as* $|E-E_c|^{-1}$.

## 7 - Other Discrete Equations

In the previous Sections, we have considered only equation (1.1) but many other discrete equations can be considered and arise from physics. For example one can consider operators H of the type

$$(H\Psi)(x) = \sum_{|x-y|=1} J(x,y)\Psi(y) \tag{7.1}$$

with $J(x,y)=J(y,x)$, acting again on $L^2(\mathbb{Z}^d)$, and now where the $\{J(x,y)\}$ will be the random variables. It is also natural to consider equations of the type

$$(H\Psi)(x) = \sum_{|x-y|=1} J(x,y)\ \Psi(y) - [\sum_{|x-y|=1} J(x,y)]\ \Psi(y) \tag{7.2}$$

For large classes of such operators, the spectrum is again almost-surely exactly known [22]. Also if the random variables are independant it was proven [22] that for one-dimensional systems the spectrum is almost-surely pure point with exponentially decaying eigenfunctions. More recently, results analogous to those of [14],[15],[16],[34] have been obtained for some of these systems [46]. It was also proven the analogue of Theorems 5.1 and 5.3 above [47]. Existence of pure point spectrum in some regions of the parameters was also proven for a class of two dimensional discrete models of systems with spin-orbit interactions or in a magnetic field [48].

## 8 - Non-Independant Random Potentials

In the previous Sections we have restricted ourselves to the case of independant random variables. Some results are nevertheless proven for more general classes of potentials. The fact that the spectrum is exactly known almost-surely is a much more general fact as is seen in [6]. Concerning the nature of the spectrum, it is important to note the fact that for one dimensional systems the Lyapunov exponent can be proven to be positive as soon as the potentials is non deterministic that is as soon as the $\sigma$-algebra generated by the potential on $]-\infty,x]$ is strictly larger than the $\sigma$-algebra at $-\infty$. This remarquable result was first proven by Kotani for the continuous Schrödinger equation [12] and then extended to the discrete Schrödinger equations [13] and to a large class of discrete and continuous wave equations [49].

As for the proofs of pure point spectrum, it can be noted that the proofs of [14],[15],[16], [17] do extend readily to very large classes of non independant random potentials as mentionned in these references. Also the results of [46] mentionned in Section 7 above do extend to a very large class of distributions of the random variables.

Finally let us emphasize the following fact: potentials which are almost-periodic as a function of x can be considered in some sense as non independant random potentials, and more precisely they are ergodic with respect to translations. However most of the results described in this review do not apply to these type of potentials for which very different phenomenon occur [50]; the main difference is that almost periodic potentials are completely determined if one knows them on some interval, which is not the case for "really random" potentials.

## 9 - Continuous Random Schrödinger Equations and Other Wave Equations

In the previous Sections we have considered only discrete i.e. finite difference equations but both from the physical and the mathematical point of view, it is also natural to study continuous random sef-adjoint operators.

Let us first start with one dimensional systems. For large classes of random coefficients, it also possible to prove that the Lyapunov exponent is again strictly positive using Furstenberg theorem or many of its extensions. I have in fact already mentionned the remarquable result of Kotani [12] who proves that for Schrödinger equation with an ergodic potential, the Lyapunov exponent is strictly positive as soon as the potential is given from a non deterministic process; this result was extended to a large class of wave equations by Minami [49]. Since the Theorem of Pastur 2.2 in fact is also valid for continuous systems, we get a large class of one dimensional random operators which have almost surely an empty absolutely continuous spectrum. In fact such results have been extended to even a wider class of random potentials since it was proven that a class of intuitively random but nevertheless deterministic potentials do exhibit also a positive Lyapunov exponent and thus do not have almost surely an absolutely continuous spectrum [51]. For a large class of distributions of potentials, one can go farther and prove that the spectrum is almost surely pure point with exponentially decaying eigenfunctions [46],[52]. Finally an interesting question arises if one adds an elctric field that is a potential of the type -F.x to the random Schrödinger equation: if the random potential is sufficiently smooth, e.g. has two bounded derivatives, then for any non zero F, we have only purely absolutely continuous spectrum [53] and in fact some square integrable singularities can also be accomodated [54]. In contrast in the case of a random Krönig-Penney potential that is a potential composed of $\delta$-functions with random strengths then one can prove [27], [28] a transition from pure point to continuous spectrum, the spectrum being almost surely pure point for small field F, and (except of course for F=0) with power law decaying eigenfunctions, and for large enough field F the spectrum is almost surely purely continuous. Finally let us also note that, as already mentionned in Section 3, it is possible to construct various examples exhibiting simultaneously pure point and continuous spectrum [24] e.g. with appropriate random potentials on the left and periodic on the right.

Let us turn now to multi-dimensional systems: the property of exponential decay of the Green's function analogous to Theorem 5.1 has been proven for a class of continuous random potentials in [55] and one can prove in the corresponding regions of the parameters that almost surely the spectrum is pure point [41].

## References

[1] Simon B, in these Proceedings.

[2] Thouless DJ, in *Ill-Condensed Matter*, Proceedings Les Houches, North-Holland, Amsterdam (1979).

[3] Souillard B, in *Common Trends in Particle and Condensed Matter Physics*, Proceedings Les Houches, Phys. Rep. 103 (1984) 41.

[4] Lee P and Ramakrishnan TV, Rev. Mod. Phys. 57 (1985) 291.

[5] Souillard B, Fractals and Localization, to appear in the Proceedings of the NATO Institute on Amorphous and Liquid Materials held in Bolzano (1985).

[6] Kunz H and Souillard B, Commun. Math. Phys. 78 (1980) 201.

[7] Kirsch W and Martinelli F, J. reine angew. Math. 334 (1982) 141.

[8] Pastur L, Preprint Kar'kov (1974) and Commun. Math. Phys. 75 (1980) 179.

[9] Fürstenberg H, Transactions of the American Math. Soc. 108 (1963) 377.

[10] Casher A and Lebowitz JL, J. Math. Phys. 12 (1971) 1701.

[11] Ishii K, Prog. Theor. Phys. Supp. 53 (1973) 77.

[12] Kotani S, Proc. Taniguchi Symp. Katata (1982) 225.

[13] Simon B, Commun. Math. Phys. 89 (1983) 227.

[14] Delyon F, Lévy Y and Souillard B, Phys. Rev. Lett. 55 (1985) 618.

[15] Delyon F, Lévy Y and Souillard B, J. Stat. Phys. 41 (1985) 375.

[16] Simon B, Taylor M and Wolff T, Phys. Rev. Lett. 54 (1985) 1589.

[17] Simon B and Wolff T, Singular continuous spectrum under rank one perturbations and localization for random Hamiltonians, Commun. Pure Appl. Math. to appear.

[18] Goldsheid Ya, Molchanov S and Pastur L, Funct. Anal. Appl. 11 (1977) 1.

[19] Molchanov S, Math. USSR Izvestija 42 (1978).

[20] Carmona R, Duke Math. J. 49 (1982) 191.

[21] Royer G, Bull. Soc. Math. France, 110 (1982) 27.

[22] Delyon F, Kunz H and Souillard B, J. Phys. A16 (1983) 25.

[23] J. Lacroix, Ann. Inst. E. Cartan, Nancy 7 (1983).

[24] Carmona R, J. Funct. Anal. 51 (1983) 229.

[25] Kotani S, Proceedings of the AMS meeting on "Random Matrices", Brunswick, 1984, J. Cohen Editor.

[26] Craig W and Simon B, Duke Math. J. 50 (1983) 551.

[27] Delyon F, Simon B and Souillard B, Phys. Rev. Lett. 52 (1984) 2187.

[28] Delyon F, Simon B and Souillard B, Ann. Inst. H. Poincaré 42 (1985) 283.

[29] Simon B, Commun. Math. Phys. 87 (1983) 253.

[30] Delyon F, J. Stat. Phys. 40 (1985) 621.

[31] Kotani S, unpublished.

[32] Guivarc'h Y and Raugi A, Z. Wahrschein. u.Verw. Geb. 69 (1985) 187.

[33] Lacroix J, Ann. Inst. H. Poincaré 38A (1983) 385.

[34] Simon B, Localization for one dimensional random systems I: Jacobi matrices, Commun. Math. Phys. to appear.

[35] Golds'heid Ya, Dokl. Akad. Nauk. SSSR, 225 (1980) 273.

[36] Lacroix J, Ann. Inst. H. Poincaré 40A (1984) 97.

[37] Fröhlich J and Spencer T, Commun. Math. Phys. 88 (1983) 151.

[38] Spencer T, in Proceedings Les Houches 1984, North Holland to appear.

[39] Martinelli F and Scoppola E, Commun. Math. Phys. 97 (1985) 465.

[40] Delyon F, Lévy Y and Souillard B, Commun. Math. Phys. 100 (1985) 463.

[41] Fröhlich J, Martinelli F, Scoppola E and Spencer T, Commun. Math. Phys. 101 (1985) 21.

[42] Jona-Lasinio G, Martinelli F and Scoppola E, Multiple tunnelings in d-dimensions: a quantum particle in a hierarchical potential, J. Phys. A17 (1984) 73.

[43] Golds'heid Ya G, announcement at the Tachkent Conference on Information Theory, September 1984.

[44] Kunz H and Souillard B, J. Phys. (Paris) Lett. 44 (1983) 411.

[45] Kunz H and Souillard B, to appear.

[46] Delyon F, Simon B and Souillard B, Exponential localization for a class of discrete and continuous random wave equations, preprint Polytechnique.

[47] Faris W, Localization for multi-dimensional off-diagonal disorder, to be published.

[48] Bellissard J, Grempel DR, Martinelli F, Scoppola E, Localization of electrons with spin-orbit or magnetic interactions in a 2-D disordered crystal, preprint Marseille.

[49] Minami N, An extension of Kotani's theorem for random generalized Sturm-Liouville operators, preprint.

[50] Simon B, Adv. Appl. Math. 3 (1982) 463.

[51] Kirsch W, Kotani S and Simon B, Absence of Absolutely Continuous Spectrum for Some One Dimensional Random but Deterrministic Schrödinger Operators, Ann. Inst. H. Poincaré to appear.

[52] Kotani S and Simon B, Localization for one-dimensional continuous sytems, to appear.

[53] Bentosella F, Carmona R, Duclos P, Simon B, Souillard B and Weder R, Commun. Math. Phys. 88 (1983) 387.

[54] Ben-Artzi M, J. Mat. Phys. 25 (1984) 951.

[55] Holden H and Martinelli F, Commun. Math. Phys. 93 (1984) 197.

# DISPERSIVE BULK PARAMETERS FOR COHERENT PROPAGATION IN CORRELATED RANDOM DISTRIBUTIONS

Victor Twersky

Mathematics Department
University of Illinois
Chicago, Illinois 60680

## 1. Introduction

The wave scattered by a configuration of obstacles in response to an incident wave (with propagation parameter $k$) can be represented functionally in terms of the waves scattered by the individual obstacles in isolation [1]. For a statistically homogeneous ensemble of configurations, the ensemble averaged wave (the coherent field) specifies a composite medium with bulk parameters depending on the properties of the particles and embedding medium, on the value of $k$ (hence dispersive), and on the statistics of the distribution of particles (their number concentration, pair correlations, etc.) [2].

The solution of the scattering problem for a single uniform particle depends in general on two complex relative parameters, say $B'$ and $C' = B'\eta'^2$ such that $\eta'$ is the index of refraction relative to the uniform embedding medium. The corresponding relative bulk parameters $B$ and $C = B\eta^2$, and index $\eta$, are associated with the coherent field. Explicit approximations of $\eta$ for special cases, and functional representations in terms of isolated scattering amplitude forms, have been obtained by direct and indirect methods [2-9]. However, in general, results for the bulk parameters have been based on indirect procedures (e.g., on comparison with forms of the solution at the interface of two uniform media). We consider the essential structure of a direct procedure [2], and give explicit acoustic and electromagnetic results to $O(k^3)$ for uniformly random pair-correlated distributions of identical spheres (of radius $a$) [9].

Our development is based on representation theorems for $C$ and $B$ obtained from the ensemble average of the field, and an approximation for $\eta$ obtained from the functional equation relating the multiple and isolated scattering amplitudes of particles in an arbitrary configuration [2]. The theorems follow from the first of the system of hierarchy integrals of the ensemble, and the approximation from essentially the second. Replacing the average multiple scattering amplitude with two particles fixed by that with one fixed (analogous to Lax's procedure for the effective exciting field [5]) truncates the system to provide a determinate equation for $\eta$, which together with the theorems determine C and B. Alternative [7,8] procedures provide insight for the truncation approximation, but analytical aspects, essentially as for the analogous statistical mechanics problems [10], are unresolved.

The determinate relations are applied to distributions of spheres specified by the average number $(\rho)$ of centers in unit volume and by the statistical mechanics radial distribution function (f). For radius a small compared to wavelength $(\pi 2/k)$, to third order in ka, we obtain the form

$$\Gamma = \Gamma_1(k^0) + \Gamma_c(k^2) + i\Gamma_s(k^3) \tag{1}$$

with $\Gamma = C$ or B. The leading term $(\Gamma_1)$ and the fluctuation scattering term $(\Gamma_s)$ depend exclusively on the corresponding isolated sphere parameter $(\Gamma' = C'$ or $B')$, but the dispersive correction term $(\Gamma_c)$ depends on both C' and B', i.e., both particle parameters are coupled in each bulk term $C_c$ and $B_c$. As required by the theorems, the dependence is such that if $B' = 1$ then $B = 1$ and $\eta^2 = C$, or if $C' = 1$ then $C = 1$ and $\eta^2 = 1/B$; thus, if the particles are specified by only one relative parameter, then so is the distribution.

The terms $\Gamma_c$ and $\Gamma_s$ depend respectively on the first and second moments of the direct correlation function $f - 1$. For impenetrable statistical spheres, the leading terms in powers of the volume fraction (w) occupied by spheres are obtained from the virial expansion of f, and simple closed forms (rational functions) in w follow from the Wertheim-Thiele solution [10] of the Percus-Yevick integral equation approximation for f. We regard only $w \lesssim 0.63$ as realistic for radially symmetric pair correlations. (The bound, corresponding to measured values for the densest random packing of identical spheres, represents an analog of an amorphous solid.) Our explicit results for B and C reduce to B' and C' at the unrealizable value $w = 1$ (full packing), which appears consistent with implicit statistical mechanics and multiple scattering approximations arising from the truncation procedure.

In order to introduce notation and representations, we begin with brief sketches of scattering problems for the scalar Helmholtz equation, and indicate some of the associated physics in the context of acoustics. Explicit acoustics results for $\eta$, C, and B are discussed with emphasis on the structure of the dispersive corrections. Then, modifications required for electromagnetics are noted, and corresponding explicit results are included [2,9].

## 2. Acoustics

For small amplitude acoustics and time-harmonic waves, we suppress the time dependence and represent the excess pressure p by $\psi(\underset{\sim}{r})$ and the normalized velocity by $\underset{\sim}{v}(\underset{\sim}{r})$. The observation point is specified by the vector $\underset{\sim}{r} = r\hat{r}$ with magnitude r and direction $\hat{r}(\theta, \ )$. In the embedding medium with physical parameters normalized to unity, we have

$$\underset{\sim}{v} = \nabla\psi, \qquad \nabla \cdot \underset{\sim}{v} = -k^2\psi, \qquad (\nabla^2 + k^2)\psi = 0 \tag{2}$$

where $\nabla^2$ is the Laplacian.

For a plane wave

$$\phi = e^{i\underset{\sim}{k}\cdot\underset{\sim}{r}}, \quad \underset{\sim}{k} = k\hat{k}, \quad \hat{k} = \hat{r}(\theta_0, \ _0) \tag{3}$$

incident on an obstacle with center (the center of the smallest circumscribing sphere) at $r = 0$, the field in the external region (outside the scatterer's volume v bounded by its surface S), is given by

$$\psi = \phi + u. \tag{3'}$$

The scattered wave u is the radiative function

$$u(r) = \{h(k|\underset{\sim}{r} - \underset{\sim}{r}'|), u(\underset{\sim}{r}')\} = \frac{k}{i4\pi} \int \hat{n} \cdot (h\nabla u - u\nabla h) dS(\underset{\sim}{r}'), \tag{4}$$

where $h(x) = e^{ix}/ix$, and $\hat{n}$ is the outward normal. The integral over S may be replaced by the integral over any surface that incloses S and excludes $\underset{\sim}{r}$. Asymptotically, for $r \sim \infty$,

$$u \sim h(kr)g(\hat{r},\hat{k}),$$

$$g(\hat{r},\hat{k}) = \{e^{-i\underset{\sim}{k}_r \cdot \underset{\sim}{r}'}, u(\underset{\sim}{r}';\underset{\sim}{k})\} = \{\phi(-\underset{\sim}{k}_r),u\}\ , \quad \underset{\sim}{k}_r = k\hat{r} \tag{5}$$

where $g$ is the normalized (dimensionless) scattering amplitude. Using the complex spectral representation for $h$ in (4), we obtain (at least for $r$ greater than the scatterer's projection on $\hat{r}$),

$$u(\underset{\sim}{r}) = \int_c e^{i\underset{\sim}{k}_c \cdot \underset{\sim}{r}} g(\hat{r}_c,\hat{k}), \quad \underset{\sim}{k}_c = k\hat{r}_c = k\hat{r}(\theta_c,\varphi_c), \tag{6}$$

where $2\pi \int_c = \int d\varphi_c \int d\theta_c \sin\theta_c$ and the contours for $\varphi_c$ and $\theta_c$ are each of the general Sommerfeld form for the Hankel function of the first kind.

In the scatterer's interior $v$, the pressure and velocity satisfy

$$\begin{aligned} &\underset{\sim}{v} = B'\nabla\psi, \quad \nabla \cdot \underset{\sim}{v} = -k^2C'\psi, \quad (\nabla^2 + K'^2)\psi = 0, \\ &K' = k(C'/B')^{1/2} = k\eta' \end{aligned} \tag{7}$$

such that $\psi$ is nonsingular. For the simplest acoustic problems, the physical parameters are real, and $C'$ is the particle's relative compressibility and $1/B'$ its relative mass density. More generally, the parameters are complex with $\mathrm{Im}C' > 0$ and $\mathrm{Im}B' < 0$ to account for various loss mechanisms. The relative index determines the phase velocity $(1/\mathrm{Re}\eta')$ and attenuation $(2k\,\mathrm{Im}\eta' > 0)$ within the obstacle. On S, we require continuity of $\psi$ and $\hat{n} \cdot \underset{\sim}{v}$:

$$\psi(k) = \psi(K'), \quad \hat{n} \cdot \nabla\psi(k) = B'\hat{n} \cdot \nabla\psi(K'), \tag{8}$$

where the arguments indicate external $(k)$ and internal $(K')$ forms.

From (4), (8), and (7), we represent $u(k)$ for $\underset{\sim}{r}$ outside of $v$ as the volume integral

$$\psi - \phi = [h(k|\underset{\sim}{r} - \underset{\sim}{r}'|)] = -\frac{k}{4\pi i} \int [(C' - 1)k^2h\psi - (B' - 1)\nabla h \cdot \nabla\psi]dv(\underset{\sim}{r}') \tag{9}$$

which also generates $\psi(K') - \phi(k)$ as a principle value for $\underset{\sim}{r}$ in $v$. The corresponding form of the scattering amplitude is

$$g(\hat{r},\hat{k}) = [\phi(-\underset{\sim}{k}_r), \psi(K';\underset{\sim}{k})] = g[\underset{\sim}{k}_r,\underset{\sim}{k}]. \tag{10}$$

The last enclosure of the argument of $g$ specifies the volume integral form.

Indicating the solutions corresponding to arbitrary directions of incidence $\hat{r}_a, \hat{r}_b$ by $\psi_a, \psi_b$ (each subject to the same condition on $S$ and $v$ as $\psi = \phi + u$), we have $\{\psi_a, \psi_b\} = 0$. Consequently, $\{\phi_a, u_b\} = \{\phi_b, u_a\}$, and from (5),

$$g(-\hat{r}_a, \hat{r}_b) = g(-\hat{r}_b, \hat{r}_a), \quad g(\hat{r}, \hat{k}) = g(-\hat{k}, -\hat{r}) \tag{11}$$

This reciprocity relation holds for all scattering problems satisfying $\{\psi_a, \psi_b\} = 0$ on $S$.

For a fixed configuration of $N$ obstacles (with surfaces $S_s$, volumes $v_s$, centers $\underset{\sim}{r}_s$, etc.), the solution external to all can be written as the incident wave $\phi$ plus a set of multiple scattered fields $U_s$:

$$\Psi = \phi + \sum_{s=1}^{N} U_s(\underset{\sim}{r} - \underset{\sim}{r}_s; \underset{\sim}{r}_1, \underset{\sim}{r}_2, \ldots, \underset{\sim}{r}_N). \tag{12}$$

The field radiating from particle-s satisfies $(\nabla^2 + k^2)U_s = 0$ outside of $v_s$, and

$$U_s = \int_c e^{i\underset{\sim}{k}_c \cdot (\underset{\sim}{r} - \underset{\sim}{r}_s)} G_s(\hat{r}_c), \quad G_s(\hat{r}) = \{e^{-i\underset{\sim}{k}_c \cdot \underset{\sim}{r}'}, U_s\}_s \tag{13}$$

The multiple scattering amplitude $G_s$ is in the form (5) over $S_s(\underset{\sim}{r}')$ with $\underset{\sim}{r}' = \underset{\sim}{r} - \underset{\sim}{r}_s$ as the local vector from $\underset{\sim}{r}_s$.

With reference to scatterer-t,

$$\Psi = \Psi_t = \Phi_t + U_t, \quad \Phi_t = \phi + \sum_s{}' U_s, \quad \sum_s{}' = \sum_{s \neq t} \tag{14}$$

such that $\Psi_t$, $\Phi_t$, $U_t$ satisfy the same relations on $S_t$ and $v_t$ as $\psi$, $\phi$, $u$ for scatterer-t in isolation. Thus $\{\psi_a, \Psi_t\}_t = 0$ gives $G_t(-\hat{r}_a) = \{\phi_a, U_t\}_t = \{\Phi_t, u_a\}_t$. Using (14) for $\Phi_t$ and the forms $g$, $U$, $G$ as in (5) and (13), we obtain a system of functional equations (essentially reciprocity relations),

$$G_t(\hat{r}) = g_t(\hat{r}, \hat{k}) e^{i\underset{\sim}{k} \cdot \underset{\sim}{r}_t} + \sum{}' \int_c g_t(\hat{r}, \hat{r}_c) G_s(\hat{r}_c) e^{i\underset{\sim}{k}_c \cdot \underset{\sim}{R}_{ts}}, \quad \underset{\sim}{R}_{ts} = \underset{\sim}{r}_t - \underset{\sim}{r}_s \tag{15}$$

which holds at least if the projections of scatterers $s$ and $t$ on $\underset{\sim}{R}_{ts}$ do not overlap. For a given configuration, this system of equations determines $G$ in terms of $g$ (the direct problem) or $g$ in terms of $G$ (the inverse).

The average of $\Psi$ of (12) over a statistically homogeneous ensemble of configurations of $N$ identical obstacles whose centers $\underset{\sim}{r}_s$ are uniformly distributed in a slab region $V_d$ may be written

$$\langle\Psi(\underset{\sim}{r})\rangle = \phi + \sum_s \langle U_s\rangle = \phi + \rho \int \langle U_s(\underset{\sim}{r} - \underset{\sim}{r}_s;\underset{\sim}{r}_s)\rangle_s d\underset{\sim}{r}_s, \qquad (16)$$

where $\int d\underset{\sim}{r}_s$ is the volume integral over $V_d$, and $\rho = N/V_d$. The ensemble average $\langle\Psi\rangle$, the coherent field, is independent of the configurational variables $(\underset{\sim}{r}_1,\underset{\sim}{r}_2,\ldots,\underset{\sim}{r}_N)$, and the average $\langle\ \rangle_s$ with $\underset{\sim}{r}_s$ held fixed depends only on $\underset{\sim}{r}_s$ (now a dummy). We use the radiative form $\langle U_s(k)\rangle_s$ if $\underset{\sim}{r}$ is outside of $v_s = v$, and $\langle\Psi_s(K')\rangle_s - \langle\Phi_s(k)\rangle_s$ if inside. From (15),

$$\langle G_t(\hat{r})\rangle_t = g_t(\hat{r},\hat{k})e^{i\underset{\sim}{k}\cdot\underset{\sim}{r}_t} + \rho\int d\underset{\sim}{r}_s f(\underset{\sim}{R}_{ts})\int_c g_t(\hat{r},\hat{r}_c)\langle G_s(\hat{r}_c)\rangle_{st} e^{i\underset{\sim}{k}\cdot\underset{\sim}{R}_{ts}}, \qquad (17)$$

where $\langle\ \rangle_{st}$ is the average over all variables but $\underset{\sim}{r}_s$ and $\underset{\sim}{r}_t$. The pair distribution function $f(\underset{\sim}{R}_{ts})$ satisfies $f = 0$ for $R_{ts}$ less than the minimum separation of centers, and $f \sim 1$ for $R_{ts} \sim \infty$. Eqs. (16) and (17) are essentially the first two hierarchy integral relations for the ensemble. The third involves an additional fixed center, etc.

**Bulk Parameters and Index**

The volume $V_d$, containing all centers of scattering particles, separates space into essentially three regions: an external region $V_E$ containing no scattering material, an internal region $V_I$ (the bulk region), and a nonuniform transition region $V_T$ with thickness of the order of a scatterer's diameter. The average excess pressure $\langle p\rangle$ equals $\langle\Psi\rangle$ throughout all space. In $V_E$, the average normalized velocity $\langle\underset{\sim}{v}\rangle$ is related to $\langle\Psi\rangle$ as in (2), i.e.,

$$\langle\underset{\sim}{v}\rangle = \nabla\langle\Psi\rangle, \quad \nabla\cdot\langle\underset{\sim}{v}\rangle = -k^2\langle\Psi\rangle, \quad (\nabla^2 + k^2)\langle\Psi\rangle = 0. \qquad (18)$$

For the simplest cases, we define bulk parameters associated with the coherent field in $V_I$ by paralleling the structure of (7):

$$\langle \underset{\sim}{v} \rangle = B\nabla\langle\Psi\rangle, \quad \nabla\cdot\langle \underset{\sim}{v}\rangle = -k^2\langle\Psi\rangle, \quad (\nabla^2 + K^2)\langle\Psi\rangle = 0,$$
$$K = k(C/B)^{1/2} = k\eta. \tag{19}$$

More generally, $B$ is a dyadic and the resulting composite is anisotropic. Earlier procedures replaced $V_T$ by an interface and determined parameters by comparing explicit forms of $\langle\Psi\rangle$ in $V_E$ and $V_I$ with corresponding forms of the solution for a plane wave incident on the interface of two uniform media.

For $\underset{\sim}{r}$ in $V_I$, the pressure field equals

$$\langle\Psi(\underset{\sim}{r})\rangle = \phi(\underset{\sim}{r}) + \rho\int_V \langle U_s(k)\rangle_s d\underset{\sim}{r}_s + \rho\int_V [\langle\Psi_s(K')\rangle_s - \langle\Phi_s(k)\rangle_s] d\underset{\sim}{r}_s \tag{20}$$

where $V = V_d - v$. Using (19) and (7), the associated velocity field is

$$B\nabla(\Psi(\underset{\sim}{r})\rangle = \nabla\phi(\underset{\sim}{r}) + \rho\int_V \nabla\langle U_s\rangle_s d\underset{\sim}{r}_s + \rho\int_V [B'\nabla\langle\Psi_s\rangle_s - \nabla\langle\Phi_s\rangle_s] d\underset{\sim}{r}_s \tag{21}$$

with $\nabla = \nabla_r$ acting on $\underset{\sim}{r}$ in the suppressed arguments $(\underset{\sim}{r} - \underset{\sim}{r}_s; \underset{\sim}{r}_s)$. The individual terms of (21) are velocity analogs of the corresponding pressure terms of (20).

From (20) and the divergence of (21), and from (21) and the gradient of (20), we have

$$(C - 1)\langle\Psi(\underset{\sim}{r})\rangle = (C' - 1)\rho\int_V \langle\Psi_s(\underset{\sim}{r} - \underset{\sim}{r}_s; \underset{\sim}{r}_s)\rangle_s d\underset{\sim}{r}_s \tag{22}$$

$$(B - 1)\nabla\langle\Psi(\underset{\sim}{r})\rangle = (B' - 1)\rho\int_V \nabla\langle\Psi_s(\underset{\sim}{r} - \underset{\sim}{r}_s; \underset{\sim}{r}_s)\rangle_s d\underset{\sim}{r}_s \tag{23}$$

Thus, $C - 1$ is proportional to $C' - 1$ and a mean value of average pressure fields, and $B - 1$ is proportional to $B' - 1$ and a mean value of the gradients.

For an internal plane wave component $\langle\Psi(\underset{\sim}{r})\rangle = Ae^{i\underset{\sim}{K}\cdot\underset{\sim}{r}} = \Psi_I(\underset{\sim}{r})$, the corresponding waves $\langle\Psi_s\rangle_s$, $\langle\Phi_s\rangle_s$, $\langle U_s\rangle_s$ have the translational property indicated by $\overline{\psi}(\underset{\sim}{r}')\Psi_I(\underset{\sim}{r}_s)$, $\overline{\phi}\psi_I$, $\overline{u}\Psi_I$ where the local fields $\overline{\psi}$, $\overline{\phi}$, $\overline{u}$ are related on $S$ and $v$ as for a single isolated obstacle. Similarly,

$$\langle G_s(\hat{r})\rangle_s = g(\underset{\sim}{k}_r|\underset{\sim}{K})\Psi_I(\underset{\sim}{r}_s), \quad g(\underset{\sim}{k}_r|\underset{\sim}{K}) = [\phi(-\underset{\sim}{k}_r), \overline{\psi}] \tag{24}$$

where we may use the brace instead of the bracket operation.

Writing $\langle\Psi_s\rangle_s = \overline{\psi}(\underset{\sim}{r}')\Psi_I(\underset{\sim}{r}_s) = \overline{\psi}(\underset{\sim}{r}')e^{-i\underset{\sim}{K}\cdot\underset{\sim}{r}'}\langle\Psi(\underset{\sim}{r})\rangle$, we reduce (22) and (23) to

$$C - 1 = (C' - 1)\rho\int e^{-i\underset{\sim}{K}\cdot\underset{\sim}{r}'}\,\overline{\psi}(\underset{\sim}{r}')dv(\underset{\sim}{r}') \tag{25}$$

$$\eta^2(B - 1) = (B' - 1)k^{-2}\rho\int\nabla e^{-i\underset{\sim}{K}\cdot\underset{\sim}{r}'}\cdot\nabla\overline{\psi}(\underset{\sim}{r}')dv(\underset{\sim}{r}') \tag{26}$$

where $\nabla = \nabla_{r'}$ acts on $\underset{\sim}{r}'$. Subtracting (26) from (25), we obtain

$$K^2 - k^2 = \rho\int[(C'-1)k^2\phi(-\underset{\sim}{K})\overline{\psi}(K'|\underset{\sim}{K}) - (B'-1)\nabla\phi\cdot\nabla\overline{\psi}]dv(\underset{\sim}{r}'),\ \phi(-\underset{\sim}{K}) = e^{-i\underset{\sim}{K}\cdot\underset{\sim}{r}'} \tag{27}$$

or, equivalently,

$$\eta^2 - 1 = -c[\phi(-\underset{\sim}{K}),\ \overline{\psi}(K'|\underset{\sim}{K})] = -cg[\underset{\sim}{K}|\underset{\sim}{K}],\quad c = i4\pi\rho/k^3 \tag{28}$$

with the scattering amplitude form $g[\underset{\sim}{K}|\underset{\sim}{K}]$ restricted to the volume integral representation. Alternative derivations of (25) and (26), and alternative representation of $\eta$ in terms of surface integral forms of a scattering amplitude plus additional volume integrals, and generalizations to anisotropic distributions are in the literature [2].

The analog of (28) for $B' = 1$ and $\eta'^2 - 1$ as the scattering potential was derived originally by Lax [5] who regarded $g[\underset{\sim}{K}|\underset{\sim}{K}]$ as proportional to the isolated scattering amplitude in a medium specified by $K$. The form of (28) in terms of the isolated scattering amplitude $g(\hat{k},\hat{k})$ in the embedding space was obtained earlier by Reiche [3] for dipoles and by Foldy [4] for monopoles, and Rayleigh's [11] original result corresponds essentially to approximating $\eta^2 - 1$ by $2(\eta - 1)$.

In order to use the representations (25) and (26) to construct explicit results for $C$ and $B$, we require a determinate form for $\eta$. We approximate $\langle G_s\rangle_{st}$ of (17) by $\langle G_s\rangle_s$, analogous to $\langle\Phi_s\rangle_{st} \approx \langle\Phi_s\rangle_s$ as used by Lax, but make no approximations in the unknown g-functions of (24) and (28). In terms of the radiative function

$$\mathcal{U} = \int_c g(\hat{r},\hat{r}_c)g(\underset{\sim}{k}_c|\underset{\sim}{K})e^{i\underset{\sim}{k}_c\cdot\underset{\sim}{R}} \tag{29}$$

we obtain a functional equation approximation (dispersion equation) for $\eta$:

$$g(\underset{\sim}{k}_r|\underset{\sim}{K}) = -\frac{c}{\eta^2 - 1}\{e^{-i\underset{\sim}{K}\cdot\underset{\sim}{r}},\mathcal{U}\} + \rho\int[f(\underset{\sim}{R}) - 1]e^{-i\underset{\sim}{K}\cdot\underset{\sim}{R}}\mathcal{U}d\underset{\sim}{R} \tag{30}$$

The surface integral is over the exclusion surface determined by the minimum separation of the centers of nearest scatterers, and the volume integral is over all space depleted by the exclusion volume.

For impenetrable spheres of radius $a$, the exclusion surface which contains the center $(R = 0)$ of only one particle, is $R = 2a$. We expand the scattering amplitude forms in (5) and (24) as series of Legendre polynomials

$$g(\hat{r},\hat{k}) = \sum a_n P_n(\hat{r} \cdot \hat{k}), \quad g(\underset{\sim}{k}_r|\underset{\sim}{K}) = \sum A_n P_n(\hat{r} \cdot \hat{k}) \tag{31}$$

where the coefficients (multipoles) $a_n$ are known, and the $A_n$ are unknown. Substituting into (29) and using (30) we obtain a system of algebraic equations for the $A_n$ in terms of $a_n$, $\eta$, and the correlation integrals

$$\mathcal{H}_n = 4\pi\rho\int_0^\infty [f(R) - 1] j_n(KR) h_n^{(1)}(kR) R^2 dR \tag{32}$$

where $f(R)$ is the radial distribution function, and $j$ and $h$ are the spherical Bessel and Hankel functions. The homogeneous system determines $\eta$ and all but one of the coefficients $A_n$. To obtain the bulk parameters, as well as the remaining coefficient, we expand the integrands of (25)-(27) as Bessel-Legendre series in order to integrate, and recast the results as

$$C - 1 = c\sum A_n d_n^C, \quad \eta^2(B - 1) = c\sum A_n d_n^B, \quad \eta^2 - 1 = c\sum A_n d_n, \tag{33}$$

where $d_n = -d_n^C + d_n^B$, and the $d$'s are known functions of $\eta$, $\eta'$, and $ka$. We may renormalize the coefficients to obtain a new set, say $A_n'$. The coefficients $A'$ satisfy a more convenient inhomogeneous system of algebraic equations, and the sum of the $A_n'$ determines $\eta$. See Ref. 2 (1977) for details.

**Explicit Illustrations**

Using $x = ka$, we keep the known isolated scattering coefficients $a_0$, $a_1$, $a_2$ (monopole, dipole, quadrupole) to $O(x^6)$, and display $C$, $B$, $\eta^2$ in the form (1), e.g.,

$$C = C_1 + C_c(x^2) + iC_s(x^3), \tag{34}$$

etc. From $n^2 = C/B$, we require

$$n_1^2 = \frac{C_1}{B_1}, \quad \frac{n_s}{n_1^2} = \frac{C_s}{C_1} - \frac{B_s}{B_1}, \quad \frac{n_c}{n_1^2} = \frac{C_c}{C_1} - \frac{B_c}{B_1}. \tag{35}$$

and express the results in terms of

$$\gamma = C' - 1, \quad \delta = B' - 1, \quad D = 1 + \delta(1 - w)/3, \qquad w = \rho 4\pi a^3/3, \tag{36}$$

where $w$ is the volume fraction.

The leading terms equal

$$n^2 = (1 + w\gamma)/(1 + w\delta/D) \tag{37}$$

$$C_1 = 1 + w\gamma, \quad B_1 = 1 + w\delta/D \tag{38}$$

Rewriting the relative parameters as $C' = C_p/C_e$ and $C_1 = C_{b1}/C_e$ etc., (with p, e, and b indicating values for the particle, embedding space, and bulk region), we have

$$C_{b1} = C_e(1 - w) + C_p w, \qquad \frac{B_{b1} - B_e}{B_{b1} + 2B_e} = w\,\frac{B_p - B_e}{B_p + 2B_e} \tag{39}$$

The $C$ form (the representation of the bulk value as the volume-weighted mean of the values of the components) goes back to Archimedes. The $B$ form (attributed to Clausius and Mossotti) was obtained originally by Maxwell [12].

The remaining terms involve one of the two correlation integrals:

$$W = 1 + 4\pi\rho \int_0^\infty [f(R) - 1]R^2 dR = 1 - 8w + 34w^2 + \ldots \approx \frac{(1 - w)^4}{(1 + 2w)^2}, \tag{40}$$

$$N = -4\pi\rho a \int_0^\infty [f(R) - 1]R\,dR = 6w - \frac{66w^2}{5} + \ldots \approx \frac{6w}{1 + 2w}\left(1 - \frac{w}{5} + \frac{w^2}{10}\right) \tag{41}$$

The series follow from the viral expansion of $f$, and the closed forms are based on the Werthein-Thiele [10] solution of the Percus-Yevick integral equation approximation for $f$. The function $W$, the low frequency limit of the structure factor, is proportional to the fluctuation (the variance) in number concentration; the closed form was obtained originally from the scaled particle approximate equation of state [13] by applying statistical mechanics theorems.

The scattering terms corresponding to (37) and (38) are

$$\eta_s^2 = \eta_1^2 \frac{x^3 wW}{9} \left( \frac{3\gamma^2}{C_1} + \frac{\delta^2}{B_1 D^2} \right) = \frac{x^3 wW}{9B_1} \left( 3\gamma^2 + \frac{\eta_1{}^2 \delta^2}{D^2} \right), \tag{42}$$

$$C_s = \frac{x^3 \gamma^2}{3} wW, \quad B_s = -\frac{x^3 \delta^2}{9D^2} wW. \tag{43}$$

The first equality in (42) and the values in (43) indicate that the relation required by (35) is satisfied.

These values for $\Gamma_1$ and $\Gamma_s$, and generalizations for aligned ellipsoids with dyadic B' parameter and ellipsoidally symmetric pair statistics, may be obtained directly from (30) by perturbation procedures based on Rayleigh's first approximation for scattering by a single isolated ellipsoid [14].

The results in (38) and (43) are simple in that $C_1$ and $C_s$ are independent of $\delta$ while $B_1$ and $B_s$ are independent of $\gamma$. However, the dispersive corrections $C_c$ and $B_c$ each involve both $\gamma$ and $\delta$, and thereby couple the particle parameters C' and B'. Before displaying the results explictly we consider the structure required by relations (22), (23), and (35). The constraints are satisfied by the forms

$$\eta_c^2/\eta_1^2 wx^2 = F_C/C_1 + F_B/B_1, \tag{44}$$

$$C_c/wx^2 = F_C + C_1 L = \gamma f_C, \quad B_c/wx^2 = -F_B + B_1 L = \delta f_B, \tag{45}$$

where the F's, f's, and L are functions of C', B' and w. The dispersion equation leads relatively directly to (44), but $\eta_c^2$ cannot be decomposed to obtain $C_c$ and $B_c$ because of the cancellation of the L terms that insure proportionality to $\gamma$ or $\delta$.

Explicitly, we have

$$F_C = C_1 S + T_C, \quad S = \gamma\delta N/B_1 9D,$$

$$T_C = \gamma t_C - \delta\eta'^2/15, \quad t_C = [\eta'^2 - 4 - 5\gamma(N - 1)]/15, \tag{46}$$

where S vanishes if either $\gamma$ or $\delta$ vanishes; $T_C$ consists of essentially two terms, one vanishing with $\gamma$ and the other with $\delta$. Similarly

$$F_B = B_1 S + T_B - \delta Q, \quad Q = \eta_1^2 2(3 + 2B_1)^2/75(3 + 2B'),$$

$$T_B = \delta t_B + \gamma/5D^2, \quad t_B = [9(\delta + 1) - \delta N(5 + 2\eta_1^2)]/45D^2, \tag{47}$$

where $Q$ involves quadrupole as well as monopole and dipole coupling. Using (46) and (47) in (44),

$$\frac{\eta_c^2}{\eta_1^2 w x^2} = \frac{T_C}{C_1} + \frac{T_B}{B_1} + 2S - \frac{\delta Q}{B_1}. \tag{48}$$

The $L$ term in (45) is given by

$$L = \gamma/5DB_1 + \delta\eta'^2/15, \tag{49}$$

and the corrections equal

$$C_c = x^2 w\gamma[t_C + \delta\eta'^2 w/15 + \eta_1^2(9 + 5\delta N)/45D],$$

$$B_c = x^2 w\delta[-t_B + \gamma(1 - w)/15D^2 - \gamma N/9D + \eta'^2 B_1/15 + Q]. \tag{50}$$

For the one parameter cases, if $C' = 1$ then $C = 1$ and $\eta^2 = 1/B$, and if $B' = 1$ then $B = 1$ and $\eta^2 = C$ as required by the theorems. For all cases (two or one parameter) the bulk values in terms of the closed form approximations of $W$ and $N$ reduce to the single particle values for the unrealizable bound $w = 1$.

To second order in particle parametric contrasts ($\gamma$ and $\delta$), the $w$ dependence of $r_s$ and $r_c$ terms are given by

$$S_s = wW \approx w(1 - w)^4/(1 + 2w)^2, \tag{51}$$

$$S_c = w(6 + 3w - 5N)/6 \approx w(2 - w)(1 - w)^2/2(1 + 2w), \tag{52}$$

such that $S/w$ decreases monotonically from 1 to 0 as $w$ increases from 0 to 1. The $S$ functions are zero at $w = 0$ and 1; the maximum of $S_s$ approximates 0.047 at $w \approx 0.129$, and that of $S_c$ approximates 0.083 at $w \approx 0.221$.

We have

$$C_s \approx \frac{x^3}{3}\gamma^2 S_s, \quad B_s \approx -\frac{x^3}{9}\delta^2 S_s, \quad \eta_s^2 \approx \frac{x^3}{9}(3\gamma^2+\delta^2)S_s, \tag{53}$$

$$C_c \approx \frac{x^2}{15}2\gamma(3\gamma-\delta)S_c, \quad B_c \approx -\frac{x^2}{75}2\delta(7\delta-5\gamma)S_c, \quad \eta_c^2 \approx \frac{x^2}{225}6(15\gamma^2-10\gamma\delta+7\delta^2), \tag{54}$$

$$C_1 = 1+w\gamma, \quad B_1 \approx 1+w\delta-w(1-w)\delta^2/3, \quad \eta_1^2 \approx 1+w(\gamma-\delta)-w^2\gamma\delta+w(1+2w)\delta^2/3. \tag{55}$$

To the same order in $\gamma$ and $\delta$ we may use

$$\eta \approx \eta_1 + (\eta_c^2 + i\eta_s^2)/2\eta_1 \approx \eta_1 + \eta_c^2/2 + i\eta_s^2/2 \tag{56}$$

to obtain the corresponding $K = k\eta$.

## 3. Electromagnetics

The structure is similar for the electromagnetic scattering problems of particles specified by two physical parameters $\varepsilon'$ and $\mu'$ (such that $\eta'^2 = \varepsilon'\mu'$) in v, and a pair of vector fields ($\underset{\sim}{E}$ and $\underset{\sim}{H}$) whose tangential components are continuous on S. To exploit the scalar development, we write

$$\langle\underset{\sim}{\Psi}\rangle = \begin{Bmatrix}\langle\underset{\sim}{E}\rangle\\ \langle\underset{\sim}{H}\rangle\end{Bmatrix}, \quad C = \begin{Bmatrix}\varepsilon\\ \mu\end{Bmatrix}, \quad B^{-1} = \begin{Bmatrix}\mu\\ \varepsilon\end{Bmatrix}, \quad \eta^2 = \varepsilon\mu, \tag{57}$$

where all fields are solutions of the vector Helmholtz equation (obtained on replacing $\nabla^2$ by $-\nabla\times\nabla\times$). The following indicates required modifications of the structural acoustic equations, and gives analogous explicit results for the bulk values $(\varepsilon, \mu, \eta^2)$ in terms of the correlation integrals W and N of (40) and (41).

The appropriate surface and volume integral operators are vector analogs of those considered for the scalar case. The forms (12)-(14) and (16) apply with the waves and G regarded as vector functions, but in (15) and (17) the g terms for incidence along $\hat{k}$ or $\hat{r}_c$ now represent the scalar product of a dyadic with the incident or G unit polarization vector. We obtain forms (20) and (22) for the corresponding vector fields, and (21) and (23) with the gradient replaced by the curl.

For an incident plane wave propagating along $\hat{z}$ (i.e., $\underset{\sim}{k} = k\hat{z}$) and polarized along $\hat{x}$,

$$\underset{\sim}{\phi} = \hat{x}e^{ikz} = \hat{x} \cdot \tilde{\phi}(\underset{\sim}{k}), \quad \tilde{\phi}(\underset{\sim}{k}) = (\tilde{I} - \hat{z}\hat{z})e^{ikz} \tag{58}$$

with $\tilde{I}$ as the identity dyadic, we express the appropriate versions of (25)-(28) in terms of the dyadic plane wave

$$\tilde{\phi}(-\underset{\sim}{K}) = (\tilde{I} - \hat{z}\hat{z})e^{-ikz'}, \quad Kz' = \underset{\sim}{K} \cdot \underset{\sim}{r}'. \tag{59}$$

Thus the analogs of (25) and (26) are

$$C - 1 = (C' - 1)\rho\hat{x} \cdot \int \tilde{\phi} \cdot \overline{\underset{\sim}{\psi}}\, dv, \tag{60}$$

$$\eta^2(B - 1) = (B' - 1)k^{-2}\rho\hat{x} \cdot \int (\nabla\times\tilde{\phi}) \cdot (\nabla\times\overline{\underset{\sim}{\psi}})dv, \tag{61}$$

and subtracting (61) from (60) provides the analog of (27) and (28),

$$\eta^2 - 1 = -c\hat{x} \cdot [\tilde{\phi}(-\underset{\sim}{K});\ \overline{\underset{\sim}{\psi}}(\underset{\sim}{K}'|\underset{\sim}{K})] = -cg[\underset{\sim}{K}|\underset{\sim}{K}] \tag{62}$$

in terms of the resulting bracket operation. We regard the kernel of (29) as the scalar product of dyadic and vector amplitudes, and the required dispersion equation is the form (30) with the amplitude and radiative function as vectors. The associated series decompositions involve two sets of coefficients, one corresponding to electric multipoles and the other to magnetic.

Explicit illustrations

Proceedings essentially as for (34) ff, we retain dipoles and quadrupoles to $O(x^6)$. From $\eta^2 = \varepsilon\mu$ we require

$$\eta_1^2 = \varepsilon_1\mu_1, \quad \frac{\eta_s^2}{\eta_1^2} = \frac{\varepsilon_s}{\varepsilon_1} + \frac{\mu_s}{\mu_1}, \quad \frac{\eta_c^2}{\eta_1^2} = \frac{\varepsilon_c}{\varepsilon_1} + \frac{\mu_c}{\mu_1}, \tag{63}$$

and express the results in terms of

$$\delta_e = \varepsilon' - 1, \quad \delta_m = \mu' - 1, \quad D_e = D(\delta_e), \quad D_m = D(\delta_m), \tag{64}$$

with $D$ as in (36). We include $\eta^2$ and $\varepsilon$ terms explicitly, with $\mu$ terms following from $\varepsilon$ by interchanging $\varepsilon'$ and $\mu'$.

The leading terms and scattering terms are

$$n_1^2 = (1 + w\delta_e/D_e)(1 + w\delta_m/D_m), \quad \varepsilon_1 = 1 + w\delta_e/D_e, \tag{65}$$

$$\frac{n_s^2}{n_1^2} = \frac{x^3 2wW}{9}\left(\frac{\delta_e^2}{\varepsilon_1 D_e^2} + \frac{\delta_m^2}{\mu_1 D_m^2}\right), \quad \varepsilon_s = \frac{x^3 2\delta_e^2 wW}{9D_e^2}. \tag{66}$$

Replacing $\varepsilon'$ by $\mu'$ in the $\varepsilon$ forms to obtain the corresponding $\mu$ terms, we see that the relation required by (63) is satisfied. These results, and generalizations for ellipsoids, were also derived by the perturbation procedure indicated after (43).

The dispersive corrections have essentially the structure discussed for the scalar case, but the explicit results are more symmetrical. The constraints (60), (61), and (63) are satisfied by the forms

$$n_c^2/n_1^2 wx^2 = F_e/\varepsilon_1 + F_m/\mu_1, \tag{67}$$

$$\varepsilon_c/wx^2 = F_e + \varepsilon_1 L = \delta_e f_e, \quad \mu_c/wx^2 = F_m - \mu_1 L = \delta_m f_m, \tag{68}$$

where symmetry requires that the interchange of $\varepsilon'$ and $\mu'$ in $F_e$ and $f_e$ produces $F_m$ and $f_m$, but that $L$ be antisymmetrical.

Explicitly,

$$\begin{aligned} -F_e &= \varepsilon_1 S + T_e - \delta_e Q_e, \\ S &= \delta_e\delta_m N/9D_e D_m, \qquad Q_e = n_1^2(2\varepsilon_1 + 3)^2/50(2\varepsilon' + 3), \\ T_e &= \delta_e t_e - \delta_m \varepsilon'^2/10D_e^2, \quad t_e = [9(1-\delta_e) + \delta_e N(10 + n_1^2)]/45D_e^2. \end{aligned} \tag{69}$$

Thus

$$-\frac{n_c^2}{n_1^2 wx^2} = \frac{T_e}{\varepsilon_1} + \frac{T_m}{\mu_1} + 2S - \frac{\delta_e Q_e}{\varepsilon_1} - \frac{\delta_m Q_m}{\mu_1}. \tag{70}$$

The $L$ term equals

$$L = \frac{\varepsilon'\delta_m}{10D_e} - \frac{\mu'\delta_e}{10D_m} = L(\varepsilon', \mu') = -L(\mu', \varepsilon') \tag{71}$$

and consequently

$$\varepsilon_c = -x^2 w\delta_e[t_e - \frac{\delta_m \varepsilon'(1-w)}{15\eta_e^2} + \frac{\delta_m \varepsilon_1 N}{9\eta_e\eta_m} - \eta_e] \tag{72}$$

with $\mu_c$ following on interchanging $\varepsilon'$ and $\mu'$.

For the one parameter cases, if $\delta_e = \varepsilon' - 1 = 0$, then $\varepsilon = 1$ and $\eta^2 = \mu$, and if $\delta_m = \mu' - 1 = 0$ then $\mu = 1$ and $\eta^2 = \varepsilon$. For all cases in terms of the closed forms of W and N, the bulk values reduce to the single particle values for the unrealizable bound $w = 1$.

To second order in particle parametric contrasts $\delta_e = \delta_m$, in terms of $S_s$ and $S_c$ of (51) and (52), we have

$$\varepsilon_s \approx \frac{x^3}{9} 2\, \delta_e^2 S_s, \quad \eta_s^2 \approx \frac{x^3}{9} 2(\delta_e^2 + \delta_m^2)S_s, \tag{73}$$

$$\varepsilon_c \approx \frac{x^2}{75} 2\varepsilon_e(11\delta_e + 5\delta_m)S_c, \quad \eta_c^2 \approx \frac{x^2}{75} 2[11(\delta_e^2 + \delta_m^2) + 10\delta_e\delta_m]S_c, \tag{74}$$

$$\varepsilon_1 \approx 1 + w\delta_e - w(1-w)\delta_e^2/3, \quad \eta_1^2 \approx 1 + w(\delta_e+\delta_m) + w^2\delta_e\delta_m - w(1-w)(\delta_e^2+\delta_m^2)/3. \tag{75}$$

We obtain $\eta$ as in (56).

**Acknowledgement**

This work was supported in part by grants from the National Science Foundation and the U.S. Office of Naval Research.

**References**

1. V. Twersky, J. Math. Phys. 3, 83 (1962); 8, 589 (1967).

2. V. Twersky, J. d'Analyse Math. 30, 498 (1976); J. Math. Phys. 18, 2468 (1977); 19, 215 (1978).

3. F. Reiche, Ann. Phys. 50, 1, 121 (1916).

4. L.L. Foldy, Phys. Rev. 67, 107 (1945).

5. M. Lax, Rev. Mod. Phys. 23, 287 (1951); Phys. Rev. 88, 621 (1952).

6. V. Twersky, J. Math. Phys. 3, 700, 724 (1962).

7. J.B. Keller, AMS Proc. Symp. Appl. Math. 13, 227 (1962); 16, 145 (1964).

8. V. Twersky, AMS Proc. Symp. Appl. Math. 16, 84 (1964).

9. V. Twersky, J. Acoust. Soc. Am. 77, 29 (1985); J. Math. Phys. 26, 2208 (1985).

10. M.S. Wertheim, Phys. Rev. Lett. 10, 321 (1963); E. Thiele, J. Chem. Phys. 39, 474 (1963); H.D. Jones, J. Chem. Phys. 55, 2640 (1971).

11. Rayleigh, Phil. Mag. 47, 375 (1899).

12. J.C. Maxwell, A Treatise of Electricity and Magnetism (Cambridge, 1873), Sec. 314.

13. H. Reiss, H.L. Frisch, and J.L. Lebowitz, J. Chem. Phys. 31, 369 (1959).

14. Rayleigh, Phil. Mag. 44, 28 (1897).

RANDOM RAYS AND STOCHASTIC CAUSTICS

Benjamin S. White
Exxon Research and Engineering Co.
Route 22 East
Annandale, NJ 08801

**Abstract**

I review a theory of high frequency random wave propagation which is obtained by applying a stochastic limit theorem to the equations of ray theory in a random medium. The method is applicable in the "strong fluctuation region" when large amplitudes occur randomly. If index of refraction fluctuations are small, of order $\sigma$, large amplitudes are accompanied by the occurrence of caustics which appear after long propagation distances of order $\sigma^{-2/3}$ correlation lengths of the random medium. Under broad hypotheses it is shown that the region of random caustics is given, up to a single distance scale parameter, by a universal curve for the probability of caustic formation along a ray as a function of propagation distance. In this region wavefront geometry is in some sense preserved despite the order 1 random wanderings of the rays, since to leading order the random displacement of the rays is along the wavefronts.

On this scale limit equations can be derived for the joint distribution of an arbitrary number of ray positions and associated raytube areas. The transformation from ray coördinates to physical coördinates is effected by deriving expressions for quantities which are transported along a ray, when summed over the random number of randomly chosen rays which pass through a fixed point in physical space. Application is given to computation of the correlation function of intensity.

## 1. Introduction

We consider solutions of the Helmholtz equation

$$\Delta u + k^2 n^2(x) u = 0 \tag{1}$$

where n(x) is the index of refraction and k is the wave number. It is assumed that n(x) is composed of small smooth random variations from a constant. Thus the normalized propagation speed is written as

$$C(x) = 1/n(x) = 1 + \sigma \hat{C}(x) \tag{2}$$

where $0<\sigma<<1$ is a small parameter and $\hat{C}$ is a mean zero homogeneous and isotropic random field.

The equations of ray theory are obtained from (1) in the high frequency limit $k\to\infty$ [1] by inserting into (1) the W.K.B. ansatz

$$u \sim e^{ik\phi(x)}(V + (1/ik)\, V^{(1)} + \ldots) \quad . \tag{3}$$

Equating the coefficient of $k^2$ to zero then gives the eikonal equation for the phase $\phi$

$$(\nabla \phi)^2 = \frac{1}{C^2} \quad . \tag{4}$$

Equation (4) may be solved along its bicharacteristic curves, or rays, using Hamilton-Jacobi theory for a first order nonlinear partial differential equation. If s is arclength along a ray X(s) and U(s) is the unit tangent then the ray equations are

$$\begin{aligned} \frac{d}{ds} X &= U \\ \frac{d}{ds} U &= -\frac{1}{C} (I - U U^T) \nabla C \end{aligned} \tag{5}$$

where I is the identity.

If the phase $\phi$ is known at some point along a ray, it can be determined at any other point by appending to (5) the equation

$$\frac{d}{ds} \phi = \frac{1}{C} \quad . \tag{6}$$

In particular, a surface of constant phase is a wavefront, and (5),(6) determine the evolution of wavefronts. To find $\phi$ at a fixed point x of space one must solve a two-point boundary value problem to determine the ray which passes through x.

Similarly, it can be shown that the amplitude V can be determined by integrating various quantities along a ray. In particular, V is determined by the "raytube area" A, defined as the Jacobian of the mapping from an initial wavefront at X(0) to the one at X(s).

$$V(s) = V(0)\sqrt{\frac{C(X(s))\ A(0)}{C(X(0))\ A(s)}} \tag{7}$$

For $x \in \mathbb{R}^2$ equations may be written along a ray for A and its derivative with respect to arclength, B. Let $U^{\perp}$ be the unit vector orthogonal to U and let $\nabla\nabla C$ be the matrix of second derivatives of C. Then

$$\begin{aligned} \frac{d}{ds} A &= B \\ \frac{d}{ds} B &= -\frac{1}{C}\left(U^{\perp}\right)^T \nabla\nabla C\, U^{\perp} A + \frac{1}{C} U^T \nabla C\, B \end{aligned} \tag{8}$$

For $x \in \mathbb{R}^3$ there are 2x2 matrices A,B satisfying equations analogous to (8), with the determinant of A corresponding to the raytube area.

Various authors have made use of the assumption (2) of small random fluctuations in index of refraction. Chernov[2], Keller[3], and Tatarski[4] applied regular perturbation theory to expand all relevent quantities as power series in $\sigma$. They thereby derived a number of fundamental results. This theory, however, is necessarily limited to weak fluctuations in the solution, an assumption inherent in a regular perturbation scheme. Secular terms in the expansion preclude the use of this method over long propagation distances.

For propagation over longer distances, Chernov made the *ad hoc* assumption ("dishonest" in the sense of Keller) that the ray direction U is a Markov process, and determined the statistics of the position of a single ray up to an unknown diffusion constant. The constant was subsequently supplied by Keller, who matched to regular perturbation theory. More recently, a rigorous derivation of this limit result was given by Kesten and Papanicolaou[5], who moreover determined the distance scale, of order $\sigma^{-2}$ correlation lengths of the random medium, for which the result is valid. Limits have not been derived on this scale for raytube areas, and the transformation from rays to physical coördinates has not been attempted.

I review here a theory[6]-[9] for an intermediate range, when the propagation distance is of order $\sigma^{-2/3}$ correlation lengths. It is shown that this is the scale on which large amplitudes first occur, accompanied by the vanishing of the raytube area, *i.e.* the occurence of caustics. At such points the amplitude as given by (7) is infinite, signalling the breakdown of the W.K.B. expansion (3). A modified expansion in the neighborhood of a caustic can be effected, *e.g.* by the method of Ludwig[10]. While the amplitude is finite for finite frequency it becomes infinite as $k \to \infty$. For propagation beyond the caustic another difficulty occurs. After the vanishing of the Jacobian A(s) the mapping from wavefront to wavefront is no longer 1-1 and a single point of space may be hit by more than one ray. Thus (4) may cease to have a solution. In general the ansatz (3) must be modified to include a sum of terms each of W.K.B. form corresponding to contributions from each ray which passes through the point x. For the stochastic problem we must consider random quantities which are transported along rays summed over the random number of randomly-chosen rays which pass through a given point x.

The various scales can be understood through a simpler problem modelling equation (5), a second order equation for X with small right hand side

$$\frac{d^2}{ds^2} X(s) = \sigma f(s) \qquad (9)$$

Here f is simply a mean zero stationary random function of s. Letting $t=\sigma^2 s$, $\tilde{X}=\sigma^2 X$ yields

$$\frac{d^2}{dt^2} \tilde{X}(t) = \frac{1}{\sigma} f(t/\sigma^2) \qquad (10)$$

The right hand side of (10) is scaled for a white noise limit. That is under wide hypotheses on f one can show the functional central limit theorem (weak convergence theorem) that

$$\frac{1}{\sigma} \int_0^t f(\tau/\sigma^2)\, d\tau \underset{\sigma\downarrow 0}{\Rightarrow} \text{Brownian motion} \qquad (11)$$

What is needed for (11) is that f have an appropriate "mixing" property, roughly, that $f(t_1)$ $f(t_2)$ are asymptotically independent as $|t_1-t_2| \to\infty$.

Note that the limit in (11) is not for X but for $\sigma^2 X$, and this is the case in the Kesten - Papanicolaou result.

To determine when deviations in X become order one, we do not scale X, but scale $t=\sigma^\nu s$ with $\nu$ to be determined. The unique choice of $\nu=2/3$ produces an equation of the form (10), but with $\sigma$ replaced by $\sigma'=\sigma^{1/3}$. Since order one deviations in X produce order one deviations in A, A=0, and thus caustic formation, becomes possible on this scale.

## 2. A Markovian Limit

Let $x\in\mathbb{R}^2$ and consider a wave which is initially plane. Let $\alpha$ be arclength along the initial wavefront, which is then the locus of rays for s=0. Denote by $X^{\sigma}(s,\alpha)$, $U^{\sigma}(s,\alpha)$ the ray and tangent vector which emanate from position $\alpha$ along the initial wavefront. Then denoting by $\phi^{\sigma}(s,\alpha)$, *etc.* the associated quantities we have, for a convenient choice of coördinate system

$$\begin{aligned} X^{\sigma}(0,\alpha) &= \begin{pmatrix}0\\ \alpha\end{pmatrix} \\ U^{\sigma}(0,\alpha) &= \begin{pmatrix}1\\ 0\end{pmatrix} \\ \phi^{\sigma}(0,\alpha) &= 0 \\ A^{\sigma}(0,\alpha)=1, &\qquad B^{\sigma}(0,\alpha)=0. \end{aligned} \tag{12}$$

Let

$$t = \sigma^{2/3}s \qquad \hat{\phi}(t,\alpha) = \frac{1}{\sigma^{2/3}}\left[\phi^{\sigma}(s,\alpha) - s\right]$$

$$\begin{aligned} \hat{X}(t,\alpha) &= X^{\sigma}(s,\alpha) - s\begin{pmatrix}1\\ 0\end{pmatrix} \\ \hat{U}(t,\alpha) &= \frac{1}{\sigma^{2/3}}\left[U^{\sigma}(s,\alpha) - \begin{pmatrix}1\\ 0\end{pmatrix}\right] \\ A(t,\alpha) &= A^{\sigma}(s,\alpha) \qquad B(t,\alpha) = \frac{1}{\sigma^{2/3}} B^{\sigma}(s,\alpha). \end{aligned} \tag{13}$$

Substitution of (13) into (5), (6), (8) yields, after using (2) and dropping terms of order $\sigma^{1/3}$

$$\begin{aligned} \frac{d}{dt}\hat{X} &= \hat{U} \qquad \frac{d}{dt}\hat{U} = -\frac{1}{\sigma^{1/3}}\begin{bmatrix}0\\ \hat{C}_{,2}\end{bmatrix} \\ \frac{d}{dt}\hat{\phi} &= -\frac{1}{\sigma^{1/3}}\hat{C} \\ \frac{d}{dt}A &= B \qquad \frac{d}{dt}B = -\frac{1}{\sigma^{1/3}}\hat{C}_{,22}A \end{aligned} \tag{14}$$

Here $\hat{C}_{,2}$, $\hat{C}_{,22}$ are first and second derivatives of $\hat{C}$ with respect to the second Cartesian coordinate, and are evaluated at $X^{\sigma}$. Initial conditions are

$$\hat{X}(0,\alpha) = \begin{pmatrix} 0 \\ \alpha \end{pmatrix} \qquad \hat{U}(0,\alpha) = \begin{pmatrix} 0 \\ 0 \end{pmatrix}$$

$$\hat{\phi}(0,\alpha) = 0 \tag{15}$$

$$A(0,\alpha) = 1 \qquad B(0,\alpha) = 0 \quad .$$

Note that from (14), (15) the first components of $\hat{X}$ and $\hat{U}$ are zero. Thus the distance travelled along the ray, the phase, and the distance travelled in the direction of initial propagation are all the same, $s=t/\sigma^{2/3}$, to leading order. To leading order, then, the wavefront is where it would be in the absence of random fluctuations even though the rays composing the wavefront have wandered by order one from their unperturbed positions.

These facts can be related to propagation of pulses in the time domain if equation (1) is considered as the Fourier transform of the time-dependent wave equation. A signal reaches a receiver at a time which is close to that predicted by ignoring the randomness but with large amplitude variations. In addition, instead of a single pulse some random number of pulses may be received, but differences in their arrival times, given by $\sigma^{2/3}\hat{\phi}$, are small compared to the time to traverse a typical random inhomogeneity.

Let y be the second component of $\hat{X}$ and v the second component of $\hat{U}$. Then we have, after eliminating first components

$$\frac{d}{dt}\begin{pmatrix} y(t,\alpha) \\ v(t,\alpha) \\ \hat{\phi}(t,\alpha) \\ A(t,\alpha) \\ B(t,\alpha) \end{pmatrix} = \frac{1}{\sigma^{1/3}}\begin{pmatrix} 0 \\ -\hat{C},_{2}(t/\sigma^{2/3},y(t,\alpha)) \\ -\hat{C}(t/\sigma^{2/3},y(t,\alpha)) \\ 0 \\ -A(t,\alpha)\ \hat{C},_{22}(t/\sigma^{2/3},y(t,\alpha)) \end{pmatrix} + \begin{pmatrix} v(t,\alpha) \\ 0 \\ 0 \\ B(t,\alpha) \\ 0 \end{pmatrix} \tag{16}$$

The initial condition for (16) is

$$\begin{pmatrix} y(t,0) \\ v(t,0) \\ \hat{\phi}(t,0) \\ A(t,0) \\ B(t,0) \end{pmatrix} = \begin{pmatrix} \alpha \\ 0 \\ 0 \\ 1 \\ 0 \end{pmatrix} \quad . \tag{17}$$

Consider a system of N rays parametrized by $\alpha_i$ for i=1,2,...,N, with associated quantities $y_i=y(t,\alpha_i)$ *etc.* Let $\hat{X}$ be the 5N dimensional vector containing $(y_i,v_i,\hat{\phi}_i,A_i,B_i)$ i=1,...,N as its components. Then letting $\epsilon=\sigma^{1/3}$, it is apparent from (16) that $\hat{X}$ satisfies an equation of the form

$$\frac{d}{dt}\hat{X} = \frac{1}{\epsilon} F(t/\epsilon^2,\hat{X}) + G(\hat{X}) \qquad (18)$$

Furthermore for fixed t, **X**, **F**(t,**X**) has mean zero. Therefore for small $\epsilon$ equation (18) is scaled for a white noise limit. That is , application of the limit theorem of Papanicolaou and Kohler[11] yields a diffusion Markov process. To describe this process, let the correlation function of $\hat{C}$ be $R(y,z) = E[\hat{C}(0,0)\,\hat{C}(y,z)]$ and define

$$g(z) = -\int_0^\infty R_{zz}(y,z)\,dy \qquad (19)$$

We can then write a forward Kolmogorov (Fokker-Planck) equation. In particular, if $P(t,y_1,\ldots,y_n,v_1,\ldots,v_n,A_1,\ldots A_n,B_1,\ldots B_n \mid \alpha_1,\ldots,\alpha_n)$ is the joint density of $y(t,\alpha_1),\ldots,y(t,\alpha_n)$ *etc.*, then P satisfies[8]

$$\frac{\partial}{\partial t} P = -\sum_{j=1}^{N}\left( v_j \frac{\partial P}{\partial y_j} + B_j \frac{\partial P}{\partial A_j}\right)$$

$$+\sum_{j=1}^{N}\sum_{l=1}^{N}\left[ g(y_j-y_l)\frac{\partial^2 P}{\partial v_j \partial v_l} - 2A_l\, g'(y_j-y_l)\frac{\partial^2 P}{\partial B_l \partial v_j} - A_j A_l\, g''(y_j-y_l)\frac{\partial^2}{\partial B_j \partial B_l}\right] \qquad (20)$$

Analogous equations can be derived in three dimensions.

### 3. Probability of a Caustic

For a single ray it can be shown from (20) that, since $g'(0)=0$, the two-dimensional process $(y(\bullet),v(\bullet))$ is statistically independent of $(A(\bullet),B(\bullet))$. v may be recognized as Brownian motion and y is its integral.

The (A,B) process depends on a single parameter, $g''(0)$, which can be eliminated entirely by scaling. Thus the statistics of A, B are governed by a canonical process without parameters. For $\beta$ a standard Brownian motion this process is characterized by the Ito equation

$$dA = B\,dt$$
$$dB = A\,d\beta \qquad (21)$$

Caustics, or the zeroes of A, occur when the wavefront curvature $Z=B/A$ is infinite. Z satisfies the one dimensional Ito- Riccati equation

$$dZ = -Z^2 dt + d\beta \quad . \qquad (22)$$

Equations (21) and (22) have been studied to determine the universal curve for the probability that a caustic first occurs in scaled distance $(t,t+dt)$ along a ray, both for an initially plane wave, and for a line source[6],[9]. These curves have an initial flat portion where caustics are unlikely, and are transcendentally small for small t, of the form $(a_1/t^{5/2})\ \exp(-a_2/t^3)$. They rise rapidly to a peak and decay exponentially at large distances, implying that caustics occur along every ray with probability one. The plane wave case has been verified by the Monte-Carlo experiments of Hesselink and Sturdevant[12].

In three dimensions the caustic theory is similar[7],[9]. The principal normal curvatures of the wavefront are eigenvalues of a symmetric 2x2 matrix R which satisfies the matrix Ito-Riccati equation

$$dR = -R^2 dt + \sqrt{2}\begin{bmatrix}1 & 0\\ 0 & 1\end{bmatrix} d\beta_1 + \begin{bmatrix}1 & 0\\ 0 & -1\end{bmatrix} d\beta_2 + \begin{bmatrix}0 & 1\\ 1 & 0\end{bmatrix} d\beta_3 \qquad (23)$$

The universal curves for caustic occurrence both for an initially plane wave and for a point source are qualitatively similar to those in two dimensions, with a small t expansion of the form $(a_3/t^4)\ \exp(-a_4/t^3)$, and exponential decay for large t.

## 4. Transformation from Ray to Physical Coordinates

Let $x \in \mathbb{R}^2$ and for $j=1,2,\ldots,m$ let $h^{(j)}(t,\alpha)$ be m quantities transported to distance $t/\sigma^{2/3}$ along the ray parametrized by $\alpha$. Using the equivalence to leading order of arclength and distance along the ray in the original propagation direction, we can write the expression for quantity $h^{(j)}$ at a fixed point $x=(t/\sigma^{2/3},\bar{y})$ in space, when summed over all rays which pass through x

$$H^{(j)}(t,\bar{y}) = \sum_{\{\alpha_1:\ y(t,\alpha_1) = \bar{y}\}} h^{(j)}(t,\alpha_1) \qquad (24)$$

It can be shown that to leading order $A(t,\alpha)=\partial y(t,\alpha)/\partial\alpha$, *i.e.* the raytube area is approximately the Jacobian of the one-dimensional mapping $y(t,\bullet)$. Then the following expression results from (24) and a $\delta$-function identity

$$H^{(j)}(t,\bar{y}) = \int_{\infty}^{\infty} d\alpha \mid A(t,\alpha)\mid \; h^{(j)}(t,\alpha) \quad . \tag{25}$$

Using (25) we can write a useful expression for expected values of products of expressions $H^{(j)}$ evaluated at different points in space

$$E\left[\prod_{j=1}^{m} H^{(j)}(t_j,\bar{y}_j)\right] = \int d\alpha_1 \int d\alpha_2 \ldots \int d\alpha_n E\left[\prod_{j=1}^{m} \mid A(t_j,\alpha_j)\mid h^{(j)}(t_j,\alpha_j)\right] \quad . \tag{26}$$

Note that the integrand in (26) can be computed in ray coordinates where the appropriate probability densities can be characterized by partial differential equations such as equation (20).

Using this method, an equation has been derived for determining the correlation function of intensity or of energy flux between two points $(t/\sigma^{2/3},\bar{y}_1)$, $(t/\sigma^{2/3},\bar{y}_2)$ separated by distance $y=\mid \bar{y}_1-\bar{y}_2 \mid$ along a line parallel to the initial wavefront. Let $\rho(t,y)$ denote this correlation function, and let $Q(t,y,v)$ be the solution of

$$\begin{aligned} \frac{\partial Q}{\partial t} &= -v\frac{\partial Q}{\partial y} + 2(\, g(0) - g(y)\,)\frac{\partial^2 Q}{\partial v^2} \\ Q(0,y,v) &= \delta(v) \quad . \end{aligned} \tag{27}$$

Then

$$\rho(t,y) = \int_{\infty}^{\infty} Q(t,y,v)\, dv \quad . \tag{28}$$

Analogous results hold in three dimensions.

For $y\neq 0$, $\rho(t,y)$ as computed by (27),(28) is well behaved and agrees well with the results of Monte-Carlo simulations of rays in a random medium[8]. For small t it matches regular perturbation theory. It has, however, a logarithmic singularity at $y=0$ which results from the lack of caustic corrections in the present theory.

To assess the effects of caustic corrections we use Ludwig's[10] theory for the smooth caustic. As $k\to\infty$ the amplitude at such a caustic is of order $k^{1/6}$, and corrections are necessary only in a small boundary layer of order $k^{-2/3}$. We thus take the probability that a fixed point in space will fall within the caustic boundary layer to be of order $k^{-2/3}$. Since the intensity at a point in space is the squared magnitude of the field u, $\rho(t,y)$ is a fourth moment involving a single point when $y=0$, but two distinct points when $y\neq 0$.

For a moment of order m involving the field at a single point of space the caustic correction will be of order $k^{-2/3}\,(k^{1/6})^m$, which is negligible for $m<4$ but will contribute to the fourth moment $\rho(t,0)$. However for $y\neq 0$ we have two distinct points of which one or both may be in a caustic boundary layer. For one point in the layer the contribution is of order $k^{-2/3}\,(k^{1/6})^2=k^{-1/3}$, while for both points in layers the contribution is of order $(k^{-2/3})^2\,(k^{1/6})^4=k^{-2/3}$, so that in either case the effect of the correction is negligible as $k\to\infty$.

Clearly proper account of caustic corrections will induce a boundary layer for finite k at $y=0$ in $\rho(t,y)$, and enable accurate calculation of $\rho(t,0)$, which is related to the "scintillation index". An extension of the present theory to include uniform asymptotics is currently under investigation.

## References


[1] R.M. Lewis and J.B. Keller, "Asymptotic methods for partial differential equations", Research Report EM-194, AFCRL, Air Force Contract #AF 19(604)5238 (1964)

[2] L.A. Chernov, *Wave Propagation in a Random Medium*, Trans. R.A. Silverman McGraw-Hill, New York (1960)

[3] J.B. Keller, "Wave propagation in random media", *Proc. Sympos. Appl. Math. Vol 13*, Amer. Math. Soc., Providence, RI (1960)

[4] V.I. Tatarski, *Wave Propagation in a Turbulent Medium*, Trans. R.A. Silverman, Dover, New York (1967)

[5] H. Kesten and G. Papanicolaou, "A limit theorem for stochastic acceleration", *Comm. Math. Phys.* 78, 19-63 (1980)

[6] V.A. Kulkarny and B.S. White, "Focussing of waves in turbulent inhomogeneous media", *Phys. Fluids* 25 (10), 1770-1784 (1982)

[7] B.S. White, "The stochastic caustic", *SIAM J. Appl. Math.* 44 (1), 127-149 (1984)

[8] D.I. Zwillinger and B.S. White, "Propagation of initially plane waves in the region of random caustics", *Wave Motion* 7, 207-227 (1985)

[9] B.S. White, B. Nair, and A. Bayliss, "Random rays and seismic amplitude anomalies", *Geophysics*, submitted

[10] D. Ludwig, "Uniform asymptotic expansions at a caustic", *Comm. Pure Appl. Math*. 19 (2) (1966)

[11] G. Papanicolaou and W. Kohler, "Asymptotic theory of mixing stochastic ordinary differential equations", *Comm. Pure Appl. Math*.27, 641 (1974)

[12] L. Hesselink and B. Sturdevant, "Propagation of shock waves through random media" *J. Fluid Mech*., submitted